THE SCIENCE OF WAR

DONALD H. AVERY

The Science of War: Canadian Scientists and Allied Military Technology during the Second World War

UNIVERSITY OF TORONTO PRESS
Toronto Buffalo London

© University of Toronto Press Incorporated 1998
Toronto Buffalo London

Printed in Canada

ISBN 0-8020-5996-1 (cloth)

Printed on acid-free paper

Canadian Cataloguing in Publication Data

Avery, Donald, 1938–
  The science of war : Canadian scientists and allied military technology during the Second World War

  Includes bibliographical references and index.
  ISBN 0-8020-5996-1

  1. World War, 1939–1945 – Science – Canada.   2. World War, 1939–1945 – Canada – Technology.   3. Military research – Canada – History.
  4. Military weapons – Research – Canada.   5. World War, 1939–1945 – Canada.   I. Title

D810.S2A93 1998   940.54'8   C98-930276-8

This book has been published with the help of a grant from the Humanities and Social Sciences Federation of Canada, using funds provided by the Social Sciences and Humanities Research Council of Canada.

The author acknowledges the assistance of the J.B. Smallman Publication Fund, Faculty of Social Science, The University of Western Ontario.

University of Toronto Press acknowledges the financial assistance to its publishing program of the Canada Council for the Arts and the Ontario Arts Council.

For my wife, Irmgard, my sons, Richard and Bruce,
and my grandsons, Brendon and Alexander

Contents

PREFACE ix
ABBREVIATIONS xiii

Introduction 3

1 Canada's Defence Scientists: Organizing for War, 1938–1940 14

2 Building the Defence Science Alliance, 1940–1943 41

3 Radar Research and Allied Cooperation, 1940–1945 68

4 Weapons Systems: Proximity Fuses and RDX 96

5 Chemical Warfare Planning, 1939–1945 122

6 Canadian Biological and Toxin Warfare Research: Development and Planning, 1939–1945 151

7 Atomic Research: The Montreal Laboratory, 1942–1946 176

8 Secrets, Security, and Spies, 1939–1945 203

9 Scientists, National Security, and the Cold War 228

Conclusion 256

APPENDIX 1: Major Military, Political, and Scientific Events   267
APPENDIX 2: Brief Biographical Sketches   272

NOTES   281
BIBLIOGRAPHY   367
ILLUSTRATION CREDITS   387
INDEX   389

Illustrations follow page 112.

Preface

The Science of War has many dimensions. First, it is an analysis of a unique aspect of Canada's military and foreign policy during the Second World War and the early years of the Cold War. While a number of excellent accounts have examined the wartime role of Canada's political leaders, diplomats, and generals within the North Atlantic military alliance, this book is the first comprehensive study of the activities and contributions of the country's defence scientists, particularly the key administrators. While the names of C.J. Mackenzie, Otto Maass, and E.G.D. Murray might not be familiar to Canadian readers, they, along with Frederick Banting, formed the core group of military science mandarins who successfully mobilized the country's scientific resources behind the war effort. This book assesses their contribution within the general parameters of alliance warfare and, more specifically, within the context of major weapons systems.

A second major theme of the book is how the war changed Canada's scientific community. Here the emphasis is on the critical role that the National Research Council assumed in meeting the weapons requirements of the Canadian armed forces, as well as the part played by the country's universities and their scientists. The establishment of the Defence Research Board in 1946 represented a continuation of this partnership between Canada's military and the academic community.

Another important subject is the debate over the social function of science, and whether scientists should be involved in the creation of new, potentially destructive weapons. This moral dilemma is traced from the antiwar *Zeitgeist* of the early 1930s through the war years, when combating totalitarian adversaries became a higher calling. Particular emphasis is placed on the postwar activities of the Canadian Association of Scientific Workers (CAScW) and its imaginative but ultimately unsuccessful attempt to champion the cause of scientific activism and international cooperation among scientists.

The spy scares of 1946 contributed most to the CAScW's demise, especially since a number of its leaders were deemed guilty by the Canadian Royal Commission on Espionage. Whether the sentences imposed on these men, for transgressions that occurred within the context of an Anglo-Soviet-Canadian alliance, were appropriate remains a subject of much historical debate, particularly after the 1991 collapse of the Soviet Union.

With the end of the Cold War, the ongoing debate among scientists about whether their expertise should be used for humanitarian or military purposes assumes new importance. It is hoped that this book will contribute to this discourse.

While preparing this study for publication I have incurred many debts. I have received scholarly assistance from a number of friends and colleagues. In particular, I would like to acknowledge the unique assistance and guidance provided by six exceptional scholars: John English of the University of Waterloo, Wesley Wark of the University of Toronto, Hector Mackenzie of the Department of Foreign Affairs, John van Courtland Moon of Fitchburg College, Roy MacLeod of the University of Sydney, and the late John Holmes, director of the Canadian Institute for International Affairs.

While researching this book, I visited a number of archives in Canada, Great Britain, and the United States. I would like to give special thanks to the staffs of the National Archives of Canada, the University of Toronto Archives, the McGill University Archives, the National Archives in Washington, the British Public Record Office, and Churchill College, Cambridge. I obtained additional information through personal interviews with former defence scientists, who spent hours sharing their wartime experiences with me. I am deeply grateful to them.

The broad range of my research would not have been possible without the generous financial support I received from the Social Sciences and Humanities Research Council, the University of Western Ontario, the Department of National Defence, and the Osgoode Society. In addition, assistance in the publication of this book has been provided by the Social Science Federation's Aid to Scholarly Publication Program, and by the J.B. Smallman Fund. I would also like to thank Gerry Hallowell, Emily Andrew, and the staff of the University of Toronto Press for their patience and professionalism, and the anonymous reviewers for their useful comments.

Special praise is reserved for Glenda Hunt, who made my computer disks legible, and for my son Bruce and Carrie Jeneroux, who prepared the book's fine index.

Above all, I would like to acknowledge the intellectual and emotional encouragement of my wife, Dr Irmgard Steinisch, who actively participated in this project throughout its many stages, despite her own demanding professional career. For this, I am deeply grateful.

Abbreviations

AAScW	American Association of Scientific Workers
ACE	Associate Committee on Explosives
ACMR	Associate Committee on Medical Research
ADRDE	Air Defence Research and Development Establishment (U.K.)
AI	Air to Air Radar
ASV	Air to Surface Vessel Radar
BAScW	British Association of Scientific Workers
BCSO	British Commonwealth Scientific Office
BTW	Biological and Toxin Warfare
CACDS	Commonwealth Advisory Committee on Defence Science
CAScW	Canadian Association of Scientific Workers
CARDE	Canadian Armament Research and Development Establishment
CCOS	Combined Chiefs of Staff
CCRD	Cabinet Committee on Research for Defence
CD	Coastal Defence Radar
CDC	Cabinet Defence Committee
CGS	Chief of General Staff
CID	Committee of Imperial Defence
CIL	Canadian Industries Limited
CJSM	Canadian Joint Staff Mission
CPC	Combined Policy Committee
CWISB	Chemical Warfare Inter-Service Board
CWS	Chemical Warfare Service (U.S.)
DCW&S	Directorate of Chemical Warfare and Smoke
DIL	Defence Industries Limited
DMS	Department of Munitions and Supply

DND	Department of National Defence
DRB	Defence Research Board
DSIR	Department of Scientific and Industrial Research (U.K.)
EW	Early Warning Radar
FAS	Federation of American Scientists
FBI	Federal Bureau of Investigation
GL	Gun-Laying Radar
GRU	Glavnoye Razvedyvatelnoye Upravlenie (Red Army Intelligence)
ICI	Imperial Chemical Industries
IFF	Identification, Friend or Foe Radar
ISSCBW	Inter-Services Sub-Committee for Biological Warfare
JCNW	Joint Committee on New Weapons (U.S.)
JCS	Joint Chiefs of Staff (U.S.)
JSM	Joint Staff Mission (U.K.)
JTWC	Joint Technical Warfare Committee
LORAN	Long-Range Navigation System (Radar)
LREW	Long-Range Early Warning Radar
MAP	Ministry of Aircraft Production (U.K.)
MED	Manhattan Engineer District
MEW	Microwave Early Warning Radar
MIT	Massachusetts Institute of Technology
MRC	Medical Research Council (U.K.)
NCCSF	National Council for Canadian–Soviet Friendship
NCU	National Conference of Universities
NDRC	National Defense Research Committee (U.S.)
NKGB	Norodny Kommissariat Gosudarstvennoy Bezopasnosti (People's Commissariat of State Security)
NRC	National Research Council
OSRD	Office of Scientific Research and Development (U.S.)
OSS	Office of Strategic Studies (U.S.)
PF	Proximity Fuse
PJBD	Permanent Joint Board on Defence
RAF	Royal Air Force
RCAF	Royal Canadian Air Force
RC	Royal Commission
RCMP	Royal Canadian Mounted Police
RCN	Royal Canadian Navy
RDF	Range and Direction Finding (Radar)
RDX	Research Development Explosive

REL	Research Enterprises Ltd.
SHAEF	Supreme Headquarters Allied Expeditionary Forces
SES	Suffield Experimental Station
TRE	Telecommunications Research Establishment (U.K.)
USAF	United States Army Air Force
USBWC	United States Biological Warfare Committee
USN	United States Navy
WRS	War Research Services (U.S.)
WTSCD	War Technical and Scientific Development Committee
ZEEP	Zero Energy Experimental Pile Reactor

THE SCIENCE OF WAR

Introduction

On 24 March 1947, the American defence scientist Vannevar Bush wrote C.J. Mackenzie, president of the Canadian National Research Council (NRC), congratulating him for being awarded the Medal of Merit of the United States for his accomplishments during the Second World War. 'One of my fine memories of the war period,' Bush stated, 'lies in our association ... [and] I hope and trust that the years to come will see closer and closer interchange between our two countries on every matter in which we have common concern.' In his reply, Mackenzie was equally complimentary: 'I have appreciated the many courtesies and kindnesses which I received from you personally and your office during the war. It seems a long time since the cooperation between our countries in the war effort started, but when one reflects on what has happened scientifically and technically during the past seven years, one wonders how so much was done in such a short time.'[1]

Nine months earlier, in June 1946, Mackenzie had received a much less pleasant message from Washington. It was an inquiry from U.S. military intelligence, requesting a detailed assessment of any top-secret information Soviet agents operating in Canada had obtained about American weapons, including the atomic bomb.[2]

These two incidents reveal much about the importance of Canada's involvement in Allied defence science during the Second World War, and about the implications of such involvement for the emerging Cold War. They also raise a number of significant questions. How did Canadian scientists come to gain access to such top-secret information in the first place? What contributions did they make to the development of these weapons systems? And why did Soviet intelligence target Canadian defence and scientific institutions? These and other questions form the basis of this study.

The Second World War gave new meaning to the concept of total war, both in the mobilization of national resources and in the utilization of science. 'Waged on the back of industrial and technological resources, the terrifying wonder weapons produced by the scientists on both sides, and the mushroom clouds soon to rise over devastated Japanese cities brought home to everyone the military and economic potential of scientific research.'[3] The Second World War differed substantially from the First World War, in which strategists and technicians 'were still operating with weapons they could understand ... in accordance with the hallowed military principle of the economy of force ... and balanced weapon systems.'[4] In contrast, the global conflict of 1939–45 required large numbers of scientists to create and produce new weapons and techniques, often with little awareness of their operational purpose, or control over their final use. In the case of radar, for example, 'civilian scientists had to be brought into the military machine ... [and] they had to become operational researchers or operations analysts, in order to see how the new apparatus could be fitted into a military context.'[5]

The United States, Britain, and Germany took the lead in mobilizing their scientific communities behind the war effort. The results were most spectacular in the case of the United States, where the Second World War brought about fundamental changes in the relationship between science and government, 'whether measured by dollar amounts of federal sponsorship, the dominance of national security as an engine of research and development, the appearance of Big Science, or the enthusiasm with which many in the nation's universities and engineering schools welcomed the new cooperation.'[6] The United States, of course, possessed a number of crucial advantages in developing a substantial weapons factory, notably 'powerful private sector capabilities for R & D in both university and industry beyond anything available during the First World War.'[7] And by the end of the war, the Office of Scientific Research and Development (OSRD), which coordinated most of U.S. defence science, had allocated over half a billion dollars in grants to more than 320 industrial and 142 academic institutions.[8] In one way or another, almost all of the country's academic scientists were mobilized for war, and the country's major universities were transformed.[9] This new state sponsorship, in turn, 'brought in its wake ... the militarization of science ... and a major breach in the wall of suspicion that historically had separated the academy from the state.'[10]

Patterns of Defence Science during the Second World War

With the outbreak of war in Europe, defence science was organized quite differently in each of the three North Atlantic democracies. In 1939 the statutory responsibility for scientific military research and development in Britain rested with two traditional bureaucracies: the Department of Scientific and Industrial Research (DSIR) and the ordnance branches of the three military services – the Army, Navy, and Air Force.[11] In the early stages of the conflict it was generally assumed that this decentralized approach gave the United Kingdom an advantage over other countries, a view shared by many scientists. By the spring of 1940, however, many within Britain's scientific establishment had become alarmed by the conservatism of the armed services in technological matters, by the ineffectiveness of British technical intelligence, and by the lack of coordination in defence science planning. They were also concerned about the growing evidence that Prime Minister Winston Churchill did not take Britain's scientific elite seriously but relied instead on the advice of the brilliant but unpopular F.A. Lindemann.[12]

Quite a different attitude prevailed in the United States, where President Franklin Roosevelt opted in June 1940 to mobilize the scientific resources of the nation through a newly created civilian organization, the National Defense Research Committee (NDRC). The decision of this body to concentrate on academic rather than industrial scientists reflected the backgrounds of its principal organizers – Vannevar Bush and James Conant.[13] In June 1940, Bush and Conant contacted the heads of some 725 colleges and universities asking for information about their laboratories and scientists, particularly in the areas of physics, chemistry, engineering, and mathematics. This research revealed that the United States had approximately 6800 physicists, 60,000 chemists, 3400 chemical engineers, 57,800 electrical engineers, 2750 radio engineers, 55,000 mathematicians, and 3400 psychologists.

Yet for all their academic elitism, Bush and his associates understood from the beginning that they must maintain close contact with major defence contractors such as Bell, Du Pont, and General Electric, while securing the trust of the generals and the admirals. Bush proved equal to this challenge and was particularly adept at balancing the scientists' quest for unfettered and open research and the military's obsession with secrecy.[14]

In Canada, the National Research Council (NRC), founded in 1916, had the major responsibility for organizing military research during the

Second World War. In some ways it was rather ill-suited for this purpose. Throughout the interwar years the NRC had functioned almost exclusively as a civilian agency with a mandate to foster scientific research in Canada, and to become an internationally recognized research centre. Although the Department of National Defence had gradually become the largest government contractor for the NRC's services, the lines of communication between the two remained rather tenuous. In 1939 it was not out of choice but out of necessity that the NRC was transformed into a defence research board, charged with creating sophisticated weapons for the Canadian Armed Forces.

Canada's Defence Science Mandarins

Four administrators dominated Canadian defence science between 1935 and 1945: General A.G.L. McNaughton, C.J. Mackenzie, Frederick Banting, and Otto Maass. All had been strongly influenced by the trauma of the First World War. During the interwar years all were highly regarded in their fields of specialization and were prominent members of the National Research Council. Although the four men shared many characteristics, they followed quite different career paths.

Before the Great War, McNaughton was recognized as one of the most brilliant graduates of McGill University's engineering faculty. His reputation as a scientifically minded officer was enhanced in 1917, when he was given the responsibility of directing the Canadian Corps artillery barrages during the decisive battles at Vimy Ridge and Amiens. Even more impressive was his rapid rise to the position of Chief of the General Staff in 1929 'in a military force bound up in seniority and red tape.'[15] In 1934 he stated: 'We have entered on one of those phases of history marked by the most intense application of human knowledge to the development of machines for war.'[16]

C.J. Mackenzie[17] was born and raised in New Brunswick and studied civil engineering at Dalhousie and Harvard. During the First World War Mackenzie served in the Canadian Army with great distinction, winning the Military Cross in 1918. After the war, he accepted a position with the fledgling engineering faculty at the University of Saskatchewan, serving as dean from 1921 to 1939. As a university administrator, Mackenzie gained a reputation for his organization and interpersonal skills.

While Otto Maass was about the same age as Mackenzie, he came from quite a different background. Born in New York City of German-speaking Swiss parents, Maass moved with his family to Montreal when he was a teenager.[18] After completing his undergraduate training at McGill, Maass pursued an advanced degree in chemistry, receiving his M.Sc. in 1913, and then studied organic chemistry in Berlin under Walter Nernst. There he met two young British scientists, Henry Tizard and Frederick Lindemann, future mandarins of British defence science. The outbreak of war brought Maass back to North America, first to Montreal, where he spent the war years teaching chemistry at McGill, and then to Harvard, where he obtained his Ph.D. in 1919. After his return to McGill, Maass quickly established an impressive reputation, and during the next two decades he was undoubtedly the most influential chemist in Canada.

Frederick Banting, who was born into a middle-class farm family in rural Ontario, began his medical training at the University of Toronto just before the Great War.[19] On graduating in 1916, Banting enlisted in the Royal Canadian Army Medical Corps, serving as a medical officer in France.[20] He was wounded in action, and subsequently received the Military Cross for valour. During the interwar years Banting gained international acclaim and the Nobel Prize (1923) for his co-discovery of insulin, and he directed one of Canada's most prestigious medical research laboratories.

The wartime experiences and subsequent academic achievements of Banting, Maass, and Mackenzie were paralleled by other Canadian scientists, who also assumed a major role during the Second World War. Some of the more prominent were bacteriologists Everitt George Dunne (E.G.D.) Murray and Guilford Reed; surgeons Duncan Graham, W.E. Gallie, J.C. Meakins, and Wilder Penfield; and physicists R.W. Boyle and J.S. Foster.[21]

During the interwar years Canada's scientific elite operated almost exclusively within a male cultural context. Historian Michael Bliss ably described this phenomenon in his outstanding biography of Frederick Banting:

He liked to drink and smoke and swap stories with his medical or army buddies ... It was a way of belonging that filled an intense need in Banting and countless men of his and most other generations. A little frightened of the complexities of life, not at all sure how to behave in the presence of women, a little lonely, you found security and acceptance and could come out of your shell when you got together with the fellows, the boys, the old gang.[22]

Otto Maass shared many of these traits. According to his biographer, Maass was 'never happier' than when he was the centre of a congenial group engaged in lively conversation, drinking, or playing poker.[23] C.J. Mackenzie also enjoyed a good drink, especially after his Saturday afternoon golf game with his wartime boss, C.D. Howe.[24]

Some Canadian women scientists were involved in war-related research between 1939 and 1945, but almost exclusively at the junior or support level. This reflected the gender profile of science and medical research during the interwar years, in which masculinity was equated with authority. Although some women scientists, particularly in medicine, managed to challenge male domination of the scientific field, obtaining a full-time academic position or a scientific job with the NRC remained extremely difficult for them.[25] The situation further deteriorated during the Depression years, when professional opportunities for women with universities and government agencies virtually disappeared.[26]

Scholarly Context

While American and British historians have extensively explored the impact of the Second World War on academic scientists, little has been written on this topic from the Canadian perspective. What has been published tends to be either institutional in focus, or deals with specific scientific disciplines. Some of the more notable works are Wilfrid Eggleston's two survey accounts, *National Research in Canada: The NRC 1916–1966* and *Scientists at War*, as well as Yves Gingras's perceptive *Physics and the Rise of Scientific Research in Canada*. Robert Bothwell's *Nucleus: The History of Atomic Energy of Canada Limited* effectively analyses wartime nuclear research. Canada's involvement in biological and chemical warfare research is discussed in John Bryden's *Deadly Allies: Canada's Secret War, 1937–1947* and Michael Bliss's *Banting: A Biography*.

The best single account of the role of Canada's armed forces in the development of military technology remains Charles Stacey's monumental study *Arms, Men and Government*.[27] It provides an excellent précis of the many projects carried out at the Army's six major research centres and at Canadian universities. These included the RDX explosive, GL Mark IIIC radar equipment, the proximity fuse, the Sexton self-propelled 25-pounder gun, and important work in the fields of biological and chemical warfare at the Suffield Experimental Station. The activities of the RCAF, with its five research centres, is briefly mentioned, as is the role of the RCN's Research Establishment in Halifax.[28] Stacey's conclusions about the contribution of Canadian defence scientists to Allied military technology bear repeating:

> It is difficult and indeed impossible to produce any definite quantitative assessment of the value of Canadian research and development as a contribution to Allied victory. A major difficulty is simply the fact that it *was a contribution* – a share, and necessarily in most cases not a major share, in a great and complicated joint effort. Many Canadian projects were closely related to British ones and were essentially adaptations or development of ideas or devices on which much British work had already been done. To arrive at a definitive evaluation of the relative importance of contributions in cases of this sort seems out of the question.[29]

Since the publication of Stacey's massive tome, Canadian military historians have explored various dimensions of defence science by focusing on the experiences of the different military branches. Included in this list are the official histories of the Royal Canadian Air Force – W.A.B. Douglas, *The Creation of a National Air Force*, and Brereton Greenhous, Stephen Harris, William Johnson, and William Rawling, *The Crucible of War, 1939–1945* – both of which provide interesting insights into aerial operational research and problems of incorporating new forms of technology. The history of naval research and development is analysed quite effectively in Marc Milner's two books *North Atlantic Run* and *U-Boat Hunters*; Michael Hadley's *U-Boats against Canada* and David Zimmerman's *The Great Naval Battle of Ottawa* provide interesting interpretations on the same topic. So too does W.A.B. Douglas's fine article 'Conflict and Innovation in the Royal Canadian Navy, 1919–1945' in Gerald Jordan's anthology *Naval Warfare in the Twentieth Century*. Additional critiques of the role of key wartime military officials are provided by John Swettenham's three-volume study of General A.G.L. McNaughton,[30] by the memoirs of General Maurice Pope,[31] and by J.L. Granatstein's collective biography of Canada's leading Second World War generals.[32] The problems of alliance warfare, particularly Anglo-Canadian military relations, are also treated in these books, as well as in Jack Granatstein's imaginative 1989 synthesis, *How Britain's Weakness Forced Canada into the Arms of the United States*. Indeed, Granatstein's central thesis provides a useful intellectual model for this study:

10   The Science of War

The British ... were not deliberately trying to force Canada into American hands during the two world wars and the Cold War ... Britain's immediate and only intent during the world wars and after was to do everything it could to keep itself out of enemy control, and its own economy functioning. A great power that properly put its own interests first, and one of utter ruthlessness *in extremis*, Great Britain did only what it had to do. The effects of British action on Canada's course, nonetheless, were pronounced.[33]

The Science of War, while utilizing the historiography cited above, also draws extensively on the work being carried out on defence scientists and military technology by American, British, and Australian specialists. These studies have reinforced the diversity of conceptual approaches that *The Science of War* has adopted, ranging from military and foreign policy analysis to elements of the history of science and technology.[34] In addition, since this book emphasizes a comparative approach, extensive reference is made to the historiography that examines major defence science trends in Britain and the United States during the war years.

This book could not have been written without access to the wartime records of Canadian, British, and American scientific and defence agencies. Of particular value are the exhaustive records of the NRC and the Department of National Defence, and the files of the Office of Scientific Research and Development. The correspondence of many of the leading Allied scientists has also been critical for elucidating the complex problems in developing sophisticated weapons systems and in coordinating the scientific resources of Canada, Britain, and the United States.

Terminology and Format

The term 'scientist' means many things to many people. For some, a scientist is anyone who has published 'at least one paper in a recognized scientific journal,' or 'is eligible for membership in a recognized scientific body.'[35] For others, a scientist is someone who has an advanced university degree. As one critic put it, 'A scientist without a Ph.D. is like a lay brother in a Cistercian monastery. Generally he has to labour in the fields while the others sing in the choir.'[36]

This book focuses on academic scientists, the vast majority of whom had a Ph.D. and either taught at Canadian universities or worked for the National Research Council. That is not to say that laboratory technicians did not assume important roles – indeed, they were indispensable to

Canada's scientific war effort. But *The Science of War* is most concerned with Canada's scientific elite, and how they interacted with similar elites in Great Britain and the United States.

Another important qualification is whether to differentiate between scientist and engineer. This study assumes that defence science, even in peacetime, combines the talents of scientists and engineers. Such collaboration was particularly pronounced during the Second World War, when the pressure to create and produce new and improved weapons was intensified.[37] C.J. Mackenzie said it best in a June 1941 memorandum: 'We all realize that in modern warfare based on science, technology and mass production, there are three stages – research and development, production and use in the field.'[38] It should be noted that Mackenzie, who ran the NRC during the war years, was a mechanical engineer, as was his boss, C.D. Howe, Minister of Munitions and Supply.

Because of the complexity of its subject, *The Science of War* does not deal with all aspects of Canadian defence science during the Second World War. The most notable gaps are in the fields of aeronautical engineering, operational research, code breaking, and military medicine. There are good reasons for these omissions. In the case of aeronautical research, almost all of the advanced design work and developmental research was carried out in Britain and the United States, not in Canada. That is not to deny the importance of the Canadian industrial effort in producing more than 16,000 military aircraft through the efforts of over 116,000 workers.[39]

Operational research was another field in which Canada borrowed extensively from British and American defence science. Described as a scientific method of developing new operating techniques, and of improving 'the uses to which equipment is put,' operational research assumed many different forms within Canada's war effort. Between 1939 and 1945 most NRC and university scientists were involved in activities with applied research, devising and testing new weapons in cooperation with the Canadian Armed Forces. In addition, about sixty scientists were assigned to the operational research branches of the Army, Navy, and Air Force and were primarily involved with logistical and strategic planning. These scientists are not the focus of this study.[40]

Another group of scientists, the mathematicians who worked for Canada's cypher and code-breaking facilities between 1941 and 1946, are also excluded from this study. One reason for this omission was the limited scope of the Canadian cryptographic unit (code-named the Examination Unit), which involved only a handful of scientists, and was almost totally dependent on the sophisticated British facilities at Bletchley Park, and on American experts at the Office of Naval Intelligence, the U.S. Signals Branch, and U.S. Army Intelligence. Another reason was that intelligence gathering really does not fit into the category of weapons, at least as defined by this book.[41] As for wartime medical research, Canadian scientists made significant contributions in such fields as aviation medicine, blood substitutes, shock research, motion sickness, nutrition, and penicillin. But to analyse and discuss these complex undertakings would require many more chapters. This important subject awaits its historian.[42]

The Science of War is arranged along thematic and chronological lines. The first chapter provides an overview of major trends in Canadian defence science during the interwar years; reference is also made to significant developments in Canadian chemistry, physics, and medicine in this period. Controversies involving the social function of science, the relationship between scientists and the military, and the changing impressions Canadian scientists had of the Soviet Union between 1935 and 1946 are also discussed.

The remainder of the book concerns the war years. First, is an analysis of how the National Research Council under C.J. Mackenzie organized Canadian scientists behind the war effort, and how the NRC responded to the ever-increasing demands of the Canadian Armed Forces and the British government. Canada's involvement in alliance warfare, both before and after the Tizard mission of August 1940, is another important subject discussed at length. The context of alliance warfare is also essential for understanding the role Canadian scientists played in creating new weapons, which is the central concern of the next five chapters.

A unique feature of *The Science of War* is its detailed and comparative analysis of the most important weapons systems of the Second World War. Radar and the proximity fuse, for instance, represented challenging new technologies that rapidly became essential to allied military victory. Meanwhile, older technologies from the First World War, such as RDX explosives, chemical weapons, and biological warfare, were upgraded in the Second World War, giving them even greater destructive power. The Montreal atomic laboratory represented Canada's involvement with the most destructive military technology in the history of mankind – the nuclear bomb.

The final two chapters of the book deal with Canadian attempts to protect highly classified defence secrets. Special attention is devoted to how the Soviet Union took advantage of its status as a wartime ally to recruit a number of 'progressive' Canadian scientists into its spy network and the implications of the 1946 Royal Commission on Espionage are assessed.

Canada's involvement in the science of war does not end with the defeat of the Axis powers. While this detailed analysis extends only to 1946, the book does provide a brief description of how Canadian defence science was organized during the immediate postwar years, in the context of new defence commitments and the continuing involvement of Canadian scientists in creating weapons of mass destruction.

1

Canada's Defence Scientists: Organizing for War, 1938–1940

> In 1914, the ideological gap between pure and applied science seemed as wide as that between Darwinian scientists and fundamental theologians. By the end of World War II, this was no longer the case. The line between pure and applied science had narrowed and, in many, sophisticated projects, had almost reached the vanishing point ... In the six years of war, Canadian science and technology, led by the initiative of NRC, had come of age. By war's end, its standing was high amongst scientists in our allied countries and, of most importance, was recognized at home.
>
> C.J. Mackenzie, cited in Thistle, ed., *The Mackenzie–McNaughton Wartime Letters*, 146–8

Canadian defence science was virtually nonexistent during the interwar years, since the Department of National Defence (DND) had neither the scientific personnel nor the financial resources to undertake even rudimentary investigations of weapons systems. The NRC, given its own underfunding, was unable to provide much assistance to the DND, at least until after 1935, when General A.G.L. McNaughton became president. This meant that the Canadian armed forces were almost totally dependent on data and equipment that the British government, usually through the Committee of Imperial Defence, decided to share with its dominions.

During the 1930s Canadians were adversely affected by the devastating downturn in the Canadian economy, with its accompanying widespread unemployment, poverty, and social dislocation. While the 'Dirty Thirties' had a disproportionate impact on the unemployed and working poor, middle-class Canadians, including scientists in the universities and in government agencies, were also affected. In addition to their difficulties in finding and keeping jobs, many scientists were aware of the controversy surrounding the social function of science that marked that decade. While the most intense debates occurred in Britain and in the United States, they also found expression in Canada. One of the most contested issues was whether the enormous scientific and technological advances of the twenties had enhanced or threatened the quality of life in capitalistic societies. Did science for profit subvert the principles of science harnessed for the improvement of humanity?

By the mid-1930s Canadian scientists also shared concern over how science in the Third Reich was being manipulated and utilized by the Nazi war machine. Their own role in military preparedness received new emphasis as the British and Canadian governments began to prepare for war. In Canada, most of this activity was coordinated by the NRC, which by 1939 had already acquired some expertise in radar, explosives, war medicine, and chemical and biological warfare.

The Legacy of the Great War

Between 1914 and 1918 many Canadian scientists became aware, either as members of the Armed Forces or as wartime researchers, of how science could transform military conflict. Some may have shared the optimistic view of W.C. Murray, president of the University of Saskatchewan, that the war had ushered in a new era of scientific research: 'The nations of the world today,' he wrote in 1917, 'have come to see what the scientists long preached, that in science they have the most potent of instruments for extending human power.'[1] But others must have reflected on the fact that modern science had also created the powerful explosives and the poison gases that were killing thousands of Canadian soldiers. The words of Ernest Rutherford, the world-famous experimental physicist, would have caused many to reflect on how military officials of all countries had wasted the talents of so many brilliant young scientists: 'It is a national tragedy that our military organization at the start was so inelastic as to be unable, with few exceptions, to utilise the offers of services of our scientific men except as combatants in the firing line.'[2]

Chemists were undoubtedly even more aware of the implications of total war, since they were deeply involved in designing some of the most destructive weapons. Indeed, by 1918, over 5400 allied and enemy chemists were involved in chemical warfare research, with thousands more in the munitions field.[3] Wartime chemical warfare research was unique in that it demonstrated the advantages of organizing the most brilliant and prestigious chemists in centralized research facilities to develop more deadly gases.[4] The fact that over 124,200 tons of poison gas were used, causing over a million casualties, was a grim reminder of the terror of modern warfare.[5] Events after the war confirmed that the major powers regarded poison gas as a devastating strategic weapon that had to be controlled. This was reflected in the provisions of the Versailles Treaty, which specifically prohibited Germany from possessing chemical weapons,[6] and by the June 1925 Geneva Protocol, held under the auspices of the League of Nations, which banned the offensive, or first, use of chemical and biological weapons, and was signed by thirty-eight countries including the United States, Britain, Germany, France, Italy, Japan, and Canada.[7]

Canada's endorsement of this ban was not surprising, since its soldiers had been among the first victims of gas warfare, during the 1915 Battle of Ypres. Canadian chemists had also assumed an important role in helping Britain develop new chemical weapons during the remainder of the war, while others worked on new forms of explosives. Meanwhile, Canadian physicists made important contributions in the fields of artillery sounding, and submarine detection,[8] and medical researchers provided valuable sources of tetanus, small pox, and typhus vaccines.

Nevertheless, Canada's First World War defence science role was limited, since the number of its scientists was small and their resources meagre. Indeed, a 1917 survey of Canadian scientific resources showed that only 160 industrial scientists were operating out of thirty-seven corporate laboratories. Nor was the situation any better within academe: the survey showed that only fifty scientists were doing pure research, and most of them were located at McGill, Queen's, and the University of Toronto.[9] The pressures of war made a bad situation worse. Many British scientists working in Canada had left their academic jobs to serve the Empire; scientific equipment, materials, and journals were in short supply; and the Dominion government increasingly commandeered the services of university personnel.[10]

Indeed, one of the reasons for the creation of the National Research Council in 1916–17 was to more effectively mobilize Canada's wartime scientific resources. Given Canada's close relationship with Great Britain, it was not surprising that the NRC was originally modelled on that country's Department of Scientific and Industrial Research, established in July 1915.[11] The engine driving the Canadian campaign for such an organization was a coalition of prominent scientists, university presidents, and industrialists. Their efforts to convince the Borden government that Canada desperately needed this type of institution were rewarded on 23 May 1916, when Minister of Commerce Sir George Foster persuaded his 'indifferent or antagonistic' Cabinet colleagues to give the Honorary Advisory Council on Scientific and Industrial Research (later called the National Research Council) an opportunity to carry out a comprehensive survey of scientific and industrial research in Canada, to help organize and mobilize existing research facilities in discovering new industrial processes, to link more effectively science and technology with labour and capital, and to suggest ways of expanding the size and scope of Canada's research talent. In August 1917, legislation officially creating the NRC was endorsed by all political parties and received royal assent.[12]

Many details about the NRC's operations remained unresolved at first. One of these was whether the NRC's laboratories would be based in Ottawa, or scattered among the major universities of the country. This debate, which lasted two years, highlighted the different perceptions of how government science should be organized and the relationship between pure and applied science. Those calling for central laboratories stressed the need for a Canadian version of the U.S. Bureau of Standards, in which the NRC would assume a key role in providing essential quality control for technologically advanced Canadian industries. Despite the endorsement of most university scientists and administrators, attempts to establish central laboratories were dealt a serious blow in May 1921, when the Senate rejected that section of the legislation. This was not the first, or last, time that Canadian scientists would feel betrayed by their legislators.

The situation did not improve during the next two years. Underfunding, poor morale, and lack of leadership prevailed until April 1923, when the Liberal government of William Lyon Mackenzie King finally decided to infuse new life into this scientific endeavour. The most important move was the appointment of Henry Marshall Tory, president of the University of Alberta, as NRC president. Under Tory's direction the NRC gradually began to provide some badly needed leadership and support for Canadian scientists.[13] In 1935 Tory was replaced by another powerful figure who would shape the NRC in his own image: General A.G.L. McNaughton.

Canadian Defence Science: Organization and Problems, 1930–1939

Throughout the 1930s Canada's defence establishment was constrained by rigid budgetary limits imposed by the federal government and by the reluctance of the Bennett and King governments to accept imperial military obligations. The severe problems facing the Armed Forces were outlined in a secret May 1935 memorandum by General A.G.L. McNaughton, Chief of the General Staff, just before his transfer to the NRC. According to McNaughton there was 'not a single modern anti-aircraft gun of any sort in Canada; ... available stocks of ammunition for field artillery represented only 90 minutes' fire at normal rates; ... coast defence armament was obsolescent, when not defective; and ... there was not one service aircraft of a type to employ in active operations.'[14]

Prime Minister Mackenzie King was stunned by these revelations and immediately appointed a Cabinet Defence Committee to investigate the situation. Yet despite this flurry of activity, little real progress was made in upgrading Canada's military capabilities,[15] and the appropriations for the Department of Defence for the fiscal year 1937–8 were only $36,194,839.[16] Nor was King prepared to alter his no-commitment defence policy even when pressured by British and Commonwealth officials at the 1937 Imperial Conference.[17] Although he eventually agreed to participate in the British Commonwealth Air Training Program, which was designed to train 30,000 British, Canadian, Australian, and New Zealand air crew annually for service in the RAF, only after prolonged and acrimonious negotiations did it became operative, in the winter of 1939.[18]

Canada was also a latecomer to the field of defence science. What was achieved was largely through the efforts of General McNaughton after he was appointed president of the NRC in 1935 by Prime Minister R.B. Bennett.[19] Although McNaughton was not initially keen for the job, he eventually accepted the challenge, reasoning that he had already taken an active part 'in promoting scientific developments in the armed forces.'[20] He also sought to overcome the concerns of some NRC scientists that he was merely a political puppet by expanding the scope of the Council's fundamental and applied research with projects carried out by the divisions of Chemistry, Engineering, Biological Sciences, and Physics and Electrical Engineering.[21] In addition, he increased the number and importance of the NRC's associate committees by judiciously recruiting Canada's most outstanding academic and industrial scientists by providing them with prestige and patronage, an attractive inducement in a country whose scientific reward system was underdeveloped.

One NRC member who caught McNaughton's attention was Chalmers Jack (C.J.) Mackenzie, dean of the University of Saskatchewan's engineering faculty, who was appointed to the Council in 1935 and who subsequently served as chair of the NRC Review Committee. McNaughton and Mackenzie quickly took to each other, and their friendship and mutual respect helped ensure continuity in NRC administrative priorities during the next ten years. An admiring Mackenzie described his boss as follows: 'He understood pure science – he understood the environment you needed. He was of the type of scientific engineer who have [sic] made industrial countries efficient – a man well versed in pure science but interested in the application.'[22]

During his four years as NRC president, McNaughton encouraged a variety of war-related projects. The ballistics group, under physicist D.C. Rose, carried out studies for the Army on a type of portable photoelectric cell that could be used to measure the velocity of projectiles fired from guns. The aeronautical branch conducted wind tunnel tests of air frames and engines, and physicist L.E. Howlett perfected the technique of making maps from photographs. Equally important was the work on the use of metallic magnesium for airframes, preliminary research on the cathode ray oscilloscope for direction finding, and the development of the cathode-ray direction finder for marine craft and airplanes.[23] In 1938, physicist J.T. Henderson convinced McNaughton of the value of establishing a permanent Radio Field Station, just outside Ottawa, for experiments in range and direction finding (RDF), short-wave transmitters, short-wave receiving arrays and beam aerial systems, long-wave transmitters (requiring large antenna systems), field strength measurements, and measurements in radio interference.[24]

While McNaughton felt that the NRC was making an important contribution to Canadian defence, he was often frustrated by the lack of encouragement he received from the Chiefs of Staff.[25] One explanation for this lack of support was that all three armed services were underfinanced, ill-equipped, and understaffed. As late as March 1939, the Permanent Force numbered only 4169 of all ranks, with a meagre 446 officers, many of whom 'were too old for active service or were in ill-health.'[26] Moreover, most of the senior officers, still rooted in their First World War experiences, remained 'out of date in their military thinking after twenty years of peacetime soldiering in a country that cared nothing for its armed forces and vigorously practised military retrenchment.'[27] Another problem was the Canadian military's tendency to imitate policies and practices of the British Armed Forces, a symptom of residual colonialism, social conditioning, and professional training. In the case of the Canadian Army, for example, more than half of the Permanent Force officers who achieved general officer rank before or during the Second World War had received British Army Staff College training. These included such officers as T.V. Anderson (in 1920), E.L.M. Burns (1928), Harry Crerar (1924), Charles Foulkes (1939), A.G.L. McNaughton (1921), J.C. Murchie (1930), George Pearkes (1919), Maurice Pope (1925), Guy Simonds (1938), and Georges Vanier (1924). A number of these men also attended the prestigious Imperial Defence College in London.[28]

While the Canadian Army was strongly influenced by British strategic thinking, tactics, and equipment, it was at least determined to maintain an identity separate from the War Office. The same could not be said for the Royal Canadian Navy, which, by 1939, was regarded as little more than an auxiliary squadron of the Royal Navy, 'relying on the British for almost all its logistic support, including weaponry, electronics such as asdics and radios.' Because of this dependency, Canadian Naval Staff Headquarters generally followed the lead of the Royal Navy in terms of technological change. This meant that it remained relatively indifferent to the possibilities of scientific assistance from the NRC until the outbreak of war.[29] In contrast, the Royal Canadian Air Force was prepared to innovate, recognizing that the next war would require new weapons and tactics, and it closely monitored British radar research and the RAF plans to develop new fighters and bombers. In addition, the King government's rearmament priorities after 1937 meant more generous funding for the RCAF.

Department of National Defence: Planning and Priorities, 1935–1939

Throughout the late 1930s Canadian and British staff officers who were interested in the application of science to war maintained close contact. The most important Canadian spokesman was General McNaughton, both in his capacity as NRC president and as the 'godfather' of ambitious and creative younger officers such as colonels Harry Crerar, Maurice Pope, and E.L. Burns. Canada's role in Commonwealth rearmament was the subject of much interest for these military planners, who shared a sense of frustration about Canada's vacillating and unimaginative defence policies. Crerar was particularly well placed, serving as Director of Military Operations and Intelligence between 1934 and 1938 – the critical period when the British Commonwealth moved toward war.[30]

Some of Crerar's most useful information came from Colonel W.W.T. Torr, British Military Attaché in Washington. Theirs was an amicable relationship, since Torr appeared to appreciate Crerar's complaints that few of the senior officers in the three British Services had any firsthand knowledge 'of the constituent parts of this Empire of ours,' and that the situation would be greatly improved if these brass hats would 'only get around more, and look at defence problems from the Melbourne, Pretoria or Ottawa points of view, instead of from an office chair in Whitehall.'[31] Torr was also willing to facilitate Canadian efforts in obtaining information about U.S. defence science developments in general and the War Department's Industrial Mobilization Plan in particular. Crerar pursued this issue when he visited Washington in October 1937, meeting with a number of 'friendly ... and informed U.S. Army officers,'[32] notably Major General Embick, Deputy Chief of Staff, and future American representative on the Permanent Joint Board on Defence. He was also impressed by evidence that President Roosevelt was moving the United States away from isolationism, and hoped that this would change the attitude of the King government.[33]

By 1938 one of Crerar's greatest concerns was whether, in the event of war, Britain would be able to supply Canada with munitions and essential military equipment, or, alternatively, whether British firms would 'set up branch plants in this country to look after part of your requirements, and all of ours.' If neither of these choices were viable, Crerar believed that Canada would be forced to consider other options. One of these would be to follow the example of Australia and place defence orders with American companies, a policy he believed would inevitably 'result in a breakaway from that long established and very important Imperial principle of similarity of equipment (and organization and training).' But modern arms and equipment of dissimilar pattern were better, he reasoned, 'than none at all ... and that is what we are apparently up against in these days of hectic re-armament.'[34]

Crerar and Pope were also fully aware that for all major weapons systems it was the British not the Canadians who determined the character and progress of the projects and that, in most instances, British policies were only decided after furious infighting among the Services and among rival groups of scientists. The most celebrated confrontation took place between Henry Tizard, Britain's foremost defence scientist, and Frederick A. Lindemann during the late 1930s over the issues of radar development and Britain's air defence priorities.[35] This dispute had two major issues. First was the argument set forth by Tizard, A.V. Hill, and Patrick Blackett that Britain's radar system (RDF) must first determine the location, direction, and height of enemy aircraft, and then coordinate plans for their destruction. In contrast, F.A. Lindemann called for a variety of detection systems, including infra-red and radar, and a defensive system of aerial mines and balloon barrages. In addition Tizard believed that the Air Ministry's Committee for the Scientific Study of Air Defence (CSSAD), which he chaired, was only an advisory body, and felt that it was improper to engage in political lobbying. Lindemann had no such qualms, and fully utilized the influence of his patron, Winston Churchill, during the bitter 1936–8 air defence debates.[36] Once Churchill became prime minister in the spring of 1940, it was Lindemann who became the dominant scientific adviser.

Canadian defence scientists were affected by this controversy in several ways. Because of Lindemann's objections, there was a delay in developing effective radar equipment, and in sharing this technology with the dominions. Part of the blame, however, must be attributed to the attitude of the British Armed Forces, who jealously guarded their defence secrets – even from each other.

By 1938 Canada was having its own defence controversies, especially after George Drew, a prominent Ontario Conservative, blasted the King government's rearmament policies in a 16 March article in the *Financial Post* entitled 'Canada's Defence Farce.' According to Drew, the Canadian Armed Forces were woefully equipped with modern weapons, largely because of the government's lack of vision and confidence in its own business community: 'If we are to have a capable defence force we must make what we require in Canada.' Significantly, General McNaughton was among those who congratulated Drew for his sweeping critique: 'Every word is true ... and I do hope that the result will be beneficial in awakening the people of Canada to the ridiculous and dangerous position in which we now find ourselves.'[37] McNaughton's endorsement was not, however, extended to Drew's next campaign: allegations of corruption associated with the lucrative Bren gun contract of 31 March 1938. Under this arrangement the John Inglis company of Toronto was responsible for producing 7000 Bren light machine guns for the Canadian Army, as well as another 5000 guns for the British War Office. While McNaughton shared Drew's arguments that an expanded Dominion Arsenals, not a private company, should have been entrusted with this important task, his position as NRC president mitigated against any active involvement in this controversy.[38]

Scientific Missions to Britain, 1937–1939

Although McNaughton was a Canadian nationalist, he recognized that

Canada's future role in scientific warfare would depend largely on Britain's needs and priorities. Being cast in a support role, however, had some advantages. In many cases British scientists and technicians had already carried out the preliminary research and design of new weapons, and this information was usually shared with the NRC, especially in the case of chemical and biological warfare.

As early as 1935, when he was appointed president, McNaughton had asked the deputy minister of National Defence to keep him posted about 'secret information which you receive from time to time on the developments which are taking place in [chemical warfare] protective clothing,' in which case qualified NRC scientists would 'give advice or ... conduct research if and when required.'[39] The threat of chemical and biological warfare loomed large after 1935, particularly after the Italian Armed Forces used mustard gas against poorly equipped Ethiopian troops and received no censure from the international community for such tactics. As a result, British military planners became increasingly concerned that their chemical warfare preparation lagged far behind that of both Italy and Nazi Germany. In July 1936 the Committee of Imperial Defence (CID) prepared a report for the War Office on 'The Possible Use of Gas as a Retaliatory Measure in War,' which warned that 'foreign countries are busily preparing for offensive chemical warfare,'and called for an immediate increase in Britain's stockpiles of HS and HT mustard gas for insurance.[40]

Although McNaughton carefully monitored technical chemical and biological warfare developments through the reports of Canadian representatives on the Committee of Imperial Defence, he wanted his own firsthand assessments. In February 1937 he made arrangements for E.A. Flood to visit British chemical warfare facilities ostensibly as part of the NRC's efforts to improve the quality of Canadian Army gas masks.[41] In an interview before Flood's departure, Colonel Harry Crerar advised him: 'Don't forget, the best defence is offence, and if you don't come back with knowledge on how to make offensive chemical weapons, I won't think you've done a proper job.' While Flood tried to meet Crerar's expectations, he was hampered by the fact that British chemical warfare preparations were woeful, with 'a near complete lack of chemical munitions at the onset of war.'[42]

An even worse situation prevailed in the area of biological warfare, a situation of much concern to Frederick Banting, who believed that Germany would not hesitate to use these weapons against Britain or its dominions. Banting first expressed his concerns in an October 1938 memorandum he prepared at the behest of General McNaughton. And in his December 1939 mission to England he tried to convince the Special Committee on Biological Warfare that a crash program in both defensive and offensive biological warfare research was essential. His efforts, however, were unsuccessful.[43]

The experiences of Flood and Banting demonstrate that Canadian scientists, although free to comment on British defence policies, had little opportunity to influence the decision-making process. This ineffectiveness was even more pronounced in the case of British radar policies, which, until mid-1939, remained a carefully guarded secret, even from the Dominions. Only in March 1939, with war imminent, did the Air Ministry announce that they were prepared 'to release information respecting a most secret device which they had adopted for the detection of aircraft.' This information was channelled through Air Vice Marshal Croil, Chief of the Canadian Air Staff, to General McNaughton, who made arrangements for John Henderson of the Radio Branch to represent Canada at the April 1939 meeting of Dominion radar specialists. It was the beginning of Canada's involvement in Britain's important radar war.[44]

But why did it take Britain so long to inform the NRC and the Department of National Defence, even though it continually sought Canadian military assistance and commitments through the CID? One reason was British concern for secrecy. Obviously the Air Ministry wanted to limit the distribution of vital technical information about their defensive weapons system when the stakes were so high. Second was the question of whether Canadian physicists were sufficiently talented to appreciate the complexities of radar equipment, a negative assessment that NRC physicists soon refuted. Then there was Mackenzie King's belated response to CID requests for assistance because of his own political calculations and his determination to avoid long-term defence commitments, an approach that British military officials deeply resented.

Canada's Scientific Resources on the Eve of War

By 1939 Canadian military planners such as Crerar, Pope, and McNaughton were aware of the importance of new weapons in the forthcoming conflict. But they were uncertain how Canada would acquire this new technology and what role the country would play in the research and development process. One option was the transfer of British military technology to the Canadian Armed Forces, either through the CID or through the separate British service. For example, under this type of arrangement the Royal Canadian Navy would obtain advanced radar and sonar equipment from the Royal Navy after it had met all its operational needs and determined that the Canadians could make proper use of this technology. Another approach, which became increasingly viable, was to mobilize Canadian scientists working through the auspices of the NRC into research teams that would concentrate on problems related to the major weapons systems – radar, explosives, proximity fuses, and chemical and biological warfare. Under this scheme Canadian defence scientists would participate in the development of new military technology, along with their British counterparts, and the new weapons would be shared by both countries. Not surprisingly, the idea of being partners rather than supplicants had much greater appeal for the Canadian Armed Forces and the National Research Council. Such a commitment was, however, based on the premise that, unlike the situation in 1914–18, Canada would have sufficient scientific talent to carry out this daunting task.

During the interwar years Canada's scientific community increased both in size and expertise. While all fields shared in this expansion, the changes were particularly important in the disciplines of chemistry, physics, and medicine. These were also the fields that contributed the highest number of defence scientists in 1939. By 1943, the National Research Council alone had a scientific staff of 1300, with hundreds of other university scientists carrying on war-related research in their laboratories.[45]

Chemistry in Canada, 1918–1939

In 1918 Canadian chemists took a major step toward professionalization with the formation of the Canadian Institute of Chemistry. Its first president was Watson J. Bain of the University of Toronto's Chemical Engineering Department. By 1930 chemists enjoyed increased opportunities to publish scholarly papers in national and international journals. Canadian publications included the *Transactions of the Royal Society of Canada*, and *The Canadian Journal of Research*, which was created by the NRC in 1929. For those seeking a wider audience, the *Journal of Physical Chemistry*, the *Journal of American Chemical Society*, the *Journal of Chemical Physics*, or the *Proceedings of the Royal Society* (London) were also available.

During the twenties most Canadian universities offered honours degrees in the major branches of the discipline: organic, inorganic, and chemical physical.[46] Some of the most promising new research was carried out in physical chemistry, especially chemical thermodynamics, spectroscopy, chemical kinetics, and theoretical chemistry.[47] Physical chemists, particularly at McGill and the University of Toronto, also had close connections with their colleagues in chemical engineering. In addition, strong links existed between academic and industrial chemists, especially in the fields of fuels, metallurgy, hydroelectricity, and pulp and paper.

Each major university had its distinguished chemists. Queen's University's small chemistry department was headed by A.C. Neish and J.A. McRae,[48] while the most productive chemist at the University of Western Ontario was Christian Sivertz.[49] Well-established chemists in western Canadian universities included Matthew Parker and John Shipley at the University of Manitoba;[50] John Wesley Shipley and Osman James Walker at the University of Alberta; R.H. Clark and E.H. Archibald at the University of British Columbia; and T. Thorvaldson, John Spinks, and Gerhard Herzberg at the University of Saskatchewan.[51] In Atlantic Canada, Dalhousie University boasted the largest chemistry department in the region.

In Ontario, the University of Toronto's chemistry department was dominant. Among the more prominent members were Andrew Gordon, who carried out important research on the thermodynamic properties of triatomic molecules and Frederick Earl Beamish, who established a reputation in the field of inorganic chemistry. Hugh Massey Barrett was well-known for his work on toxic gases,[52] while George Wright's research on the chemistries of explosives and organometallic compounds had great importance during the Second World War.[53] Additional expertise was provided by chemical engineer Watson J. Bain, first president of the Canadian Chemical Institute,[54] and by two German refugee chemists – Hermann Otto Fischer and Richard Baer.[55]

Toronto's great rival during the interwar years was McGill's chemistry department. It was composed of George Stafford Whitby, C.F.H. Allen, and Robert Ruttan in organic chemistry; F.M.G. Johnston in inorganic; Charles Hibbert in wood chemistry; and Otto Maass in physical chemistry. The latter was to exert the greatest influence both in the life of the department and in Canadian chemistry in general.

Throughout his long academic career Maass published over 200 papers on a wide range of topics: equations of state, viscosities, sound propagation, electrolyte phenomena, surface tension, specific heats and the physics and chemistry of hydrogen peroxide, the physical chemistry of the sulphite wood process and the surface energies of solids, notably the alkali halides.[56] According to his Royal Society biographers, 'Maass was ... more a "doer" than a "dreamer." Hence ... his main contributions to scientific knowledge are in the experimental field rather than in the field of theoretical speculation ... Maass might be regarded as Canada's Faraday ... when we compare Maass's contributions with those of other Canadian chemists of his day.'[57] Maass was frequently invited to present scientific papers at the most prestigious conferences, and became a member of both the Royal Society of Canada (1922) and the Royal Society of London (1940).[58] In addition to his scholarship, Maass was a successful teacher, his most outstanding protégé being E.W.R. Steacie, who went on to become one of Canada's most famous scientists and eighth president of the NRC.[59]

Throughout the 1930s Maass maintained a wide range of academic, government, and industrial contacts. Many of his students staffed chemistry departments across the country, and they kept in touch with their former mentor. Even more important were his strong connections with the Chemistry Division of the NRC through his former colleague George Whitby and through his close relations with 'Andy' McNaughton.[60] Maass's position as director of the Pulp and Paper Institute of Canada brought him into contact with many of Canada's leading industrialists.

On 5 May 1939 Maass gave the presidential address to the Canadian Institute of Chemistry (CIC).[61] After acknowledging the CIC's membership growth and its professional certification status, as well as improvements in the quantity and quality of Canadian chemists, Maass stressed the need for CIC members to commit themselves to Canada's forthcoming war effort:

It is a matter of keen satisfaction to all of us that the Government of Canada has recently recognized the Canadian Institute of Chemistry as a body capable of advising in matters of national import ... This is in line with the cooperation given the British Government by the Chemical Organizations of Great Britain. You will agree with me when I say that we shall not fail our government.[62]

Physics in Canada, 1920–1939

The discipline of physics also experienced growth in Canadian universities throughout the interwar years.[63] Between 1900 and 1916 thirty-three master's degrees and six doctorates were granted; in the period 1917–39, these figures rose to 208 master's and 85 doctorates. While Toronto and

McGill continued to dominate the doctoral field, over half of the master's students graduated from Dalhousie, Queen's, and the four western universities. Although many of these graduates subsequently pursued their doctoral studies at Toronto and McGill, others left for British, American, or European universities.[64]

The University of Toronto had a particularly strong physics program, due largely to the work of J.C. McLennan, a world-famous authority on atomic spectroscopy, superconductivity, and low-temperature physics. In 1920 McLennan was able to convince the university's president, Robert Falconer, to support his research on the spectra of hydrogen, helium, and nitrogen at low temperatures, for which he and his assistant Gordon Shrum earned the 1928 Gold Medal of the Royal Society of London. McLennan's international prestige was reflected in his election to the Royal Society of London in 1915 and his knighthood in 1935. In addition to his scholarship, McLennan was a successful promoter of his graduate students, who obtained half of all the scholarships awarded by the NRC in physics between 1918 and 1932.[65] These McLennan graduates strongly influenced the development of other Canadian university physics departments throughout the early twentieth century. Among the more prominent were Raymond Dearle at the University of Western Ontario, J.F.T. Young at the University of Manitoba, and Gordon Shrum at the University of British Columbia. During his long tenure at Toronto, McLennan built a strong experimental physics department, with assistance from Eli F. Burton, whose research projects included radioactivity of crude petroleum and the physical properties of colloidal solutions.[66] McLennan's attempts to create a chair in theoretical physics were never realized, largely because of the university's decision to establish the department of applied mathematics in order to convince John L. Synge to return to the fold.[67] In turn, Synge recruited the Polish theoretical physicist Leopold Infeld, a friend and collaborator of Albert Einstein.[68]

Physics at Queen's,[69] Dalhousie,[70] and McGill was highly influenced by the work of Nobel laureate Ernest Rutherford. Although he stayed in Montreal for only nine years (1898–1907), as Macdonald Professor of Physics, Rutherford transformed physics not only in Canada but throughout the world. Carried out in 'the most costly, and perhaps the largest ... institute in the world when it opened in 1893,'[71] Rutherford's scientific achievements during these years are legendary: the discovery of thorium emanation; the transformation theory of radioactivity (along with chemist Frederick Soddy); deflection of alpha rays and their recognition as charged atomic particles; and detection of alpha ray scattering.

By 1908, the year he was awarded the Nobel prize, Rutherford was regarded as one of the greatest experimental physicists of all time.[72]

Rutherford's prestige attracted to McGill a promising group of Canadian physicists, such as R.K. McClung, Harriet Brooks, S.J. Allen, A.G. Grier, and R.W. Boyle, and an impressive collection of foreign scholars, notably Tadeusz Godlewski from Poland and Otto Hahn from Germany.[73] However, this flow was curtailed after 1907, when Rutherford accepted a position as head of the University of Manchester physics laboratory. In 1919 he went on to become the director of the famous Cavendish Laboratory at Cambridge University. Despite his departure, Rutherford retained a proprietary interest in the well-being of physics in Canada. This was demonstrated in 1912 when he sent one of his more promising Manchester doctoral students, Joseph A. Gray, to Canada for what would become a long and distinguished scientific career.[74] Gray, in turn, assumed an important role in recruiting promising young physicists for the Cavendish, particularly after 1924 when he was appointed to the Chown Research Chair at Queen's University, a position he held until his retirement in 1952.[75]

Although his early research had concentrated on the scattering of Y-rays of radium and X-rays by gases, during the interwar years Gray turned his attention to the field of atomic physics.[76] And it was in this area that his scientific protégés – Harold W. Cave, Donald C. Rose, Bernice W. Sargent, and W.J. Henderson – chose to pursue their academic careers.

The individual and collective experiences of these young Canadian physicists reveal much about scientific networking. All, for instance, were admitted to Cavendish because of Gray's personal connection with Rutherford; this was also an important asset in their ability to obtain British Exhibition of 1851 scholarships, one of the few forms of stable funding. For their part, they kept Gray informed about the major debates within the British physics community, especially during the early 1930s when research carried out by Cavendish physicists garnered three Nobel awards.[77] Being at Cavendish also meant being exposed to the ideas of leading European nuclear scientists such as Niels Bohr and Albert Einstein, who would periodically appear at the meetings of the Cambridge Physical Society.[78]

For his part, Rutherford maintained a lively interest in his Canadian students even after they returned home and accepted jobs at the NRC or with a Canadian university. He was particularly well connected with the NRC, since Robert W. Boyle, who had studied with him at McGill, was director of the NRC's Physics Branch and regularly consulted with Rutherford on appointments.[79] By 1931, two of the eight scientists in the division – George Lawrence and Donald C. Rose – were Cavendish graduates, with John T. Henderson coming from King's College, London.[80] Rutherford was also instrumental in getting J.S. Foster, an experimental nuclear physicist, appointed in 1932 to McGill's prestigious Macdonald research chair in physics, which he had occupied twenty-five years earlier.[81]

Foster, unlike his McGill colleagues, was much more involved with American-based nuclear research than with British,[82] taking special interest in the work of Nobel laureates Ernest Lawrence,[83] Harold Urey, and Arthur Compton.[84] Indeed, one of Foster's greatest goals after 1932 was to convince his university, and the Rockefeller Foundation, to provide the necessary funds for him to acquire a powerful particle accelerator, or cyclotron, in order to emulate the work of the famous Lawrence Berkeley Laboratory. Although Foster came close to realizing his dream, his timing was bad. With the outbreak of war in September 1939, only projects directly connected with military preparedness were to receive university funding.[85]

By the 1930s many other Canadian atomic physicists sought to advance their careers in the United States, attracted by opportunities in the rapidly expanding research laboratories at Columbia, Chicago, Harvard, Princeton, and Berkeley. Some, such as Foster's students Robert Thornton and Arthur Snell, went on to the Lawrence Laboratory, while others, such as Harry Thode, Walter Zinn, and George Volkoff, gravitated toward other U.S. nuclear laboratories. Thode accepted a three-year postdoctoral job at Harold Urey's Columbia laboratory to work on experiments separating uranium isotopes using heavy water as a moderator, 'where you couldn't tell a chemist from a physicist.'[86] In this creative environment Thode met some of the world's most outstanding nuclear scientists before beginning his long career at McMaster University in 1939.[87] Another Canadian student at Urey's lab was Walter Zinn, a former protégé of J.A. Gray; Zinn obtained his Columbia Ph.D. in 1934.[88] George Volkoff studied with Robert Oppenheimer at the University of California, Berkeley.[89] He remembers his three years studying theoretical physics with Oppenheimer as a 'magical experience'; through Oppenheimer he met many of the leading U.S. and European physicists. In 1941 Volkoff returned home to the physics department of the University of British Columbia.

During the Second World War these Canadian-born, American-trained physicists were to play a key role in developing nuclear weapons. Thode and Volkoff, for example, were members of the Anglo-Canadian Montreal nuclear laboratory, while Arthur Snell and Walter Zinn worked on the U.S. Manhattan Project.

While the connections between Canadian and American atomic scientists were important, they were not as strong as Canada's links with the Cavendish connection.[90] Throughout his long career, Ernest Rutherford had groomed many Canadian scientists who absorbed the Cavendish 'spirit and good humour' that produced so many outstanding scientific achievements.[91] Canadian physicists also established lifelong contacts with the elite of British science: John Chadwick, John Cockcroft, P.M.S. Blackett, Mark Oliphant, E.D. Ellis, P.B. Moon, F.H. Fowler, Norman Feather, and William Lawrence Bragg. To be a Rutherford man meant instant recognition and prestige.

Canadian Medical Researchers, 1918–1939

Another group of scientists who were crucial to Canada's war effort were medical researchers. Again, the country was fortunate in being able to draw upon its universities for specialists to provide the necessary expertise in fields as diverse as aviation medicine and biological warfare. Indeed, the interwar years witnessed an enormous growth in the number and quality of Canadian medical researchers. Some of the most exciting work was done at the University of Toronto and McGill. But Queen's, Laval, Dalhousie, the University of Western Ontario, and the western Canadian universities had their share of promising medical researchers.[92]

The 1923 Nobel Prize for physiology or medicine, awarded jointly to Sir Frederick Banting, Charles Best, J.J.R. Macleod, and James Collip for their discovery of insulin, was the most spectacular achievement of Canadian medical science in the interwar period. Banting was rewarded with an appointment as permanent Professor of Medical Research at the University of Toronto at an annual salary of $5000, a lifetime annuity of $7500 from Parliament 'sufficient to permit Dr. Banting to devote his life to medical research,' and a knighthood in 1934.[93] With these assets, Banting was able to groom an impressive team of medical researchers, such as biochemists E.J. King, Colin Lucas, and George Hunter, and physiologists Wilbur R. (Bill) Franks, D.A. Irwin, George Manning, and G. Edward Hall. Hall, after obtaining his M.D. and Ph.D. degrees at Toronto, became Banting's chief lieutenant prior to the war. Historian

32 The Science of War

Michael Bliss has aptly described the ambience of Banting's laboratory: 'To the young researchers in the department, Banting's attitude was refreshing and liberating. They would always get the credit they deserved from a chief who stressed teamwork and collaboration ... It was an ultra-egalitarian, democratic approach ... not commonly followed in Banting's time or since.'[94]

By 1939, other research 'empires' had been built at the University of Toronto. Duncan Graham was the Sir John and Lady Eaton Professor of Medicine and head of the department; in 1940 he became president of the Canadian Medical Association. Kenneth McKenzie, a world-class neurosurgeon, attracted a large number of outstanding medical students.[95] On another front, R.D. Defries, John G. FitzGerald, and Charles Best directed the fortunes of the successful Connaught Laboratory, which after 1922 turned its attention from developing rabies and diphtheria vaccines and antitoxins to the large-scale production of insulin.[96] Best, who had been involved with Connaught since 1922, succeeded Macleod as professor of physiology at the University of Toronto in 1929. Best also attracted his own loyal group of talented researchers, most notably Donald Solandt, who trained at Cambridge under Archibald V. Hill, the 1922 Nobel laureate for physiology or medicine.[97]

McGill was Toronto's great rival in medical research. Neurosurgeon Wilder Graves Penfield, biochemist James Bertram Collip, surgeon J.C. Meakins, and bacteriologist E.G.D. Murray all had international reputations. The most impressive medical operation was the Montreal Neurological Institute (MNI), which opened its doors in 1934. Under its director, Wilder Penfield, and his team of surgeons and researchers, the MNI soon gained considerable professional prestige.[98]

E.G.D. Murray had similar ambitions to elevate his department of Bacteriology and Immunology to world-class standards. The thirties was an exciting period for microbiologists, who made great strides in their understanding of bacteria and viruses. Of immense importance was the discovery and development of the sulfonamides, which forced scientists 'to change their mental gears.'[99] Murray, along with his close friend Guilford Reed, a bacteriologist at Queen's University, was in the forefront of this research.

By the late 1930s medical scientists were conducting important research at other Canadian universities and hospitals. These included Donald Maitland, Professor of Anatomy, Dalhousie University; J.E. Gendreau, Director of the Radium Institute, Laval University; C.L. Masson, Professor of Pathological Anatomy, University of Montreal; P.H.T.

Thorlakson, Professor of Surgery, University of Manitoba; and J.C. Paterson, Director of the Pathological Department, Regina General Hospital. These researchers, together with Banting, Duncan Graham, V.E. Henderson, Wilder Penfield, Grant Fleming, and E.H. Ettinger, were founding members of the NRC Associate Committee on Medical Research (ACMR), which was created in 1938.[100]

Given Banting's national prestige, his power base at the University of Toronto, his close association with Canada's medical elite, and his excellent relationship with General McNaughton, it is not surprising that he became the ACMR's first chairman, a job he accepted with enthusiasm and vigour.[101] Banting's most ambitious undertaking was an extensive survey of Canada's medical research facilities, in which he visited medical schools, major hospitals, and provincial public health laboratories, and interviewed some 300 medical researchers.[102] Banting was also in contact with the major British and American medical organizations, which, by 1938, were increasingly interested in war-related research.

Canadian Scientists and the Social Issues of the 1930s

Canadian scientists, like other social groups, were strongly affected by the many changes that occurred during the 1930s. They were forced to confront disturbing political and social problems: the devastating impact of the Great Depression could not be ignored; nor could the rise of fascism and the possibilities of another world war. Unfortunately, unlike the situation in Britain and the United States, debates about scientific activism and the social function of science were not institutionalized in Canada. There was no Canadian equivalent in these years to the British Association of Scientific Workers or its U.S. counterpart. As a result, it is necessary to examine a wide range of scientific and political organizations that discussed these issues, in order to assess the extent to which Canada's scientific community, on the eve of war, had been affected by the major ideological debates of the 1930s, and how it responded to the threat of fascism.

Whether the great scientific and technological advances of the twenties had created massive structural unemployment was a subject of much debate.[103] Indeed, many academic and industrial scientists were themselves facing unemployment or underemployment. To deal with these issues U.S. scientists formed organizations such as the Committee on Unemployment and Relief for Chemists and Chemical Engineers and the Professional Engineers' Committee on Unemployment. In 1936, more politically minded scientists established the American Association of Scientific Workers (AAScW). The AAScW was modelled on the British Association of Scientific Workers (BAScW), which had been founded in September 1918 by a group of elite British scientists who were concerned about the social status and economic well-being of the country's scientific community. They claimed that scientific workers did not 'exercise in the political and industrial world an influence commensurate with their importance.'[104]

For BAScW militants such as J.D. Bernal,[105] Hyman Levy, J.B.S. Haldane, Joseph Needham, Lancelot Hogben, and Patrick Blackett,[106] commitment to the social function of science also meant an acceptance of socialist goals and policies.[107] The British scientific Left viewed the Soviet Union as a model of how science and Marxism could converge and provide the necessary institutional structures and encouragement for a dynamic and caring society. Such views intensified in the 1930s, particularly after the 1931 visit of the Soviet scientific delegation, under Nikolai Bukharin, to the London meetings of the International Congress of the History of Science and Technology.[108] No sooner had the Congress ended when a group of twenty prominent British scientists – including J.D. Bernal, J.B.S. Haldane, John Cockcroft, and Julian Huxley – accepted an invitation to visit the Soviet Union that fall. All were impressed by what they saw and by how they were treated. Julian Huxley captured their enthusiasm in his (1932) personal account *A Scientist among the Soviets*.[109] Bernal went even further in his praise of Soviet society, both in his writings at the time, and in his seminal work *The Social Function of Science*, published in 1939. 'It is to Marxism,' he wrote, 'that we owe the consciousness of the hitherto analyzed driving force of scientific advance, and it will be through the practical achievements of Marxism that this consciousness can become embodied in the organization of science for the benefit of mankind.'[110]

Many North American scientists shared Bernal's and Huxley's positive views of the Soviet Union, especially those who attended the International Congress of Physiology held in Moscow in 1935. They were particularly inspired when the eighty-six-year-old Ivan Pavlov officially welcomed them as scientific comrades for 'the rational and final unity of mankind.'[111] According to A.V. Hill, one of Britain's most eminent physiologists, 'We came away filled with affection and regard for our Russian colleagues and ... moved by the ardour and enthusiasm of the army of young scientific workers.'[112]

Frederick Banting, another delegate at the Congress, was also impressed by the achievements of Soviet science, and by the quality of Pavlov's laboratory and residence: 'It is a great tribute to science,' he wrote, 'that such honour should be paid by a government which the world considers so harsh and inhuman.'[113] Like A.V. Hill, Banting was profoundly moved when briefed about the great advances in Soviet medicine: 'Infant mortality has been cut in half. The general death rate has been reduced one third. Research institutions are attacking tuberculosis, cancer, malaria ... social insurance, sick benefits, provision for maternal aid, rest homes and sanatoriums all aim at the health of the worker.' Banting speculated that if he were younger, 'it would be one of the highest and finest things in life to be permitted to take some part in this glorious, changing scene.'[114] On the other hand, he was sufficiently discerning to notice 'the abject stupidity of some of the minor officials,' and the 'peculiar psychology' of Russian leaders, as well as the tendency of Soviet scientists to 'work too much and think too little.'[115]

Norman Bethune, the controversial Montreal surgeon, also attended the 1935 Moscow Congress and was deeply impressed with the Soviet system of hospitalization, welfare, and social medicine.[116] Bacteriologist Guilford Reed of Queen's University found much to admire in Soviet medicine,[117] writing a glowing account of his visit in *Queen's Quarterly*. In his article Reed reserved special praise for the Soviets' provision of 'preventive and medical services alike for all' and the emphasis on social values in a country where the doctor 'was rated as one of the most important members of society ... [since] his professional work demanded intelligence, training and skill of the highest order.'[118]

By 1935 this positive image of the USSR was further enhanced when that country supported the Spanish Republican government against Franco militarism and fascist intervention. In fact, until the Nazi–Soviet Pact of August 1939, most Canadian 'progressives' held a favourable view of the Soviet Union, especially those who were involved in Popular Front organizations such as the Canadian League against War and Fascism (changed to the League for Peace and Democracy, in 1937) or the Committee to Aid Spanish Democracy.[119] The Canadian Communist party also gained new adherents because of its involvement with the Mackenzie-Papineau Battalion, which fought alongside antifascist volunteers to defend Spanish democracy, and by its identification with Norman Bethune, whose outstanding blood transfusion work on behalf of the antifascist forces attracted international praise.[120] In contrast, many Canadian intellectuals and students were disillusioned not only by the failure of the King government to assist Spanish democracy, but also by its harassment through the punitive Foreign Enlistment Act,[121] of those who served in Spain.

Canadian civil liberty organizations were even more critical of the government of Maurice Duplessis for its support of the Spanish Nationalists, and for its iniquitous Padlock Law, which curtailed freedom of expression and association. McGill students were particularly affected by these controversies either through a direct association with the Communist party, or through popular front organizations. McGill chemist Jack Edward, reflecting on his own student experiences during the late 1930s, captures a sense of this period:

> You had to have been a leftist in the thirties to know how powerful the feeling was that Marxism was the key which somehow unlocks this tangle of the world ... Being a Communist then was ... enormously attractive both for its intellectual pretensions – because it provided one with a ready made philosophy by which you could judge everything, including movie pictures, art, architecture – the good guys were clearly distinguished, and also because of the human fellowship which one gets in any religious community.[122]

With the advent of the Cold War, much would be made of the pro-communist inclinations of some McGill students of this era, particularly those named in the 1946 Report of the Royal Commission on Espionage: Eric Adams (B.Sc. 1929), Frank Chubb (B.Sc. 1935), John (Jack) Edward, Harold Gersen (M.Sc. 1929), Matt Nightingale (B.Sc. 1928), David Shugar (Ph.D. 1940), and Raymond Boyer (Ph.D. 1930).[123]

The activities of Raymond Boyer during the late 1930s are particularly intriguing, given his social background and his prominent role as one of Montreal's most outspoken scientific 'progressives.' The son of wealthy and influential French-Canadian parents, the 'tall, dark, charming and good looking' Boyer had given little attention to social issues as an undergraduate student, concentrating instead on being a playboy. After graduation, while his McGill classmates scrambled for jobs, Boyer was able to leisurely continue his studies at universities in Boston, Vienna, and Paris.[124] However, with the triumph of National Socialism in Germany, Boyer became deeply disturbed by the spread of the fascist threat in Europe and the outbreak of the Spanish Civil War. He was also shocked, on his return to Montreal in 1937, with the growth of right-wing extremism and anti-Semitism in Quebec.[125] He soon became an active member of the Canadian Committee to Aid Spanish Democracy, the Canadian League for Democracy, and the

Canadian Civil Liberties Union, and an outspoken critic of French-Canadian nationalists. They, in turn, denounced him as a traitor and social deviant who 'should be burned at the stake.'[126] In September 1939 Boyer was among the first scientists to offer his services to the Canadian war effort.

Scientists Prepare for War, 1936–1939

Japan's 1931 assault on Manchuria, Italy's 1935 attack on Ethiopia, and fascist involvement in the Spanish Civil War provided growing evidence that the world was speeding toward another world war. As a result, by 1936 many antifascist organizations were prepared to discard their previous commitment to pacifism and accept the need for collective security military measures as a means of national self-defence. In Britain, this transition was evident on many fronts. But of particular importance to this study was the growing enthusiasm the BAScW adopted toward their country's rearmament policies during the late 1930s, including an acceptance that Britain's scientific resources would have to be mobilized behind the war effort. The June 1936 edition of *The Scientific Worker* summed up how the BAScW felt about the potential destructiveness of modern warfare:

Bravery, physical fitness, unselfishness and nobility of character are of no avail against the methods of large scale mechanical slaughter developed by engineers, of burning and maiming developed by physicists, and of asphyxiation and blinding developed by the chemists. Soon the bacteriologists may surpass the other scientific workers and produce methods of 'destroying the morale of the civilian populations.'[127]

Further deterioration in international relations, especially growing evidence of Germany's hostile intentions, led BAScW members at Cambridge, under the direction of J.B.S. Haldane, to publish, in 1937, an indictment of the inadequacy of British civil defence, and of the Baldwin government's duplicity in concealing this situation from the British public.

The Munich crisis of October–November 1938 further mobilized the BAScW, to consider how British scientific talent could best be utilized in case of war. Although some members opposed direct involvement in the development of weapons, most followed the lead of BAScW chairman W.L. Bragg, who argued that the British armed forces 'should be given the opportunity of direct contact with university laboratories.'[128] In February 1939 the BAScW Council, in an official statement, agreed to allow its 1600 members, in the event of war, to contribute their scientific expertise to the British war effort. It stated:

While we regard war as the supreme perversion of science, we regard anti-democratic movements as a threat to the very existence of science. Hence we are prepared to organize for defence in a democratic cause and would actively resist any attempts to introduce fascism or any other anti-democratic system into this country either from inside or outside.

The BASW stipulated, however, that the British government reorganize the structures and procedures of British defence science, recognizing that the organization and control of scientific work 'should be in the hands of scientists.' The BAScW concluded 'that the mistakes of the last war shall not be repeated. Our best brains must not be wasted in the trenches.'[129]

Similar debates about the wartime role of scientists were occurring in the United States, especially after the December 1938 formation of the American Association of Scientific Workers (AAScW). One of the most active branches was in Cambridge, Massachusetts, where scientists from Harvard and MIT predominated,[130] including the Toronto-born mathematician Israel Halperin and his brother-in law, physicist Wendell Furry.[131] Not surprisingly, the AAScW shared many of the principles and policies of the BAScW, namely, to encourage a wider application of science and the scientific method for the welfare of society, to combat fascist misuse of science, and to assist Jewish scientific refugees.

The AAScW was not the only U.S. scientific organization to become involved with social issues. The prestigious American Association for the Advancement of Science (AAAS)[132] also made a commitment to examine 'the profound effects of science on society,' and invited 'all other scientific organizations ... to cooperate, not only in advancing the interests of science, but also in promoting peace among nations and intellectual freedom in order that science may continue to advance and spread more abundantly its benefits to all mankind.'[133] As part of this campaign, the AAAS leadership stressed the importance of establishing closer relations with the British Association of Science, including reciprocal membership.[134]

In 1938 the AAAS also made an overture to Canadian scientists[135] when, for the fifth time, the organization held its annual meeting north

Organizing for War, 1938–1940   39

of the forty-ninth parallel.[136] According to its organizers the conference was a great success: 'Science was not Canadian or United States; it was just science ... and it was conducted in the spirit of perfect harmony and good fellowship.'[137] Between 27 June and 2 July, delegates heard papers from some of Canada's best scientists.[138] But the highlight was a series of medical biochemistry papers from the laboratories of Banting, Best, and Collip.[139]

At another memorable session, NRC scientist F.E. Lathe and Principal R.C. Wallace of Queen's University presented papers in the special series on 'Science and the Future,' along with Nobel laureates chemist Harold Urey and physicist Arthur Compton.[140] But it was the grim message of the latter two U.S. scientists that gripped the audience. Urey, for example, noted how modern science had intensified the savagery of war: 'In the efficient destructive machines which are in use to-day the most important agent is a chemical substance, an explosive, an incendiary mixture or a poison gas ... war chemistry will ... destroy people, their material possessions, and will dissipate and destroy the resources of the earth.'[141] While Compton was somewhat less pessimistic in tone, he too stressed the fact that the world was at a crossroads: 'The growth of physics is ... intimately bound to the future of civilization ... the great power given to man by his new knowledge of the world may be used either to his good or to his harm ... it may become terribly destructive.'[142] Ironically, within a year, Compton and Urey would themselves be carrying out experiments to confirm Otto Hahn's discovery of the artificial fission of uranium 235 – the basis for the atomic bomb. They would also be key members of the U.S. National Defense Research Committee, created in June 1940 under the chairmanship of Vannevar Bush.[143]

Conclusion

In 1938, while American scientists were deeply divided over whether they should become involved in military research,[144] most Canadian and British scientists were now convinced that war was imminent, and that they must offer their services to the armed forces. This commitment became a necessity during the spring of 1940, when Nazi Germany, triumphant over its continental enemies, was poised to invade Britain.

With the enactment of the National Resources Mobilization Act in June 1940, Canadian scientists, after years of isolation from military concerns, were now expected to commit their professional services to the state in the struggle against fascism. The National Research Council was given the task of organizing Canada's defence science, and of creating sophisticated weapons for the Canadian and British Armed Forces. Fortunately, the country's scientific resources had greatly expanded and improved since the Great War, and the NRC was able to recruit experts, primarily from the universities, to participate in the development of radar, new explosives, proximity fuses, as well as chemical and biological weapons. However, these were uncharted waters for an organization that had previously concentrated almost exclusively on civilian scientific projects, despite the efforts of General McNaughton during the late 1930s to establish closer links with the Canadian Armed Forces. C.J. Mackenzie expressed his own anxiety, as Acting Head of the Council, in a June 1941 letter to General McNaughton: 'We all feel keenly that unless our endeavours produce equipment and findings of use to the man in the field we will not be achieving our fundamental purpose, for we all realize that in modern warfare based on science, technology and mass production there are three stages – research and development, production and use in the field.'[145]

2

Building the Defence Science Alliance, 1940–1943

> Medical Science & Medical research in Canada enjoys the reciprocation of ideas & research results of England & the United States. These two countries stand preeminently above others in their contributions to the advancement of scientific knowledge during the last two decades. The agencies that are responsible for the application of the newer knowledge to the problems of warfare are alike & awake.
>
> Frederick Banting, Wartime Diary, 11 August 1940

Defence science cooperation within the North Atlantic triangle began well before the United States formally became a belligerent. The primary motivation for such cooperation was fear of Nazi Germany. In June 1940 the possibility of military defeat forced the British government to seek assistance from the United States and to call on its major military ally, Canada, to make every possible sacrifice to save the mother country. The scope and character of defence science cooperation among the three countries would, however, change dramatically during the next two years as the United States gradually became the dominant partner both militarily and industrially and in the creation of major weapons systems.

Formalized British and American military cooperation began with the 1940 bases-for-destroyers deal and the creation, in March 1941, of the ABC-1 defence arrangement. Meanwhile, Canada and the United States, through the Ogdensburg Agreement of August 1940, had established the Permanent Joint Board on Defence with a mission to cooperate 'in the broad sense [on] the defence of the north half of the Western Hemisphere.' Eight months later, the Hyde Park Declaration created a common North American market for defence production.[1]

These developments had an enormous impact on the organization and performance of Canadian defence science. During the first year of the war, the National Research Council (NRC) and the Department of National Defence (DND) had been granted limited access to top-secret information about British weapons, and had virtually no entry into the world of American military technology. All this changed after the Tizard and Conant scientific missions of 1940–1, when the British connection made it possible for Canadian scientists to become involved in the most sophisticated of American military projects. The fact that the United States was a nonbelligerent until December 1941 also worked to Canadian scientists' advantage, since information about various British and American weapons was often channelled through Canadian agencies.[2] As C.J. Mackenzie was later to admit, 'if America had declared war in September, 1939, instead of twenty eight months later, the status of Canadian science would have been quite different from what it is today.'[3]

In the early stages of the war, cooperation between the National Research Council and the three Armed Forces evolved through a series of ad hoc arrangements. By 1940, however, as the demands for more advanced weapons grew, direct institutional connections were established. These varied according to the particular requirements of each Service and the extent to which the Army, Air Force, and Navy hierarchies had confidence in the research and development capabilities of the NRC. In dealing with the military, C.J. Mackenzie had a number of assets: he had the steadfast support of General A.G.L. McNaughton, who was the country's most prestigious war hero, and he enjoyed close personal contacts with powerful Cabinet ministers such as C.D. Howe, Minister of Munitions and Supply, and C.G. (Chubby) Power, Minister of Defence for Air.[4] While Mackenzie had some rocky moments, particularly in his negotiations with the Royal Canadian Navy, he persevered.[5]

Mobilizing Canadian Universities for War, 1939–1940

During the late 1930s General McNaughton had tried to ensure that the NRC maintained close contact with Canadian university scientists through its postgraduate scholarship bursaries for 'brilliant students ... in certain fields of science.'[6] With the outbreak of war, this meant military science. But McNaughton did not continue in this coordinating role – in September 1939 he was appointed Commander of the First Division, and remained overseas until 1943. His legacy, however, continued to influence the NRC throughout the war years.

To ensure that the country's university presidents understood their responsibilities to Canada's war effort, and the impact war would have on higher education, McNaughton issued a circular letter on 16 September stressing that short-term military manpower requirements should not take precedence over the training and utilization of university science and engineering students:

Owing to the possibility of the present war extending over a very long period and the need for ... large numbers of well trained men in all branches of pure and applied science, including medicine, dentistry and agriculture ... Students now pursuing successfully in these fields will serve their country in a most valuable way by continuing their university training until graduation ... specially able students should be encouraged to continue their studies in all branches of science, especially along the lines required to meet national requirements as they develop.[7]

McNaughton had high expectations for Canadian universities, not only to undertake important military research and development, but also to provide the Armed Forces and vital war industries with the necessary specialists. The country's university presidents shared his views. In October 1939 R.C. Wallace of Queen's University, chairman of the National Conference of Universities (NCU), announced that all of his members were mobilizing 'their scientific forces and gathering information on personnel experience and other pertinent data that would be useful.'[8]

Another of McNaughton's immediate goals was to have C.J. Mackenzie, his hand-picked successor, appointed acting president for the duration of the war. It was an excellent choice, although, according to McNaughton, Mackenzie King was not initially enthused when Mackenzie's name was mentioned: 'He looked at me, knowing that my Saskatchewan antecedents were Conservative, and he said: "Oh, but he's a Conservative, too!" And I looked at Mr. King in reply, and said: "Mr. King, does that really make any difference?" And he stopped; and he said: "No Andy, it does not make any difference. It is approved and accepted."' Mackenzie, or C.J. as he was commonly known, had many qualities that made him an outstanding wartime science administrator:

He was exceptionally successful in his staff relationships, so that he obtained and developed a high degree of cooperation and affection from hundreds of colleagues throughout the war and afterwards, in Canada, the United States and the United Kingdom. He was a prodigious worker himself and he inspired a similar level of industry in others. He had an exceptionally clear mind on the essentials of administrative structure, and he was a keen judge of human capacity. He was eminently practical and yet a visionary too.[9]

In trying to meet the multitude of wartime obligations, Mackenzie had a number of advantages. By 1939 the NRC had well-equipped laboratories in Ottawa with a professional staff of 300 in the four research divisions, who were, according to Mackenzie, 'of high quality and of the proper active age group admirably suited for the demands of war research.'[10] Another asset was the existence of the various associate committees, which brought scientific, industrial, and military experts together to assist the NRC's regular staff, as circumstances required.[11] In addition, a program of extramural grants provided university scientists with thousands of dollars for war-related research, and a system of intermural projects helped graduate students at Canadian universities obtain military deferments and graduate degrees for their war-related research. Otto Maass, who designed the scheme, described its merits in a 1943 letter to C.J. Mackenzie:

I am only qualified to state my views in this connection as far as students in chemistry are concerned, although I have ample evidence that this is the case as far as students in other branches of Science, including that of Physics at Toronto University is concerned. By giving the professors at universities post-graduate students to carry out useful war research these men have been ready to stay at universities and carry on at the same time some of their teaching duties ... feeling they were useful in the war effort. As Director of Chemical Warfare and Smoke I can state that results of great importance have been obtained as a result of this extra-mural research carried out by post graduate students under the direction of the university staff. You are familiar with some of the results and for reasons of security I will not reiterate them in this letter which you may wish to pass on to others.[12]

Not surprisingly, given their size and level of expertise, McGill and the University of Toronto were most involved with NRC-directed war projects. But it was McGill that first responded to McNaughton's 16 September appeal when it created a War Advisory Board under the direction of C.F. Martin, Dean Emeritus of the Medical Faculty, and physicist David Keys.[13] One of the board's first initiatives was to prepare a comprehensive inventory of the university's manpower resources by circulating a questionnaire. The response rate was tremendous. Almost every faculty member, with the notable exception of economist Eugene Forsey, responded with an enthusiastic commitment to the war effort.[14] Chemists and physicists were particularly forthcoming, and outlined their specific skills as follows:

W.H. Barnes: X-ray diffraction studies

J.S. Foster: Nuclear physics, radiation

D.K. Fromm: Optical and electrical methods of signalling and detection

W.H. Hatcher: Organic chemistry, munitions

H. Hibbert: Cellulose chemist, expert in the manufacture of nitrocellulose and its use in explosives

Otto Maass: Physical chemistry, service in the direction and organization of war research work

R.V.V. Nicholls: Organic chemistry, synthetic resins and plastics

Norman Shaw: Thermodynamics and molecular physics, devising special physical tests for the standardization of production and quality of materials

David Shugar: 'Research work of a physical nature on problems of military and economic importance.'

W.H. Watson: Experimental and mathematical physics, artillery, ballistics.[15]

While all of the above were willing to serve, some felt that they had special qualifications. Maass and Hibbert both believed that they should assume important leadership positions – Maass because of his status in Canadian chemistry and his many powerful friends in Ottawa, and Hibbert because of his First World War experience as a scientific liaison officer between the British War Mission in Washington and the Gas Warfare Service in Edgewood Arsenal, Maryland.[16] In the end, Maass was given the critical wartime administrative positions: coordinator of Canada's explosives program, director of the chemical and biological warfare operation, and special assistant to NRC president C.J. Mackenzie.

Maass outlined how the McGill chemistry department with its 'sixty trained chemists' were prepared to make an immediate contribution to Canada's war effort: 'We have developed ... what might be called an artificial lung ... which will make it possible to test the efficacy of gas masks ... The possibility of dealing with poisonous liquid smokes (such as mustard gas) is being undertaken.' Maass also indicated that his team had developed ways of 'vaporizing materials such as diphenylchlorasine when fired from shells,' as well as providing 'improved methods for the purification of organic intermediaries required in the production of explosives of various kinds.'[17]

McGill's medical faculty were also prepared to contribute, as were the psychologists, engineers, lawyers, nurses, historians, linguists, economists, and all the other faculty. But one of the most challenging proposal, came from physicist J.S. Foster, who predicted that if McGill's administration were to proceed with his cyclotron project the university would be in a position to transform the war effort by producing artificially radioactive substances.[18] Significantly, he did not mention the possibility of creating either uranium 235 or plutonium, a subject of much interest to British, French, German, and American atomic scientists in the fall of 1939.[19]

University of Toronto scientists also responded to the patriotic bugle call. They decided, however, not to duplicate the McGill inventory model.[20] Instead, plans were made to establish a Scientific War Service with the following mandate:

1. To assist the Department of National Defence, other government departments and industry to solve problems which arise from war conditions.
2. To assist the acceleration of war-time production by aiding in the improvement of methods and equipment.
3. To give advice relating to the modification of specifications for materials or processes used in production.
4. To assist in the selection and training of scientific personnel.[21]

At Toronto, individual scientists were expected to offer their services to either the NRC or the Armed Forces, and many seized this initiative. Psychologist E.A. Bott, for example, convinced DND officials of the importance of having effective IQ and aptitude screening tests for service personnel.[22] Physicist Eli Burton created his own Sub-Committee on Physics Problems, which concentrated on 'the development, production and use of R.D.F. apparatus for the various branches of war service,' as well as sending 'suitably qualified men for R.D.F. service in Great Britain.'[23] Toronto chemists were also busy, especially in explosive and chemical warfare research. One of the most exciting projects – improving the process for manufacturing RDX – was carried out by George Wright's research team. In October 1940 Wright's colleague A.R. Gordon predicted that 'this explosive (an extremely powerful one) can be used much more extensively with possibly far reaching consequences.'[24]

The War Technical and Scientific Development Committee

With the fall of France, and the impending invasion of Britain, Canadians of all classes rallied behind their country's war effort. One manifestation of this sense of solidarity and commitment was the large number of offers and gifts of money that poured into Ottawa. In July 1940 the King government decided to channel some of these donations into scientific warfare research because of the timely intervention of C.J. Mackenzie and Frederick Banting, who had convinced J.S. Duncan, Acting Deputy Minister for Air, of the vital importance of funding Canadian defence science.[25] On 27 August the new organization, called the War Technical and Scientific Development Committee (WTSDC), was established by order-in-council PC 4260, with provision for an executive body, composed of three scientists (Mackenzie, Banting, Maass), five bureaucrats, and three businessmen.[26] It was subsequently determined that the donations, which totalled over $1.3 million, would be allocated to the NRC on the basis of specific weapons projects, but only after review by the WTSDC.

Most of the WTSDC's twenty-one meetings were dominated by C.J. Mackenzie, who outlined the merits of each project and how it related to the secret scientific work of Britain and the United States. In this endeavour, he was assisted by Maass, who provided more details about work being carried out in university and industrial laboratories, and by Banting's reports of the various medical research projects. By the end of November 1940, WTSDC commitments were close to $900,000,[27] with another $102,000 allocated to an emergency fund that Mackenzie could draw on to fund secret and high-priority undertakings.[28] The major projects are outlined in Table 1.

TABLE 1 WTSDC Funding for 1940–1941[29]

I Government Laboratories	
Project	Amount
Radar	$ 350,000
Aeronautical Engineering	128,784
Ballistics	46,000
Special Aeronautical	40,000
Explosives Testing Laboratory	40,000
Radar, 10 cm.	40,000
Optics	35,000
Naval Services	31,500
Special Aviation Medicine	25,000
Medical re Chemical Warfare	25,000
Tailless Aircraft	25,000
Experimental Flight Machines	25,000
Proximity Fuse Research	20,000
Chemical Warfare Testing	15,000
Examination Unit	15,000
Pilot Plant for butylene glycol	15,000
II University Laboratories	
Project	Amount
Chemical Researchers (chemical warfare; explosives) 34 researchers	$ 119,000
Physics and Engineering 7 researchers	60,000
Biological 1 researcher	2,000
Medical Research 4 researchers	21,743
III Emergency Fund	102,000
IV Administration	35,000[29]

The creation of the WTSDC 'Santa Claus' fund also gave Mackenzie greater clout in dealing with Canadian universities, an advantage he soon exploited. On 20 August 1940 he sent a circular letter to all Canadian universities describing both the work of the committee, and the guidelines that had been worked out as a result of consultations between the NRC, the DND, and university presidents 'to determine what needs of these Services could be met by Canada's scientific and educational institutions.' According to Mackenzie there was a consensus on the following points:

(1) The training necessary for men who would follow trades in the armed forces should normally be done in technical schools rather than in universities ...

(2) It was observed that in Canada no university offers a course in which radio engineering is the primary subject ... At present there is definite shortage of men with training in this field.

(3) In view of the importance in the war effort of men trained in radio, the National Research Council was asked to obtain, if possible, from the officials of the Canadian universities, the names of their graduates who are now engaged in radio work in the United States in order that these men may be approached to serve at home or abroad if necessary.

(4) It was agreed the training of graduate students in science should be continued, not only because there is a need now for demonstrators and research workers, but because the disruption of university work in Great Britain reduces the possibility of obtaining trained men from there and makes necessary the development of a substitute supply elsewhere in the empire.[30]

Although some university presidents complained that their scientists were often working on NRC projects 'without the knowledge of the head of the university by whom the professors were employed,' in general, the role of the NRC in coordinating and funding this research was applauded.[31] By the end of 1940 Mackenzie King was praising Canadian universities for their 'sympathetic and highly patriotic response' to Canada's war effort.[32]

Another project that brought university and NRC scientists together was the screening of applications to the Inventions Board, which had been established shortly after the outbreak of war. The Board's executive committee was composed of C.J. Mackenzie (chair) and the deputy ministers of three Services. Its mandate was to assess the merits of specific proposals that had previously been reviewed by panels of specialists from science and engineering.[33] By November 1941, over 7000 submissions had been examined, but only a small number were regarded as having wartime relevance. The reasons for this high rejection rate were described by NRC scientist S.J. Cook in a November 1942 memorandum to Colonel DesRosiers, Deputy Minister of Defence:[34]

Consideration of war inventions has its lighter moments. I recall that one man had a scheme to freeze the clouds and mount guns on them, but the Board was not very optimistic. Then another did not see why ships which are needed to carry cargoes to Great Britain should be used to transport troops. He thought it would be simpler to lay a pontoon bridge from Newfoundland to England and let the troops march across. That idea has not yet been accepted either. Quite a few still have the idea that aeroplanes could be flown with sails instead of being powered with engines. Perhaps one of the most ingenious and naïve suggestions put forward was for a gun that would shoot in two directions, either forward or backward.[35]

Mackenzie shared Cook's scepticism about the value of utilizing 'inventors,' and he generally regarded involvement with the Board as a 'troublesome' experience. His frustration was intensified when he had to deal with complaints from research officers of the Canadian Navy that they were not getting their share of proposals.[36]

Another ongoing problem was the difficulty in obtaining sufficient numbers of qualified and willing French-Canadian scientific and engineering panel members who could assess submissions from Quebec.[37] That is not to say that no important defence science research was being carried out in the francophone universities. At the University of Montreal, for example, the chemistry department used the wartime crisis to demand more money and space for research, while the École Polytechnique offered radio courses and conducted studies on the resistance of airplane wings to fire.[38] Many Laval scientists were also involved with wartime research.[39] Clearly, the old myth about the lack of French-Canadian scientific participation in the war effort was untrue.[40]

Canada-U.S. Defence Arrangements, 1940-1942

Between 1938 and 1942, the relationship between Canada and the United States was transformed from one of friendly isolationism to one of committed military and economic collaboration. The Americans realized that its northern approaches were increasingly vulnerable, particularly with the development of long-range bombers. These concerns were intensified after the British Commonwealth went to war. In addition to its strategic concerns regarding the protection of U.S. coastal installations, Washington believed that the Canadian government, in its attempts to assist Great Britain in the war against Germany, had stripped itself 'of war materials and personnel to such an extent that it finds itself in a defenceless position.' This image of Canadian weakness was reinforced in July 1940 when Ottawa pleaded with Washington to provide '200,000 rifles and ... machine guns and field pieces' for home defence.[41] And as the military situation in Europe worsened, Canada's search for additional U.S. security guarantees became more pronounced. In August 1940 the American ambassador Pierrepont Moffat reported that 'even elements which in the past have been least well disposed towards us, such as the Toronto public and the English-speaking sections of Montreal, are now out-spoken in [our] favour ... and the old fear that cooperation with the United States would tend to weaken Canada's ties with Great Britain has almost disappeared.' 'Instead,' he continued, 'Canada believes that such cooperation would tend to bring Britain and the United States closer together.'[42] On 17 August President Roosevelt and Prime Minister Mackenzie King signed the Ogdensburg Agreement, which established the Permanent Joint Board on Defence (PJBD) 'to advise on immediate needs and to constitute the permanent advisory implement for planning the defence of both countries.'[43]

Moffat's prediction that a Canada–U.S. defence alliance would be supported by all major political parties and the leading newspapers was soon vindicated.[44] He was also pleased by reports that General McNaughton had described the agreement as 'the most heartening thing that has happened in decades ... to find two nations which for a hundred years and more have settled their differences by arbitration, taking an equally practical point of view about defence.'[45] Moffat's high regard for Canadian generals did not, however, extend to Harry Crerar, who had recommended 'converting the Permanent Joint Defence Board into a triple British-Canadian-American Board, or if this were not possible to invite British representatives to sit in at some of their meetings.' Indeed, news of Crerar's suggestion sent a shudder through the U.S. State Department, and Moffat was instructed to use all his resources to quash any such initiative since it 'would virtually destroy the premise on which we have thus far worked in the Joint Board.'[46]

State Department concern over British involvement with the PJBD remained acute. In February 1941, for example, Undersecretary of State Sumner Welles prepared an assessment of future Anglo-American-Canadian military relations. On the positive side, Welles acknowledged that 'Canadian defence comes so close to our own that we have to consider Canadian needs as though they were to a considerable extent needs of the American armed forces.' This did not mean, he added, that either country should modify its defence priorities to meet British needs. He felt that Ottawa was gradually coming to this same conclusion:

Behind this is a background of steadily growing distrust between the Canadian and British officials. Bluntly, the Canadians believe that the British have stripped Canadian defence in favour of their own. They likewise feel that the British Purchasing Mission not only does not transmit Canadian representations adequately, but occasionally suppresses them in favour of British requests.[47]

Significantly, Welles's assessment occurred just before the Hyde Park Agreement of April 1941, which established a more effective and equitable distribution of war contracts, as well as the flow of essential raw materials and technology among the three allies. Ambassador Moffat was, in fact, quick to note that the agreement 'met with unanimous approval' in Canada, largely because it guaranteed that North American resources would be mobilized 'for the most efficient and expeditious aid to the British Commonwealth ... [while] the hemispheric aspect of the declaration is almost completely ignored.' He was confident, however, that, with time, the Canadian public would appreciate 'the step which has been taken in the direction of Canadian–American economic solidarity.'[48]

Building the System: The Tizard Mission of 1940

The most important catalyst in the emergence of scientific military cooperation between Canada, Britain, and the United States was the Tizard mission of 1940. This imaginative venture has been the focus of many studies, the most recent being David Zimmerman's *Top Secret Exchange: The Tizard Mission and the Scientific War*.

The decision to send a scientific mission to North America during the summer of 1940 had a long and complicated background. Until the spring 1940 British military and political officials were only marginally interested in American military resources, despite the efforts of elite scientists such as Henry Tizard and A.V. Hill, who both had a wide range of professional and personal connections in North America. Both men were also early advocates of British military preparedness, and had been key members of the 1935 Air Ministry's Committee for the Scientific Survey of Air Defence.[49] During the Munich crisis, they had taken the lead in mobilizing Britain's scientific establishment behind the government's rearmament program, most notably by preparing a register of scientists 'to ensure that scientific and technical manpower was properly employed in the war that loomed ahead.'[50] Equally important was their ability to convince Air Marshal Sir Cyril Newall that Hill should be sent to Washington as a special adviser to the RAF Attaché, Air Vice-Marshal Pirie, on the grounds that Britain desperately needed U.S. assistance and 'we need not be diffident about asking for it.'[51]

Hill's arrival in Washington was well received by the American scientific elite,[52] and he received many invitations to visit university and industrial laboratories.[53] In his campaign to exploit American scientific resources, Hill was also able to count on the support of the British ambassador Lord Lothian, who on 23 April issued a personal appeal to the British Foreign Office for a sweeping scientific exchange system with the United States: 'Britain's present bargaining power is strong owing to earlier start and service use. If we wait the U.S.A. will probably discover the essentials for themselves.'[54] This line of argument was consistent with the Hill–Tizard position that while American 'war machines' were still inferior to those of the British,[55] the United States had enormous capability for mass production given their powerful industrial sector and 'by having more energetic and scientifically minded engineers than we have.'[56]

Another aspect of Hill's North American mission were his two trips to Ottawa to consult with Canadian defence scientists.[57] The first, a five-day outing in the early part of May 1940, brought him into contact with his old friend Charles Best, with whom he discussed problems of aviation medicine.[58] Once in Ottawa, Hill conferred with C.J. Mackenzie and Otto Maass, who explained the NRC's work in radar, explosives, and chemical warfare. Hill decided to test Mackenzie's reaction to his plan for the creation of a permanent British scientific liaison office in Ottawa,[59] and was relieved when the NRC president strongly endorsed the proposal.[60] As a result, on 28 May Hill officially recommended:

1. That the British Government should ask the Canadian Government to undertake the development of certain applications of R.D.F.
2. That two experts in the work should be sent to Canada forthwith to help to start the Canadians off and give them full information.
3. That the Canadian Prime Minister should be invited to approach the President of the United States with the suggestion that, in return for full information from us, the Canadian Government should be permitted to employ the

full resources for research and development of American Companies known to be engaged in U.S. Government work on these lines.[61]

In his report, Hill also made much of Canada's 'special' opportunities for 'obtaining scientific and technical help in the United States.' This, he argued, was largely possible because 'Canadian scientists and engineers are in intimate touch with American colleagues, and many of the great companies in the U.S. have close connexions with Canadian companies,' a situation that Hill believed had already assisted the dominion's war effort.[62]

If Hill expected a sudden change in Britain's protective stance toward its defence secrets, he was soon disappointed. Secretary of State for Air, Sir Archibald Sinclair, who had originally authorized his mission, now appeared lukewarm about the proposed exchange system, while the War Cabinet decided that there would have to be 'a specific report on each list of items before it is offered for exchange.'[63] Matters improved, however, when Tizard called Sinclair's bluff and threatened to resign unless the Air Ministry adopted more imaginative policies toward exchanging defence secrets with Canada and the United States. Sinclair's change of heart was reflected in his July announcement that the distinguished nuclear physicist R.H. Fowler would be sent to Ottawa as the chief scientific representative of the Air Ministry.[64] Similar concessions were soon forthcoming from other British defence agencies. In contrast to his previously secretive approach, D.R. Pye, director of research for the Ministry of Aircraft Production, now made a point of inviting NRC scientists to Britain in order 'to acquaint themselves with the work going on at the N.P.L. and other experimental establishments.'[65] In addition, the interdepartmental planning meeting of 9 July agreed that Canadian representation on a technical mission to the United States 'would be welcome.'[66]

At this juncture, Tizard and his associates were receiving valuable information on American military and political developments from another source: Brigadier Charles L. Lindemann, Air Attaché of the British Air Mission in Washington, and brother of Churchill's scientific adviser. Prior to the German's victory over France, Lindemann had performed a critical role in coordinating British and French scientific and military cooperation in Paris. He had then been reassigned to Washington, where he was able to keep Tizard informed about how different American groups were responding to the war in Europe and to Britain's plight, in particular.[67] While Lindemann, like Tizard, appreciated the long-term capabilities of the United States to wage war, he was not impressed by its state of military readiness in the spring of 1940. 'At the present moment,' he wrote, 'I am told that the U.S. Army and Air Force is equal to that of Bulgaria and possibly a little inferior.'[68]

During the summer of 1940 Brigadier Lindemann received numerous requests from the British Armed Forces requesting information about sophisticated American weapons. In the case of the Air Ministry, the highest priority was the Norden bomb sight, a device reputedly superior to those used by the RAF, but which was, by presidential edict, kept from the British. Lindemann also listed six other important weapons that had been placed on the United States' secret list and could not 'be seen or even discussed at present.'[69] These restrictive policies were, in Lindemann's opinion, symptomatic of the bitter American political debate over rearmament and neutrality,[70] as well as the U.S. military's obsession with 'narrow security arguments,' when they should 'recognize the immense advantages associated with free and reciprocal exchange of secret military technology.'

By contrast, Lindemann, after his June 1940 trip to Ottawa, found Canadians unified in their willingness to aid Britain. He was, however, surprised that the National Research Council had not shown more initiative and 'gone ahead independently instead of waiting to be told what we would like them to do.'[71] This deficiency he attributed primarily to unimaginative and stodgy Canadian politicians and bureaucrats, who seemed unable to arrive at 'a definite decision concerning the work that they themselves will require ... planes, wooden or otherwise, and ... radio equipment.'[72] In his opinion, the problem could be dealt with in two ways: by convincing the various British departments and Armed Services to distribute instruments, reports and models to the NRC, and by utilizing the services of Ralph Fowler, who had just arrived in Ottawa, as a conduit for British radar secrets.[73]

Both Lindemann and Fowler made a good impression on their Canadian hosts. To Frederick Banting, for instance, Fowler was a gifted scientist with 'great theoretical knowledge of physics in relation to war,' while Lindemann fit his image of the typical British Secret Service agent: 'He gives little information that has not appeared in magazines, newspapers etc., but he listens well & has a grasp of the whole situation such as few people I have ever met.' Banting's only reservation was whether he and other Canadians scientists were really being taken seriously: 'When I finished talking with Lindemann & Fowler I wonder whether I was a totally uneducated colonial or a sort of average individual who had a vague knowledge of things – just enough to ask questions. The funny thing about an Englishman is that he is so polite that one never knows whether he is pulling one's leg.'[74]

Meanwhile, the gravity of Britain's military situation had forced the Churchill government to consider bold new policies. On 1 August a surprised and elated Henry Tizard was summoned to Whitehall by his old foe, Prime Minister Winston Churchill. The North American scientific mission was on, but would Tizard lead it? 'I asked if he [Churchill] would give me a free hand, and would rely on my discretion. He said "of course"–and would I write down exactly what I wanted. So I said I would go.'[75] During the next two weeks Tizard assembled his scientific liaison team. The civil departments were represented by two outstanding physicists: John Cockcroft (Supply) and E.G. Bowen (Aircraft Production). Captain H.W. Faulkner (RN), Lt.-Colonel F.C. Wallace (War Office), and Group Captain Pearce (RAF) spoke for the three Services; all three combined technical expertise with operational experience.[76] Armed with Churchill's endorsement, Tizard insisted that the Armed Forces provide information to give to the United States about a wide range of weapons: gun turrets for aircraft, rocket-defence of ships, new types of armour plate, radar-assisted anti-aircraft guns, RDX explosives, aviation medicine, and chemical and biological warfare. To complement the technical reports, films of British equipment in action, as well as actual equipment, were included. But the most important secret of all was radar – or more specifically the intricacies and potential of the cavity magnetron.

In planning his itinerary, Tizard first decided to visit Ottawa and consult with the NRC and the Canadian military; he would then proceed to Washington just before the arrival of the official mission, which was steaming toward North America on the *Duchess of Richmond*.[77] On 15 August Otto Maass greeted Tizard at the Montreal airport, and together they proceeded to Ottawa. The next day Tizard met with Mackenzie, Maass, and Banting, as well as C.D. Howe, Minister of Munitions and Supply, Colonel J.L. Ralston, Minister of National Defence, and James Duncan, Acting Minister of National Defence for Air. This was followed by a tour of NRC laboratories and briefings about major military projects. Tizard was particularly impressed by the aeronautical high-speed wind tunnel, and with Banting's work on aviation medicine. During the evenings he and his old friend Ralph Fowler had their intimate briefing sessions over drinks at the Château Laurier Hotel, where they discussed how they could obtain vital military assistance for Britain's

Building the Defence Science Alliance, 1940–1943 57

beleaguered armed forces.⁷⁸ Along with Air Commodore Pearce and High Commissioner Sir Gerald Campbell,⁷⁹ they mapped out the immediate and long-term strategies of how Canadian resources could be best utilized, and how Canada's unique relationship with the United States could facilitate an Anglo-American defence science pact.⁸⁰

Tizard earned high marks from his Canadian contacts. In his meeting with Mackenzie King, Tizard impressed the prime minister when he discussed Canada's important contribution in assisting Britain's radar research and development, as well as 'becoming the great air training centre.'⁸¹ Tizard was particularly pleased that the Canadian prime minister had shown considerable enthusiasm for the cause of Anglo-American cooperation, and had even 'pounded the table and said ... he would speak to Roosevelt, which he did.'⁸²

On the scientific front, Tizard and C.J. Mackenzie quickly developed an excellent rapport. The NRC president valued Tizard's willingness to appreciate 'all our difficulties,' while at the same time 'never ... underestimat[ing] our potential.'⁸³ He also felt that they 'seemed to talk the same language and reacted to most situations alike.'⁸⁴

On 21 August Tizard was ready to undertake the most crucial part of his Mission: negotiations in Washington. After being briefed by Charles Lindemann and Lord Lothian, Tizard met with a number of key American political and military figures, including President Roosevelt, Secretary of War Henry Stimson, and General George Marshall, Chief of the General Staff. This was followed by a series of intense sessions with General Mauborgne, Chief of the Army's Signals Division, and Admiral Furlong, Chief of Naval Ordnance.⁸⁵ Finally, there was his long-awaited encounter with Vannevar Bush, head of the newly created National Defense Research Committee (NDRC), whom Hill had once described as 'the finest kind of yankee.'⁸⁶

On 8 September the other members of the mission arrived in Washington, and the substantive negotiations now began. Between 9 September and 31 December, approximately 207 meetings took place between British and American scientists and military officials. While radar was the most absorbing topic, proximity fuses, fire control systems, explosives, sonar, aircraft design, and chemical and biological warfare⁸⁷ were also discussed extensively.⁸⁸ In the case of chemical weapons, for example, it was agreed that Otto Maass should act as liaison between the U.S. and British research communities to facilitate closer cooperation and planning.⁸⁹

By this stage, C.J. Mackenzie had arrived in Washington with an official scientific and military delegation that included Lieutenant-General Kenneth Stuart and Air Vice-Marshal E.W. Stedman. One of the Canadian mission's primary concerns was the exchange of secret radar information and equipment with the Americans, in keeping with Mackenzie's earlier commitment to manufacture ASV radar sets for Britain through the auspices of the NRC.[90] The actual details of this arrangement were worked out when Tizard returned to Ottawa on 25 September for a whirlwind two-day tour before his return to Britain.[91]

His mission completed, Tizard now faced the challenge of convincing the British War Cabinet that an extensive exchange arrangement with the Americans and Canadians was critical for Britain's survival. On 14 October the first report of his mission was presented to a special meeting of representatives from the departments of Supply and Aircraft Production, the War Office, the Air Ministry, and the Admiralty.[92] While they disagreed on specific issues,[93] on one item there was a consensus: the advantage of having 'a single permanent mission for all three Services which should be centralized in Canada to liaison with both the Canadian and American War departments.' They also supported Tizard's proposal to convey British expertise in operational research to the American military, and to achieve this by sending 'our scientists over there with a view to introducing our requirements into the equipment which they were developing themselves.'[94]

Tizard's second presentation occurred on 4 November, when he appeared before the Defence Services Panel of the newly formed Scientific Advisory Committee (SAC) of the Cabinet.[95] Although it was only an advisory body, the SAC had the advantage of having as its chairman Lord Hankey, a veteran of many political wars and a staunch Tizard ally.[96] Such a connection was fortunate, since Tizard had a number of contentious recommendations: that British radar experts immediately be sent to Canada; that samples of all new radar equipment be sent to both Canada and the United States; that Canadian scientists and military personnel be asked to test all new radar equipment under operational conditions; and that British technical experts in the United States arrange for the testing in Canada 'of any new device which might be of use to the British forces.'[97] Tizard was pleased that the committee endorsed his recommendations, but was dismayed when the War Cabinet still refused to make any long-term commitments.[98] In part, this reflected Tizard's diminished status within British defence science, and the elevation of his archrival F.A. Lindemann (Lord Cherwell), who had direct influence on major policy decisions as Churchill's personal scientific adviser.[99] The long power struggle between these two talented scientists, which had begun in the 1930s, was about to be settled in Lindemann's favour.

While C.J. Mackenzie was kept informed of Tizard's struggles, he was more affected by another event: the death of Fredrick Banting. On 20 February 1941, Banting's plane crashed, shortly after its takeoff from Gander, Newfoundland, while en route to Britain.[100] For the NRC president it was not only the loss of a close friend, it was also the removal of a gifted administrator whose scientific intuition and tireless energy on behalf of the NRC had been so crucial during the early stages of the war. But Mackenzie was fortunate in having other dedicated advisers: Otto Maass, who had assumed control of chemical warfare and explosives research; J.B. Collip, who replaced Banting as chairman of the Associate Committee for Medical Research (ACMR);[101] and Ralph Fowler, who, despite failing health, carried out his dual functions as scientific adviser to the NRC and as British liaison officer.[102]

Fowler's advice was particularly useful when Mackenzie was trying to decide whether to support the establishment of a defence science liaison office in Washington. 'I told Fowler,' he wrote, 'that we would not waive our right to negotiate and exchange information directly with the Americans as we had built up contact over many years which we did not propose to drop, and that we would not accept the position that our dealings with the Americans had to be through any British organization.'[103] Mackenzie's position was, however, about to be challenged by important developments in the Anglo-American scientific exchange system.

The Conant Mission and Anglo-American Exchange, 1941–1943

During the latter part of 1940, American defence scientists tried to convince their political masters that scientific cooperation with the British was crucial. On 20 November, for instance, Vannevar Bush wrote a letter to Secretary of State Cordell Hull, arguing that the dispatch of NDRC scientists to Britain was essential for American national security:

I think it has been amply demonstrated through the experience of the British Technical Mission to this country that only through direct contact of experienced scientific personnel with their counterparts in another country can the most effective exchange of information be arranged. It is inevitable that the military attachés are more concerned with the operational or service characteristics of methods and equipment than in the basic scientific devices comprising such equipment. Moreover, through the extensive personal friendships and contacts between British and American scientists which have existed over a period of years there is ready acceptance and cooperation to such persons who may be engaged in such a mission.[104]

Once Bush received a formal invitation from the British government in January 1941, he sought the endorsement of President Roosevelt.[105] Bush also accepted the passionate request of James Conant, president of Harvard, to head the mission: 'I would say that my going would have certain small advantages from the point of view of my moral leadership in this academic community ... plus my general desire for interesting and adventurous experiences that really motivates me in making this long plea.'[106] Despite the criticisms of isolationist groups, Conant plunged ahead with his mission, confident that Britain was the front line of defence against Nazism and should be supported at all costs.

In the end, the mission turned out to be a great success, as an anxious Bush heard from his special assistant Carroll Wilson on 16 March:[107]

British authorities ... have indicated in every way a desire to work closely with us ... it is apparent that the choice of Dr. Conant as head of the Mission has been ideal. He has lunched twice with the Prime Minister [Winston Churchill] and been received by the King ... Perhaps the most important meeting ... was one held with the DSR's [Directors of Service Research] of the Admiralty, Ministry of Supply, Ministry of Aircraft Production and various other service officers.[108]

Bush was particularly pleased that Conant had been able to establish an Anglo-American exchange system that would allow the two countries 'to utilize to the best advantage their total scientific and technical talent.' He also liked Conant's proposal for a division of responsibility that favoured the NDRC: 'immediate battle development problems' would be left to the British, while long-term and complex new projects would be undertaken in the United States.[109] To further justify American primacy in the development of complex weapons such as the proximity fuse, Conant stressed that 'in physics and engineering (mechanical and electrical) all manpower in england [sic] has been exhausted.'[110]

How were these secrets to be exchanged? Would it be through diplomatic or military channels? Or should American and British defence science organizations have their own system? Bush and Conant preferred the latter approach, and they immediately made arrangements to establish an NDRC liaison office in London that would facilitate the visits of American scientific missions and ensure that they would be granted access to top-secret British information.[111] This U.S. initiative was strongly endorsed by Tizard, who used all his remaining influence to create the British Central Scientific Office (BCSO) in Washington, which would be the conduit for American defence secrets to reach Britain. Bush welcomed the BCSO proposal, and assured Tizard that its two key administrators, Sir Charles Darwin and W.L. Webster, a Canadian-born physicist, would be given all the assistance possible.[112] During its initial operation, the BCSO was also able to draw support from the new British ambassador, Lord Halifax,[113] and from the NRC's Washington liaison officer Alan Shenstone, as well as Brigadier H.F.G. Letson, Military Attaché at the Canadian legation.

At first the system seemed to work well.[114] But despite its early promise, the BCSO had one serious liability: neither the War Cabinet nor the British Armed Forces took it seriously.[115] This was reflected in the government's unwillingness to either increase BCSO's manpower resources or expand its mandate. Indeed, by September 1941 Webster was so desperate for a suitable technical officer that he announced his intention 'to beg, borrow or steal a suitable physicist' from the Canadians.[116] Nor was the BCSO very successful in coordinating the research priorities of the Canadian, Australian, and New Zealand scientific delegations in Washington.[117] Part of the problem was that these delegations were already in place; for example, the Australian office had been created in May 1941 by Sir John Marsden, professor of electrical engineering at Sydney University. Another contentious issue was the British insistence that the BCSO function as the central clearing-house to avoid U.S. complaints about redundant requests and security breaches. For the Dominion scientists, however, these arguments often appeared as Britain's 'thinly masked fears of losing political control'[118] and as an unwillingness on the part of its scientific elite to take their colonial colleagues seriously.[119] Although the BCSO continued to function until the end of the war, it never achieved the important role Henry Tizard had envisaged for it.[120]

The decline of the BCSO did not greatly disturb C.J. Mackenzie, who much preferred to deal with the four official British liaison officers in Ottawa[121] or with key British scientific administrators. His most active correspondence was with Henry Tizard, a communication that continued into the postwar years. Since Mackenzie appears to have viewed Tizard as his father-confessor and as a major ally in the

British defence establishment, his letters were generally deferential.[122] In August 1942, for example, he praised the contribution of Tizard's scientific allies such as Geoffrey Hill to the Canadian war effort: '[Hill] fits into our picture like a glove ... due to the fact he is the first ... real specialist in the aeronautical field.' That was not to say, Mackenzie hastened to add, that Canada had been disappointed with the work of R.H. Fowler, or Nobel laureates Sir William Lawrence Bragg and Sir George Paget Thomson.[123] 'You have,' he assured Tizard, 'sent absolutely top-notch men to us and I think it has proven to be of great mutual benefit.' Mackenzie also described the enormous changes in Canadian defence science since Tizard's mission: 'Our RDF establishment ... now takes in about three hundred scientists and engineers, and Research Enterprises ... now has somewhere in the neighbourhood of six thousand employees, with orders on hand well over one hundred million dollars.' Chemical warfare, according to Mackenzie, was another success story, having developed from a small unit 'to a large Directorate of the Department of National Defence with a budget this year of nearly fifteen million dollars, and a large experimental field for offensive inventions with its complement of airplanes, artillery, engineers etc.' Mackenzie, however, reserved his greatest praise for Tizard's ongoing efforts to facilitate defence science cooperation between the British Commonwealth and the United States: 'If the same degree of confidence, friendliness and respect existed in all the other non-scientific contacts the war effort of the United Nations would probably be more effective than it is.'[124]

Another means of obtaining information about recent developments in British defence science was through the National Research Council's permanent liaison office in London, which had been established in August 1941 under the direction of L.E. Howlett, a senior NRC physicist.[125] Howlett's operation was soon flourishing, especially when arrangements were made for 'circulating' NRC specialists in physics, chemistry, and engineering, who would visit British laboratories and installations when necessary.[126]

Canada as the Scientific Linchpin, 1940–1943

Between 1940 and 1943, Mackenzie worked amicably and effectively with British scientific administrators. However, he had one major complaint about all of them: their tendency to overemphasize the importance of the British contribution to the war effort compared with that of the United States. In Mackenzie's opinion, this was one of the 'real shortcomings of British diplomacy all through the war,'[127] and it was to have an adverse effect on the development of nearly all joint weapons systems.

By contrast, Mackenzie found American scientific mandarins easier to deal with, especially those connected with the NDRC and its successor, the Office of Scientific Research and Development (OSRD). His most agreeable relations were with James Conant and Vannevar Bush, who he felt adopted an 'extremely friendly' attitude toward the National Research Council.[128] As Mackenzie wrote in his diary,

Dr. Conant is a very attractive man. He is modest, unassuming and very courteous ... His mind is crystal clear and I understand that he can be very firm when he reaches a conclusion. He approaches things very fairly and of all the people I have met in Washington I would prefer to work with Conant as one would always know where he was at. Bush is probably a better diplomat and promoter, is more optimistic and imaginative ... They make a grand team and apparently get on very well together.[129]

Throughout the war Mackenzie was convinced that direct bilateral negotiations with American scientific and military agencies was necessary if Canada was to be a full and active partner in war research, and not merely a satellite of Britain.[130] The first NRC attempt to tap American scientific resources came in October 1940 when Vannevar Bush agreed that copies of all relevant NDRC reports would be forwarded to Ottawa, and that the NDRC would welcome 'visits both by Prof. Fowler and/or his representatives to laboratories in the U.S.'[131] Further improvements in this exchange system occurred in June 1941 with the formation of the Office of Scientific Research and Development (OSRD), with Bush as its director.[132] Caryl Haskins, one of Bush's closest advisers, visited Ottawa in September to explain the advantages of this new organization to C.J. Mackenzie.[133] Haskins enjoyed his encounter with the NRC hierarchy and was impressed by the quality of research being carried out, especially at the Radio Branch's experimental station: 'The station has been very considerably developed, with a fair sized staff predominantly composed of engineers. Most of the buildings were of a temporary character. Several G.L. sets were under construction and test. Talked at some length with Bell [J.W.] and McKinley [D.W.R.]. The station is not comparable with the Radiation Laboratory, but is nevertheless a good show.'[134]

Haskins and his OSRD colleagues liked Canada and Canadians. Moreover, they were aware that they might be able to obtain additional top-secret British information through their contact with Canadian scientific and defence agencies.[135] In December 1942, for example, the following message was sent to the Washington office of the DND Director of Staff Duties on Weapons:

Although it is our understanding that your interest is principally in the possibilities for tactical use of new developments, I am sure there will be many phases of our work which will be of interest to you. Also, I believe that some of the information available to your group would be of decided interest to various sections of OSRD ... In particular, I believe that you mentioned a regular AFV letter of British origin and I believe some of our investigators would have a definite interest in these documents.[136]

As the war progressed, Canadian and American scientists became involved in a wide range of important collaborative projects, all of which necessitated close connections between the university, military, and industrial laboratories of the two countries. Inevitably, because American resources were greater, most of the movement of personnel that occurred was southward. One of the most frequent destinations was the Radiation Laboratory at the Massachusetts Institute of Technology, with its vast and highly advanced radar research facilities. For those Canadian scientists working on antisubmarine devices, the greatest attraction was the sophisticated experiments being conducted at the Woods Hole Oceanography Institute and the New London naval testing laboratory.[137] In November 1942, for instance, a request was made by W.L. Webster, Secretary of the NRC Associate Committee on Explosives, for permission to visit these installations, and for any information, 'in the form of secret reports,' about new kinds of depth charges.[138] His appeal was favourably received as the Americans recognized Canadian operational expertise in antisubmarine warfare.[139]

The OSRD was often prepared to utilize the specific talents of Canadian scientists and, where possible, to employ them in their extensive operations. In June 1941, for example, Mackenzie was able to convince Bush that Joseph Wilson Greig, a Canadian scientist working in the United States, should be brought into the highly sensitive OSRD proximity fuse project once he had obtained clearance from U.S. Army and Navy intelligence authorities.[140] What this incident revealed, aside from

Greig's obvious talents, was that Canadian scientists were increasingly being subjected to American security standards.[141]

In trying to coordinate Canada's security procedures with those of the United States, Mackenzie often used the services of General Maurice Pope, who had been appointed head of the Washington-based Canadian Joint Staff Mission (CJSM) in March 1942.[142] Pope's mandate was sweeping and vague:[143] to maintain 'continuous contact with the U.K.–U.S. Combined staffs and the Combined Planning Committee, and to represent the [Canadian] War Committee before the Combined Staffs when questions affecting Canada were under consideration.'[144]

Fortunately, Pope had a solid military reputation, was a charming and innovative diplomat,[145] and had excellent contacts with the British Joint Staff Mission (JSM), which by 1943 had about 2650 personnel, divided among the Admiralty (800), the RAF (650), and the British Army (1200).[146] Two of Pope's most useful contacts were Lieutenant-General Sir Gordon Macreary, an old friend from his days at the War Office in the 1930s,[147] and Brigadier Vivian Dykes, British secretary of the Combined Chiefs of Staff (CCOS).[148] With their assistance, Pope was usually able to gain access to the deliberations of the CCOS, created in January 1942 with responsibility for the strategic direction of the Allied war effort.

But it was a frustrating experience for Pope, who was often enraged by the myopic and condescending views of British and American military officials. In his opinion the British still retained their 1930s imperialistic outlook and 'completely failed to comprehend that we Canadians ... were conscious of having developed a separate, though completely friendly identity.'[149] On the other hand, he was disturbed that the American Joint Chiefs of Staff had a disturbing tendency to overlook their earlier commitments to Canada under the August 1940 Ogdensburg Agreement and the subsequent joint defence plans of October 1940.[150] The problem for Pope was how to reconcile the responsibilities of the Permament Joint Board of Defence with the broader Anglo-American military plan (ABC-1), drawn up in Washington in March 1941, which bilaterally allocated strategic responsibilities. Despite an attempt to clarify spheres of responsibility in August 1941 under the ancillary Canada–U.S. joint operational plan, code named ABC-22, Ottawa was concerned that American and British military policies could be imposed on Canada.[151]

Matters came to a head in July 1941 when Pope formally objected to the Combined Chiefs of Staff's lack of consultation with his mission. His complaint received immediate attention. In August 1942 a report was prepared by U.S. Brigadier General A.C. Wedemeyer, Chief of the Strategy and Policy Group, and by Major General George V. Strong, Chairman of the U.S. Joint Intelligence Committee, which emphasized that the Canadian Staff Mission and the Canada–U.S. Permanent Joint Board on Defence were 'mutually supporting.' Their analysis concluded that there was no problem: the CJSM should be primarily concerned with 'strategic planning and operational matters,' while the PJBD would deal more with 'matters of broad policy determining the relationship and participation of the two nations in the war effort which pertain to the joint area [North America].'[152] The report ended with the recommendation that the U.S. Joint Chiefs of Staff 'should deal *directly*' with the CJSM in Washington 'on all matters which pertain exclusively to operations with the U.S. area of strategic responsibility and in which Canadian forces may be directly concerned.'[153]

During his three years in Washington, Pope often found deliberations of the Combined Chiefs of Staff acrimonious and confused. In his opinion, the most contentious issues were the viability of a second front in Europe, the logistics of the North African landings, and, above all, whether the United States should devote most of its resources against Nazi Germany or Imperial Japan.[154] Pope noted: 'While the political heads of the United Kingdom and the United States never weary of reiterating their complete accord as to the objects they are determined to achieve, there can be do doubt that their respective advisers are not of one mind ... The United States Navy is ... adamant in its refusal to divert any appreciable measure of naval forces ... the United States Army has a burning desire to annihilate the Japanese.'[155] Pope was also intrigued by the shift in the relative military strength of the two allies, which had become pronounced by the time of the Quebec Conference of August 1943.[156] Pope shared C.J. Mackenzie's frustration with Britain's unwillingness to accept its diminished alliance status or to appreciate what the war meant to the United States. 'Of all the peoples of the world,' he wrote, 'the British understand the Americans the least.'[157]

Conclusion

In June 1940 the possibility of military defeat had forced the British government to seek assistance from the United States and Canada in developing vital weapons systems. While A.V. Hill and Charles Lindemann helped prepare the way, it was Henry Tizard, and his multitalented mission team, who ultimately convinced the U.S. scientific and military elites of the advantages of an Anglo-American-Canadian exchange system. The Conant mission of March 1941 confirmed the deal and established the necessary machinery for its operation. The scope of the OSRD exchange arrangement with the various British agencies is reflected in the fact that, between March 1941 and July 1945, the London office handled more than 1400 visits and 59,135 separate reports, letters, and samples from British and Canadian research agencies. In this same period the OSRD Washington offices forwarded to London some 82,153 documents.[158]

By the fall of 1942 both the Department of National Defence and the National Research Council had established a presence in Washington that was independent of the British. This was made possible by one critical factor: American support. Did it matter that these U.S. policies were primarily based on American strategic priorities? Of course not! What really counted was that the United States Joint Chiefs of Staff were favourably disposed toward the Permanent Joint Board on Defence and the Canadian Joint Staff Mission. It was also to Canada's advantage that Vannevar Bush and James Conant preferred to deal with the National Research Council directly rather than through the British Central Scientific Office.[159]

Canada's role as a scientific linchpin between the British and Americans assumed great importance during the fall of 1940, when great efforts were made to give the western allies a strategic advantage in the operational use of radar – one of the wonder weapons of the Second World War.

3

Radar Research and Allied Cooperation, 1940–1945

> About four o'clock Dr. Bowen came in ... He has been attached to the B.A.C. working half time at M.I.T ... I told Bowen that I had told Tizard, Cockcroft, etc. and my other English friends that if they had played with us instead of with the Americans on the radio game they would have had all the A.I.s and A.S.V.s they wanted ... that if they would send two or three men like Bowen over to give us the advantage of their knowledge in England, and tie in with the operational experience, that we could get along much faster than they could in the United States. My judgement has been borne out.
>
> C.J. Mackenzie Diary, 21 July 1942

From 1941 to 1945 Canada's defence science was conducted within the framework of alliance warfare, and characterized by increased quality control and long-term planning. Most weapons programs were now in place, and were advancing from the experimental to the production and operational stages. From his position as acting president of the NRC, C.J. Mackenzie remained the key coordinator between university scientists, the Armed Forces, the industrial sector, and Canada's allies. Another integral part of his job was negotiating with British defence planners, which allowed him to appreciate major changes in the dynamics of alliance warfare that occurred after Pearl Harbor, when the United States became committed to total war.

Mackenzie also appreciated the indispensable role that the U.S. Office of Scientific Research and Development (OSRD) assumed in mobilizing American industrial, scientific, and military resources behind the war effort. In this undertaking the OSRD received outstanding leadership from Vannevar Bush, who was able to deal effectively with the highest level of U.S. political and military decision makers, while still retaining the affection and respect of the vast majority of American scientists. Fortunately for Canada, Bush and Mackenzie maintained a close personal and professional relationship throughout the war years, despite some major policy disputes.

Radar cooperation between Canada and its two major allies provides a useful case study of the intricacies and dynamics of alliance warfare. Although there had been preliminary discussions between the NRC and the British Air Ministry in 1939, it was Tizard's August 1940 visit to North America that paved the way for a substantial Canadian contribution to the development of radar technology. As a result, between 1940 and 1943, NRC scientists were involved with a variety of radar projects – notably those dealing with air defence. But it was not an easy undertaking, given Canada's limited number of top-flight physicists and electrical engineers.[1]

During the Battle of Britain, radar also became associated with operational research. This meant, in the words of radar guru Robert Watson-Watt, the need

> to examine quantitatively whether the user organization is getting from the operation of its equipment the best attainable contribution to its overall objective, what are the predominant factors governing the results attained, what changes in equipment or method can be reasonably expected to improve these results at a minimal cost in effort and time, and the degree to which variations in the tactical objectives are likely to contribute to a more economical and timely attainment of the overall strategic objective.[2]

Radar was not the only NRC contribution to Britain's air war. Another important undertaking, aviation medicine, was coordinated by the NRC Associate Committee for Aviation Medicine, with many sophisticated projects being carried out at Banting's University of Toronto laboratory or at the Montreal Neurological Institute. Other useful work, conducted by the NRC's Mechanical Engineering Division, included attempts to develop wood plastic plane parts, experimental research on jet engines, cold weather testing, and work on improved de-icing techniques.

Although these activities seemed far removed from the NRC's prewar commitment to advance pure research, Mackenzie viewed this involvement in applied and operational research as both necessary and natural. In October 1940, for instance, he noted that defence science was essentially 'an engineering problem, because if developments have no chance of ending up in industrial and effective tactical use in the field, no scientist in war work is interested ... The pure scientist becomes an engineer overnight ... In a war of survival, we ... work for today.'[3]

Models of Defence Science Organization, 1942–1945

As the junior partners in the Anglo-American alliance system, Canadian scientific administrators carefully observed the different ways defence science was being organized and mobilized in Great Britain and the United States. They were particularly impressed with American developments. Under Vannevar Bush's management, the Office of Scientific Research and Development became increasingly involved in the job of coordinating the activities of academic and industrial scientists and engineers and the Armed Forces. This system was enhanced in 1942 with the establishment of the Joint Committee on New Weapons (JCNW), which had a wide-ranging mandate to coordinate the efforts 'by civilian research agencies and armed services in the development of new weapons and equipment,' and to 'maintain liaison with scientific research agencies in the United Nations through their representatives in Washington.'[4] Although the JCNW had an executive board of three permanent members, including Bush as chairman, most of the detailed work was carried out by its specialized subcommittees, the most important dealing with special projects in the fields of radar, missiles, rockets, antisubmarine weapons, and operational research.

In contrast, British defence science seemed increasingly plagued by bureaucratic obstruction, duplication of effort, problems of communication, and lack of creativity.[5] This situation was particularly galling to Henry Tizard, who in February 1942 launched a vigorous critique of how the country's scientific resources were being used. He was particularly critical of 'special' advisers, such as Frederick Lindemann, now Lord Cherwell, who appeared to be 'working in the void,' out of touch with the work of their fellow scientists and the real needs of the Armed Forces.[6] A.V. Hill shared Tizard's views, and as secretary of the Royal Society and a member of Parliament, was admirably situated to lead a public campaign for reform of British defence science. In March 1942 he circulated a memorandum in which he stated: 'The key idea, as you know, is to link technical policy right up to the top with tactical and strategic policy and to have the lines of executive authority running downwards from the top both in the services and in the supply departments, dealing with technical matters, not simply to trust to rather casual cross linkages by "co-ordination" or "liaison."'[7]

Although Hill had the enthusiastic support of most of Britain's scientific elite, and of some concerned administrators such as Lord Hankey, his proposal to create a separate Joint Technical Committee, similar to the U.S. Joint Committee on New Weapons, drew sharp criticism from Britain's military hierarchy.[8] One of his most outspoken opponents was Lord Chatfield, First Lord of the Admiralty, who told the House of Lords that he could not see why academic scientists should have any sustained involvement with weapons beyond the laboratory.[9] This comment brought a caustic response from Hill: 'It was the spirit which Chatfield so clearly expresses which sent the Prince of Wales and the Repulse to their fate, regardless of the scientific and technical conclusion, which was quite certain, that a ship cannot defend herself against really determined attack from the air, by anti-aircraft gunnery alone.'[10]

Hill's indictment of the government's 'Colonel Blimpism' would become even more strident in late May 1942, when Rommel's Afrika Corps drove the British 8th Army out of Libya and threatened the Suez Canal.[11] Under intense pressure from his critics, Winston Churchill, in his dual role as prime minister and minister of Defence, reluctantly agreed to appoint three distinguished scientists as special advisers to the minister of Production.[12] They did not, however, have direct access to Cabinet, a situation that both Tizard and Hill regarded as critical. As a result, strategic scientific planning of the war effort remained in the hands of Churchill and the scientific branches of the Armed Forces.[13]

Fortunately for the Allied war effort, by 1942 Britain's political and military leaders were interested in expanding cooperation with the United States, as was evident in the increase in the number of joint Anglo-American weapons projects. One of the most important was in the field of radar, in which the United States was gradually becoming the dominant partner. This was reflected in the 1943 agreement that stipulated that Washington would 'undertake primary responsibility' for the research, development, and production of a wide range of radar equipment with the understanding that the 'reasonable' needs of the British would be met by American production.[14] A similar trend prevailed in the case of radio countermeasures, a field in which the British had achieved an initial lead both by jamming enemy signals and by falsifying radar images through the use of Window, a tactic that entailed dropping hundreds of sheets of tin foil from planes, which gave the impression that a large-scale Allied bombing raid was occurring away from the designated target. The major use of Window occurred during the devastating July 1943 bombing of Hamburg, which destroyed most of the city and killed over 42,000 German citizens.[15] Effective jamming required exact knowledge of enemy radio frequencies and characteristics of radar beams. By April 1944 American and British scientists had developed thirteen forms of radar jamming and seven systems of radio jamming. Plans were also made to devise countermeasures against guided missiles, since it was anticipated that Germany would use these weapons in the near future.[16]

Rockets and infra-red devices were two other weapons systems developed jointly by the United States and Britain. In the rocket field the British had an early lead and by 1943 had developed forward-firing aircraft rockets for use against German U-boats and tanks; air-to-air rockets were in the planning phase. OSRD teams, especially at the California Institute of Technology, soon matched and surpassed these achievements.[17] American defence scientists also made great headway in infra-red research and favoured close cooperation with the British. This was evident during the February 1944 meetings of the JCNW, where it was proposed that a liaison subcommittee be created for the purpose of coordinating the development, production, and operational use of infra-red devices in Britain and the United States 'principally in connection with night operations of planes, boats and vehicles.'[18] Vannevar Bush strongly endorsed this proposal as long as ultimate decision making about their operational use remained in the hands of the U.S. Joint Chiefs of Staff and the Joint Committee on New Weapons.[19] Significantly, many of the guidelines for these exchange provisions had already been worked out by Allied radar scientists during the early stages of the war – before the United States officially became a belligerent.

Radar Research: Outlining the Problem

In 1940, radar, or Range and Direction Finding (RDF), consisted of five major systems: Air to Air, Air to Surface, Ground to Air, Fire Control, and Navigational. Although some systems were more developed than others, each had its own characteristics and set of problems. The need for such technology was particularly acute in Britain during the fall of 1940 because the country's scientific and military resources were stretched to the limit, and because of the crippling effects of German mass bombing attacks and the possibility of an imminent German invasion. Being able to detect, locate, and destroy enemy aircraft was, therefore, crucial for the defence of Britain and the preservation of the Commonwealth.

According to E.G. (Taffy) Bowen, one of Britain's foremost radar experts, RAF pilots during the Battle of Britain 'had no idea what to expect from the new equipment and they had no prior training in the very special technique necessary to pull off night interception.' Effective tactics of aerial engagement needed to be developed: 'First, the pilot must creep slowly up on the target, rather like a hunter stalking his prey ... then close gradually to minimum range and place himself in exactly the right position behind the target before actually opening fire.'[20] Preparing air crews for these challenging new situations meant recruiting personnel who could 'handle exotic electronic equipment,' especially since most of the radar equipment was not well engineered or capable of withstanding 'the vibration and altitude problems found in aircraft.'[21] Fortunately, scientists at the Telecommunications Research Establishment (TRE) of the British Air Ministry, and the Radiation Laboratory (Rad Lab) at the Massachusetts Institute of Technology, were able to meet this challenge with new air-to-air (AI) sets using 10 cm wavelengths, which gave RAF night fighters a decisive advantage over German bombers.[22] Another important innovation was Identification, Friend or Foe (IFF) radar technology, which allowed pilots and gun crews to differentiate between friendly and enemy aircraft.

Improving Air to Surface Vessel radar sets (ASV) was another priority for Commonwealth scientists during the perilous fall of 1940. In part, these deficiencies could be attributed to lack of foresight on the part of the British Admiralty, who until the outbreak of war gave 'little thought ... to the use of ASV radar against submarines.' In addition, there was considerable difficulty coordinating the research programs of the Admiralty and the Air Ministry,[23] which greatly complicated the development of more effective surveillance equipment and operational planning against the growing U-boat threat. Between December 1940 and March 1941, such weaknesses resulted in the sinking of 96 Allied ships without the Germans losing a single submarine.[24]

An additional task was the need to develop effective ship search radar systems, especially given complex engineering and installation factors. This meant that the radar had to be carefully integrated with other electronic devices, the communication systems, and the other types of detection apparatus in the vessel, and had to withstand the demanding conditions of warfare at sea and 'suit the tactical assignment of the type of vessel.'[25] No less challenging was the task of utilizing radar fire control systems for anti-aircraft fire, for coastal defence, for field artillery, and for naval guns. Each of these systems required different radar equipment and tactical training. For instance, while the Navy needed fixed equipment, for the heavy guns and anti-aircraft battery, Army field artillery and anti-aircraft systems had to be mobile. Significantly, the NRC was involved in the design and production of one of the more successful gun control systems – the GL Mark III – which remained a highly regarded anti-aircraft device until it was superseded by the American SCR 545 and 584 in 1944.[26]

A fifth research undertaking was the development of radar navigational systems. One of these was the Ground Control Radar (GCR) sets for blind landings. Another was a Long-Range Navigational (loran) system that 'broadcast a vast net of crisscrossed radiations in the sky so that a ship or plane a thousand miles from home could determine its position to an accuracy of 1 percent.'[27] These innovations were eagerly adopted by the Royal Canadian Air Force (RCAF), which viewed the loran pulse signal navigational system as superior to its twenty-five beacon stations that had been established earlier in the war.

Trends in Allied Radar Cooperation, 1940–1941

Canadian involvement in Britain's radar war was late in coming. It was only in March 1939 that the Air Ministry told Ottawa that it was prepared 'to release information respecting a most secret device which they had adopted for the detection of aircraft.' This overture was channelled through Air Vice-Marshal G.M. Croil, Chief of the Canadian Air Staff, who in turn relayed the message to the National Research Council. After a series of consultations, General McNaughton appointed John T. Henderson of the NRC's Radio Branch to represent Canada in the forthcoming meeting of Dominion radar specialists. Henderson was the obvious choice because of his previous work on cathode direction finders, his involvement in the establishment of the Radio Field Station, and his close working relationship with E.V. Appleton and Robert Watson-Watt, key members of Britain's radar program.[28]

On his arrival in England, Henderson was met by Squadron Leader F.V. Heakes of the RCAF, and together they attended a series of briefings and demonstrations and toured production facilities, along with scientists from Australia, New Zealand, and South Africa. In addition to these formal sessions, Henderson met with E.V. Appleton and several radar scientists who had worked with his boss R.W. Boyle during the First World War.[29] Henderson's 10 May 1939 report to McNaughton reflected his enthusiasm for the British radar operation, especially its five major British projects: the Night Watchman for guarding harbours; the radio range finder for coast defence guns; the early warning system to 'locate approaching aircraft at distances of at least 100 miles'; the GL Mark III for ranging and laying anti-aircraft gun fire; the blind landing systems for aircraft; and the airborne radar for fighter aircraft. In keeping with his assumption that British authorities were prepared to recognize the NRC as a possible partner, Henderson attempted to obtain the specifications and production data of the air-to-surface models so that they could be produced with Canadian components.[30]

Both McNaughton and the Canadian Chiefs of Staff were impressed by Henderson's report and his recommendations for an active Canadian role in the radar war. As a result, plans were drawn up to assign twenty new scientists and technicians to the Radio Branch, and a budget of $750,000 was allocated 'to purchase from Great Britain RDF equipment for different applications.' Unfortunately, the Treasury Board was not in a generous mood, and the request was denied.[31] Despite this set-back, the NRC continued to press forward with its radar agenda. In June 1939 McNaughton was able to convince the Chiefs of Staff that the NRC should become its official radar agency with a mandate 'to adapt British designs to suit Canadian practice; to undertake research work in RDF ...; to assist with the installation and calibration of new types of RDF equipment ...; and to train the nucleus of the higher technical ratings or expert operator staff.'[32] In August, McNaughton left for England on his own special mission to discuss the possibilities of producing radar equipment in Canada, a task that Robert Boyle, head of the NRC Physics Division, continued during the fall of 1939.

At the same time, Henderson was sent to the United States to inspect a new type of cathode-ray direction finder developed by the Western Electric Company, and to attend a demonstration of an innovative radio altimeter developed by Bell Laboratories. Good news greeted him on his return to Ottawa: there would be 'seed money' for the development of 'a Canadian version of ASV.'[33] However, the Treasury Board would not make a long-term commitment to the project. This penny-pinching approach enraged Boyle, who was still in England trying to arrange a deal for Canadian radar production. 'I wish to God,' he wrote in October 1939, 'they had with alacrity accepted 6 months ago the advice of a few of us ... We should then have got a start in the training of a valuable corps of men ... We shall have to make up lost time.'[34]

Meanwhile, Henry Tizard, taking full advantage of his key position with the Air Ministry and the Air Defence Research Sub-Committee, was busy promoting radar exchange within the Commonwealth.[35] In January 1940, he and Robert Watson-Watt met with Sir John Madsen, scientific liaison officer of the Australian government, and hammered out an agreement whereby Australia and Canada would be given special status in future radar development. On another front, Tizard was actively campaigning for closer relations with American radar scientists, although he was hampered by Britain's reluctance to share its military radar secrets. On 3 May 1940 Tizard called a meeting to discuss the advantages and disadvantages of an Anglo-American RDF exchange: 'I said that on general grounds I was very much in favour of telling the Americans our results. The likelihood was that they have got something good in return. In any case we should get their goodwill and cooperation which meant a great deal considering their facilities for applied research and production.' Yet, despite Tizard's eloquence and the gravity of Britain's military situation, this proposed system of radar cooperation came under attack.

Robert Watson-Watt, for instance, claimed that American scientists 'could not teach us anything ... that we should get the worst of the bargain,' while Admiral Sommerville warned that sharing radar secrets would undermine Britain's strategic position since 'anything told to the American Navy went straight to Germany.'[36] Fortunately, for the sake of future cooperation, Tizard was able to refute these charges, in large part because of the glowing reports about American radar progress he had received from two well-placed British scientific experts in the United States.

During the spring of 1940 both A.V. Hill and Mark Oliphant carried out extensive investigations of American civilian and military radar facilities, and were particularly impressed by the work being carried out by U.S. Navy scientists on shipboard gun-laying radar equipment.[37] At the same time, both commented on the fact that American scientists had not yet developed the resonant cavity magnetron, with its vast potential for microwave radar, or made much progress in airborne radar for the detection of either ships or aircraft. The British observers were, however, most concerned by the fact that powerful elements within the U.S. military remained sceptical whether radar was a significant weapons system.

The Battle of Britain in 1940 did much to convert radar's nonbelievers. Reports from Colonel Carl Spaatz of the Army Air Force in London during the 'Blitz' were full of praise about the operational effectiveness of radar-aided fighter squadrons. Vice-Admiral Harold B. Bowen of the

Naval Research Laboratory was another influential advocate, although he favoured radar research under the control of the Services.[38] The most important boosters of radar, however, were Vannevar Bush, Alfred Loomis, and Karl Compton of the newly established National Defense Research Committee (NDRC), who were impressed by the radar prototypes the Tizard mission had displayed in September 1940 and by the special briefings provided by John Cockcroft and E.G. Bowen.[39]

In late September 1940 the NDRC began making plans for a large-scale American radar undertaking, with the following research and development priorities: '(1) a 10 centimetre airborne radar, with air interception as the first objective; (2) a 10 centimetre gun-laying (GL) radar for anti-aircraft gun-laying; [and] (3) a long-range navigational system.'[40] Once it had been decided that the Massachusetts Institute of Technology (MIT) would be the site of the Radiation Laboratory,[41] Lee DuBridge of the University of Rochester was appointed director of the operation. According to Taffy Bowen, who handled the AI liaison at MIT, DuBridge was 'at once the best qualified and most pleasant scientist to lead a radar laboratory during the whole war.'[42] By the end of 1942 DuBridge's 'Rad Lab,' as the Radiation Laboratory came to be known, employed over a thousand scientists, including such outstanding physicists as Isidore Rabi, Ken Bainbridge, Edwin Macmillan, Lou Turner, Alex Allen, Louis Ridenour, Bob Bacher, and Luis Alvarez. In his memoirs, E.G. Bowen described the ambience of this unique wartime laboratory:

> Everyone worked long hours and did not spare themselves. Here was the cream of American scientists, hell-bent on doing all they could for the war effort – some fourteen months before America itself entered the war ... If there was any relaxation, it was on Friday night – in the bar of the Commander Hotel just behind Harvard Square. Inevitably, this became known as Project 4 and, although attendance was not compulsory, there was always a good turn-out and the hotel did a roaring trade.[43]

Tizard's mission had many ramifications for radar developments in Canada as well. This can be attributed to several factors. First, there was Tizard's on-site endorsement of NRC efforts and his success in convincing C.J. Mackenzie that the Radio Branch should work on a variety of systems. A second factor was the 28 September joint meeting in Washington of American radar scientists and the Tizard mission, where it was decided that the NRC would be given the task of developing a sophisticated gun-laying radar system, code named GL Mark IIIC, with the prospect that this weapon would be adopted for use by both the British and American armies.[44] On 14 October Tizard also called on the Air Ministry to make plans for 'the immediate establishment of a first-class R.D.F. staff in Canada for co-operation in the air-borne and gunnery problems where work could proceed uninterrupted by the enemy.'[45] Tizard's campaign to make the NRC an important partner in Britain's radar war continued after his return to Britain; in this endeavour he was greatly aided by positive reports from Ralph Fowler about the achievements of the Radio Branch.

During the fall of 1940 C.J. Mackenzie responded to this new challenge by providing the Radio Branch with a $350,000 allocation from the War Technical and Scientific Development Fund, by negotiating a production agreement with Colonel W.E. Phillips of the Crown corporation Research Enterprises Ltd.,[46] and by adding Colonel F.C. Wallace, of Tizard mission fame, to the staff of the Radio Branch.[47] All this was good news for John Henderson and his Radio Branch colleagues since their task of developing radar systems depended on obtaining scientific data and sophisticated components in Britain and the United States.

By October the Branch employed sixty scientists, organized into distinct research teams: D.W.R. McKinley directed the Early Warning (EW) system; F.H. Sanders handled the microwave section; H.R. Smyth coordinated naval research; H.E. Parsons was responsible for mechanical design; K.A. MacKinnon supervised aerial design; W. Happe presided over the GL IIIC gun-laying project; and W.B. Whalley and A.H.R. Smith worked on cathode-ray tubes.[48] Most of this activity was carried out at the Radio Field Station, near Ottawa, in close cooperation with scientists and engineers at Research Enterprises in Toronto. But not all projects were deemed equal. The most basic technology was the Night Watchman (NW) system, which was 'a barrier for use across harbour entrances, rivers and ship channels to detect passage of surface vessels over fixed areas.' Somewhat more sophisticated was Coastal Defence (CD) radar, 'for use with coast defence guns,' and the Early Warning (EW) system, which located approaching aircraft at distances of at least 100 miles. The other systems, however, represented much more challenging technology: this included anti-aircraft GL Mark IIIC apparatus, blind-landings radar sets, loran equipment, and, above all, airborne systems (AI), which allowed fighters 'to locate and attack enemy aircraft in the air.'[49]

The Radio Branch faced many problems in making these various systems operational. One of the most serious was the gap between British promises of technical assistance and the dearth of information and equipment coming to Ottawa.[50] Despite the best efforts of Hill, Tizard, and Fowler, Britain's defence establishment tended to remain uncooperative and condescending. In December 1940, Donald McKinley, the Radio Branch's expert on EW systems while on assignment in London, provided a report of how difficult it was to acquire advanced radar equipment from officials of the Ministry of Aircraft Production. McKinley received little satisfaction when he appealed to the minister, Lord Beaverbrook, for assistance: 'In spite of being a Canadian ... [Beaverbrook] is loath to part with any ASV until England's needs are fulfilled.' McKinley's last resort was to approach John Cockcroft, a member of the Tizard mission, who 'cuts a good deal of ice and can get things done in ways that are frowned on in official circles.'[51] McKinley ended his report with a vivid account of the Blitz and the need for accelerated efforts to develop effective air intercept systems.

London got a real plastering last night from 6:00 pm to 9:00 pm. Thousands of incendiaries lit the place up bright as daylight. They plunked a H.E. at ... Charing Cross Road ... just as I was buzzing round from Leicester Square ... Counted eighteen fires at one stage and some of them were whoppers ... At the peak of the show the A.A. quit firing and we could hear plenty of machine-gun fire from upstairs ... meaning the fast fighters with A.I. were on the job. The raiders soon went away after that. They should get this A.I. going faster; in my humble opinion it is the only thing that will stop the night raider.[52]

Despite many obstacles, Canadian radar research and production dramatically improved during the next six months. In June 1941, Sir Lawrence William Bragg, the new Air Ministry liaison officer in Ottawa, informed Tizard that the NRC was doing 'very well with their GL Mark IIIC. They are going into production soon and are talking of 50 a month by December.' Although Bragg confessed that he had not yet seen the latest British prototypes of the device, he remained confident that the Canadian version was 'very good.' Much of the credit for its operational excellence, he added, should go to Colonel Wallace, who had 'been supervising it throughout from the point of view of the gunner.'[53] C.J. Mackenzie, in an August 1941 letter to General McNaughton, provided an even more enthusiastic account of Canada's radar achievements:

Next year I am asking for one million, two hundred thousand dollars for the

Radio Section alone. As you know we have actually installed equipment on the Atlantic coast on warships, etc., but our greatest accomplishment, I think, is the construction of the GL 3 set which, as you know, is the RDF end of an anti-aircraft assembly consisting of radio locator, predictor, and guns ... I think it is safe to say that few people in Britain thought we could do anything about it at all; in fact, many such as Sir Frank Smith pooh-poohed the idea of anything coming out of Canada ... American scientists and chief engineers from Westinghouse, General Electric, Sperry, Bell Laboratories, etc., have all been up for demonstrations and have expressed amazement that we have been able to do in ten months what none of the other countries has achieved.[54]

Forging the Link: The Radiation Laboratory and the NRC, 1940–1945

By 1941 radar cooperation between Canada's National Research Council and the U.S. National Defense Research Committee (NDRC) was already well established thanks to previous negotiations between C.J. Mackenzie and Vannevar Bush. This had been initiated on 30 October 1940, when Bush invited the NRC to send some of its microwave experts to the newly established Radiation Laboratory 'to work with the scientists there.' Mackenzie, delighted by this opportunity, proposed going to Washington immediately to meet with NDRC officials so they could 'coordinate research projects.' The NRC's Radio Branch, he told Bush, had been working with British scientists for sixteen months and had 'built and put into operation certain special equipment for the Department of Defence and have now enlarged our activities by organizing special groups at the Universities of Toronto and McGill for work on the shorter wave length systems.'[55] This information was subsequently relayed to Karl Compton and Alfred Loomis, who coordinated the NDRC radar operation from its nerve centre at MIT.[56]

Mackenzie's interest in having a joint North American radar operation was further evident when Loomis and his assistant, E.L. Bowles, visited Ottawa on 25–28 November to check out NRC work on the GL Mark IIIC and the Night Watchman systems.[57] Although the British Armed Forces had priority on the delivery of both these systems, Mackenzie felt that it would be expedient to meet any requests the Americans might submit 'in view of the fact that Canada and Great Britain have made, and will continue to make, demands on the United States for equipment.'[58]

Mackenzie's strategy paid immediate dividends. In December 1940 a Radio Branch team of A.J. Ferguson, W.R. Wilson, and F.J. Heath were invited to spend three months at the Rad Lab, as well as to monitor radar development and production work being conducted at the laboratories of Bell, General Electric, RCA, and Sperry.[59] Such exposure to the most advanced forms of U.S. radar research was exciting and profitable for the NRC scientists, who felt that their newly acquired knowledge would give them sufficient expertise to construct prototypes of advanced NDRC air-to-air 10 cm sets.[60] Mackenzie was also delighted that his NRC researchers had impressed the Rad Lab hierarchy with 'their scientific capacity and energy.'[61]

This positive relationship continued throughout 1941 as equipment and data flowed between the MIT and NRC laboratories. Canadian university scientists also profited from this exchange. In January, for instance, A. Pitt of the University of Toronto physics department, was loaned an ultra-high-frequency generator 'for an indefinite period,'[62] while McGill physicist J.S. Foster received both pulse equipment and a generator.[63] In July the NRC's Fred Heath, along with Taffy Bowen and Dale Corson of the Rad Lab, transported a sophisticated air-to-surface 10 cm set to Britain for final tests before adoption. It was an elaborate device, consisting of a plan position indicator, antennae that scanned 360 degrees, and sensitive modulators and transmitters.[64] The expedition was a success, in part, because it established the Radio Branch as an important intermediary between Britain's Telecommunications Research Establishment (TRE) and the U.S. Radiation Laboratory. There was also common agreement that developmental priority should be shifted from AI equipment to air-to-surface sets, as the Battle of the Atlantic intensified.[65]

Being the junior partner in this radar exchange system had its limitations, particularly when the Radio Branch became more and more dependent on the Radiation Laboratory, with its thousands of scientists and engineers, for sophisticated electronic equipment, vital components, and technical information. The NRC, however, worked hard to appear cooperative, and took every opportunity to provide its powerful partner with either its own or British apparatus. In February 1941, for instance, it sold the U.S. War Department models of the Night Watchman and the GL Mark IIIC 'at cost.'[66] Despite such cooperation, there was always the prospect that American scientific administrators might curtail the flow of essential data and equipment either because of a dispute between Washington and Ottawa, or, more likely, because of complications in the Anglo-American radar exchange system. In January 1941, for example, A.H.R. Smith reported that his NRC team had been asked by Rad Lab scientists to explain why information about 'the long time delay phosphorescent screen cathode ray tube,' which British radar scientists had promised to deliver, had not materialized, a delay that was causing serious research problems. While the situation was soon resolved, Smith noted that the confrontation had caused 'hard feelings' toward all British Commonwealth scientists.[67]

Canadian University Scientists and Radar Development, 1940–1943

Between 1940 and 1945 the National Research Council had a variety of obligations in the radar field. One of these was a contractual commitment to produce prototypes of U.S. radar sets that would then be manufactured by Canadian companies, usually Research Enterprises, for the British Armed Forces. The Council also had the obligation, as the primary scientific agency of the Canadian Armed Forces, to design and produce specific kinds of radar for their immediate operational use. The problem was that the NRC faced serious shortages of trained radar personnel, for both their own labs and those operated by the Services.[68] This meant that almost from the beginning the NRC had to depend on the universities to provide specialized research and training of radar technicians.

In August 1940 a special NRC committee recommended two options to deal with this situation: the establishment of crash courses in radio physics and engineering at all Canadian universities and the recruitment of radar specialists from the United States. In the latter case, it was recommended that the Council obtain the names of Canadian graduates who were engaged in radio work in the United States 'in order that these men may be approached to serve at home or abroad [Britain] if necessary.'[69] This strategy was relatively effective until Pearl Harbor, when U.S. neutrality laws and security guidelines created barriers for non-nationals to participate in war-related research. It was in this context that Canadian-born A.G. Shenstone, professor of physics at Princeton, was authorized 'to visit and inspect secret work going on in the U.S.' on behalf of the NRC during the summer of 1940.[70] But not all such overtures were successful. A case in point was the negative reaction of C. Blewett, a Ph.D. student from the University of Toronto, who in December 1940 was employed at the General Electric Laboratory at Schenectady, New York. Despite the personal appeal of his former classmate A.H.R. Smith to join the Radio Branch, Blewett rejected the offer on the grounds that his research with GE 'would be of greater value to our cause than anything he could do in Ottawa ... (with) research on ultra high frequency generators of the magnetron type which ... were so promising as ... to render the best British magnetron obsolete.'[71] Blewett also contrasted his salary of $3000 (U.S.) and the stability of his position at GE with the low pay and uncertain long-term job prospects with the NRC.[72]

Another factor in the shortage of trained radar personnel was the siphoning of young Canadian physicists to the British Armed Forces. This had begun in January 1940, when the British Admiralty asked Vincent Massey, Canadian High Commissioner in Britain, whether Canada could supply forty-three physicists and electrical engineers 'with special knowledge of radio and considerable practical ability ... capable of being trained rapidly to understand, operate and control the special apparatus of this service.'[73] R.W. Boyle and J.T. Henderson of the NRC were entrusted with the recruitment campaign.[74] Most of the initial twenty university recruits were obtained through the efforts of four prominent physicists: Eli Burton of Toronto, J.A. Gray of Queen's, R.C. Dearle of Western, and David Keys of McGill.[75] However, two important administrative details had to be clarified: would the radar recruits be members of the Royal Navy for the duration of the war, and if so, at what rank? This situation was clarified. In April 1940 Admiral Percy Nelles of the Royal Canadian Navy announced that the recruits would be appointed as lieutenants or sub-lieutenants on loan to the Royal Navy 'on Canadian pay.'[76]

From all accounts, the program was a success.[77] In January 1941 one of these graduates wrote Eli Burton a glowing letter of appreciation for the quality of training he had received: 'We have had ... practically all the fundamental theory that we will require ... [and] the University taught us too to think for ourselves ... with this combination I see no reason why we should not give a good account of ourselves.'[78] Even more important was the February endorsement of C.S. Wright, Director of the Department of Research and Experiment of the British Admiralty, who praised the high performance of these young Canadian radar operators and asked for another thirty.[79] While the RCN approved Wright's request, it insisted on its right 'to bring some of these men back to Canada for service from time to time.'[80]

Although preparing young physicists and engineers for war service was an important function of Canadian universities, their major contribution was in developing new weapons, and in 1940 this usually meant advanced radar equipment. Initially, university scientists operated under General McNaughton's 30 September 1939 guidelines, 'favouring the use of existing facilities and teams of research workers ... and definitely opposing any centralization of work in the Council's own laboratories.'[81] This was welcome news to most university physicists. J.A. Gray of Queen's University, for example, made it clear in a May 1940 letter to the NRC that he expected a $5000 grant to support his short-wave radar research, since there were 'as far as ordinary radio engineering is concerned ... few, if any, better equipped (laboratories) than ours.'[82] He also pointed out that his proposal for 'ultra high frequencies' had been endorsed by Ralph Fowler, Britain's chief scientific liaison officer, as being 'of the highest importance.'[83] Gray got his grant, and his laboratory made a substantial contribution to microwave radar research during the next four years. This was, in part, due to the quality of his research team, led by R.A. Chipman, and to the support he received from two of his former students – W.J. Henderson and H.M. Cave, who were key members of the Radio Branch's microwave program.[84]

Toronto's Eli Burton was an even more influential advocate of university-based radar research. Burton's role as director of the McLennan Laboratory ensured that his views carried weight on campus, in Ottawa, and with Ralph Fowler.[85] By the summer of 1941, Burton was supervising five major projects, including work on radar and communication sets used by the British and Canadian naval forces and the top-secret proximity fuse research of physicist Arnold Pitt. Both projects profited from Burton's access to funds through the McLennan Laboratory Research Fund[86] and through the Viking Foundation, set up by Swedish industrialist Axel Wenner-Gren in 1940.[87] This latter connection worked well until January 1941, when Wenner-Gren came under attack from the *Financial Post* and the *Toronto Star* because of his alleged pro-Nazi sympathies.[88] Despite Burton's attempt to refute charges that Wenner-Gren 'might inadvertently get ... secret information' from the McLennan Laboratory, the university decided to distance itself from the Viking Foundation.[89] Fortunately, grants from the NRC and the Armed Forces more than offset this loss of funding.[90]

A more serious problem facing the University of Toronto was the siphoning of faculty physicists and engineers to the Canadian Armed Forces, the NRC, and Research Enterprises Ltd. In September 1942, for example, Burton reported that four of his colleagues – J.O. Wilhelm, C. Barnes, A.B. Misener, and D.S. Ainslie – had been seconded by the military.[91] In April 1941, the university decided to counteract this academic emigration by offering the deanship of engineering to C.J. Mackenzie, with an annual salary of $12,000, and the appointment to take effect 'whenever the war is over, or whenever you are free.'[92] Mackenzie declined on the grounds that his first duty was 'and must remain for the duration of the war to the National Research Council in its war effort and to Lieut. General McNaughton whose responsibility in that regard I assumed.' He was also concerned that acceptance of a job with Toronto would prejudice his 'standing as an independent and impartial' adjudicator in dispensing wartime grants to the country's university scientists.[93]

Indeed, many 'outback' universities had already complained that the large central Canadian institutions were receiving a disproportionate number of lucrative war projects. One of the most vigorous protests came from the president of Dalhousie University, C. Stanley, who argued that his institution had been doubly jeopardized by wartime developments: it had been forced to undertake a large number of defence contracts, and it was experiencing serious financial problems.[94] Stanley's appeal elicited a sympathetic response from C.J. Mackenzie, who had good reason to appreciate the valuable contribution of Dalhousie's scientists, particularly physicists J.H.L. Johnstone and W.J. Henderson, who had developed imaginative degaussing methods for neutralizing German magnetic mines.[95] As a result, a compensation package was arranged. Johnstone and Henderson were seconded to the RCN Halifax Research Station in May 1942,[96] while Dalhousie was compensated for this loss by a special grant of $1200 a month for a two-year period.[97]

Physicists at the University of Western Ontario (UWO) were also involved in radar research, largely through the efforts of R.C. Dearle and his assistant G.A. Woonton,[98] whose microwave research was well regarded by the NRC.[99] Although Dearle received strong support from R.W. Boyle, director of the NRC's Physics Division, he was often critical of how the NRC awarded radar contracts.[100] This brought a heated response from C.J. Mackenzie, who lectured Dearle on the realities of war-related research and emphasized that university laboratories must accommodate themselves to NRC priorities:

Our situation here is getting very serious. The urgent problem now is one of getting into actual production and manufacturing of equipment even if it is not the best and getting it into the hands of the soldiers in the field. The Chiefs of Staff of the three Services have just formed a very strong committee charged with the responsibility of seeing that this is done and there is no doubt in the world that the most effective way of arriving at this need is to concentrate the workers in the existing laboratories at Ottawa where they are in daily contact with the equipment and prototypes being made and where constant contact can be maintained with the tactical and Service personnel.[101]

Dearle was outraged by this proposal. Was Mackenzie really suggesting that the UWO physics department be closed down and its faculty transferred to Ottawa? Did this mean 'that no further war research should be done in university laboratories?' Was Mackenzie aware that UWO scientists had given their services 'freely and gladly as a contribution to our concerted scientific war effort'?[102] Confronted in this manner, Mackenzie quickly backtracked. Of course, Dearle's team should remain at Western and continue their research projects as long as the NRC was able 'to get actual prototypes made so that Research Enterprises Limited can get on with the manufacture.'[103] In the end, a compromise was reached: Dearle's team would continue their research on improving microwave antennae in London, Ontario.[104] It was a wise decision. In January 1942 Dearle's laboratory received high praise from G.P. Thomson, the most recent British liaison officer between the NRC and the Air Ministry, who commented on 'the outstanding results ... obtained with flattened wave lengths' and with the forthcoming measurements 'of radiation resistance.'[105]

But it was at McGill that the most extensive microwave research was being carried out. This was not surprising, since that university's J.S. Foster was one of Canada's most gifted and prestigious physicists, with many scientific contacts at the Radiation Laboratory.[106] In the spring of 1941 Foster was assigned to the Radio Branch's highest-priority job: helping to develop the GL Mark IIIC fire control system, a task that required close liaison with Rad Lab scientists. His visit to MIT in March was informative, revealing the enormous progress Rad Lab scientists had made in developing a precision gun-laying radar system, as well as how backward were the Canadian and British efforts by contrast.[107] After being exposed to this environment, Foster did not find it easy to return to his small McGill laboratory with its limited NRC budgets.[108] He was elated, however, when in July 1941 he received an invitation from L.A. DuBridge to join the Rad Lab 'and bring along any others who ... were free to come.'[109]

Foster's departure for the greener pastures of the Radiation Laboratory created a number of problems for Canadian radar research.[110] In the first place, McGill's microwave and gun-laying projects were left in limbo since no one else at McGill possessed Foster's scientific skills, drive, and administrative ability.[111] Mackenzie was also concerned that Foster's recruitment by the Rad Lab might encourage other talented Canadian scientists to gravitate to American research centres.[112] This fear of losing Canada's brightest researchers helps to explain why the NRC president reacted so badly when Foster tried to include one of his Ph.D. students in his entourage. 'I could not,' Mackenzie wrote, 'possibly facilitate any young man of military age leaving Canada for what amounts to a job in a foreign country, without any real authority in connection with his future service.' He also opposed Foster's appointment as the official NRC representative at MIT on the spurious grounds that the McGill physicist had 'never worked in Ottawa and is not conversant with all our problems.'[113] It was only when Rad Lab's director, L.A. DuBridge, intervened that Mackenzie backed down, and Foster was given the responsibility of 'carrying information on recent developments back and forth' between the Rad Lab and Ottawa.[114] Mackenzie's concerns, however, were duly noted by the NDRC hierarchy. Both DuBridge and Alfred Loomis made a point of assuring him that they had no interest in pirating Canadian scientists away from the NRC and that their only concern was finding 'a place where Canadian citizens could be of use in the defence effort.' DuBridge also emphasized that in the case of Foster, 'the work he does here will be of value to us and the experience he gains will be great value to you.'[115]

Throughout these complex and frustrating negotiations, Mackenzie always kept in mind one essential fact: the vital importance of the Rad Lab connection to the NRC's well-being. More specifically, he was advised that Canada's radar program could not operate effectively without MIT data and equipment and the opportunity for NRC scientists to observe research 'on both 10 cm and 3 cm sets.'[116] Mackenzie was, therefore, more than receptive when DuBridge suggested, in the fall of 1941, that the National Defense Research Committee carry out a comprehensive inventory of the radar equipment that was 'in production, test and design in the U.S. and Canada.' He readily accepted DuBridge's proposal for comparative operational radar research so that 'questions dealing with the proper design of a set to fill certain desired military characteristics ... could be handled much more satisfactorily on the basis of a complete knowledge of the comparative performance of different existing outfits.'[117] By this stage in the war, the Canadian radar group consisted of 229 scientists, technicians, and support staff, distributed among the NRC (141), the Navy (21), the Army (20), the RCAF (38), and Research Enterprises Ltd. (9).[118]

During 1942 and 1943 the Radio Branch was also affected by British and American attempts to achieve greater radar cooperation and coordination. As was so often the case, British scientists took the initiative in calling for changes. This was reflected in the formation of the British Central Radio Bureau, which sought to facilitate 'the exchange of data relating to radio research, development and manufacture, between the Governments and industrial organisations of the U.S.A., the Dominions, Great Britain and the other Allied nations.'[119] An even greater catalyst for cooperation was the establishment, under the auspices of the British Chiefs of Staff, of a special Radar Sub-Committee under the chairmanship of Sir Henry Tizard, with a mandate 'to recommend on major issues in Radio Communications, Radar and Radio Counter-Measures.'[120] On the American side, the Radar Development Planning Committee, under the chairmanship of MIT's president Karl Compton, was established in September 1942, with a mandate to work closely with the Telecommunications Research Establishment and other British radar research establishments.[121] This interaction greatly improved in April 1943, when Compton's mission to England worked out a number of important arrangements with their British counterparts, the most important being that the United States 'should generally take the lead in longer-range projects, such as pushing the development of one-centimeter radar components.'[122]

While American and British scientific mandarins debated broader radar policy, NRC scientists were primarily concerned with acquiring badly needed radar information and components from the United States. More specifically, they were apprehensive about the formation of the U.S. interservice Electronic Research Supply Agency (ERSA) in April 1943.[123] While the NRC applauded this attempt to achieve greater industrial rationalization, they were distressed by the stipulation that Canadian agencies could be placed on the ERSA list 'only if the development work is in behalf of the U.S. Armed Forces.' Such a measure, Canadian representative in Washington pointed out, was a violation of the terms of the Hyde Park Agreement of 1941, which had granted the Canadian Armed Forces procurement rights in the United States. Even worse, it was feared, such a policy would devastate the various radar projects being carried out by the NRC, by Research Enterprises Ltd., and by major Canadian electronics companies.[124] In this instance, the Canadian appeal was successful.[125]

Radar and the Royal Canadian Navy

In the years 1940 to 1945 Radio Branch scientists devoted much energy and time trying to meet the radar needs of the Royal Canadian Navy. This was a frustrating task, in part because few RCN or NRC scientists had experience at sea and were thus unable to keep the NRC informed 'on operational as opposed to strictly technical requirements.'[126] In addition, at the outbreak of war the RCN had been slow to develop its own radar technology, preferring to remain 'totally reliant on the RN for all its technical and scientific needs.'[127] Although Naval Headquarters took the initiative in 1941 in securing the services of J.R. Millard and D.C. Rose, the emergence of an effective R&D relationship between the RCN and the NRC was slow in developing.[128] Yet another problem, emphasized by historian David Zimmerman, was the poor communications between the Royal Canadian Navy and the National Research Council, a flaw Zimmerman attributes primarily to C.J. Mackenzie's determination 'to increase the power and prestige of the research council, something that put him in direct conflict with the RCN's requirements.'[129]

Yet the initial relationship between the Radio Branch and the Royal Canadian Navy seemed satisfactory to both sides.[130] In June 1940, for example, the Radio Branch devised and established a shore-based radar system, the Night Watchman, for Halifax Harbour.[131] The NRC was also involved in attempts to standardize radar operating standards between the RCN and Britain's Royal Navy. But the real problem was detecting and sinking German submarines, which by the fall of 1940 were poised to attack the convoys inching their way across the North Atlantic toward Britain. Unfortunately, the Canadian Navy was ill equipped to meet this serious challenge since its destroyers and corvettes were only belatedly provided with the British long-wave ship radar set, the 286, and its Canadian clones, the experimental CSC (Canadian Sea Control) and the operational SW1C (Surface Warning One Canadian).[132] Although this technology performed a useful role in navigation and in detecting surface vessels and aircraft, it had great difficulty in detecting trimmed-down U-boats in the turbulent North Atlantic.[133] The situation did not improve substantially during 1942, when German U-boats sank 755 Allied merchant ships between January and September, many of which were being shepherded by the RCN.[134]

The question of whether to acquire British or U.S. radar equipment or to design and produce Canadian sets, became a major issue in 1942, when the operational advantages of 10 cm ship-borne radar sets became obvious. The RCN had three choices: purchase the British 271 set, which was being used so successfully by the Admiralty; acquire an already existing American set; or manufacture the RX/C, a set designed by NRC scientists.[135] After much indecision and delay, the RCN exercised all three options; however, crews of Canadian merchant ships and RCN corvettes paid a high price for this vacillation.[136]

Sound-detection systems were another priority of the Royal Canadian Navy, and throughout the war the NRC provided expertise in the field of sonar research, or 'asdic,' as the British called it. The RCN was also fortunate in being able to draw on the expertise of A.E.H. Pew, a member of the July 1940 British Admiralty technical mission, whose primary goal was to investigate the possibilities of having asdic sets produced in Canada. Pew's mission was a success – by the end of the war 'nearly 2,600 sets were produced in Canada for the RCN, RN, USN, and other Allied navies.'[137] Scientists from the NRC and Navy also refined and tested antisubmarine weapons such as the Hedgehog and Squid, which became operational in 1942 and 1943.[138] Unfortunately, most Canadian corvettes, the mainstay of convoy service, were badly suited to deploy these weapons because of inadequate wiring and their lack of a modern gyro compass.[139]

Radar and the Canadian Army

One of the major projects of the Radio Branch in the period 1940–2 was the development of the Canadian 10 cm GL Mark IIIC for use by the Canadian Army and the British War Office. The set had two portions: the Accurate Position Finder (APF), 'providing accurate range, bearing and elevation ... housed in a rotating trailer,' and the Zone Position Indicator (ZPI), which showed the 'plan position of all aircraft within a radius of 60,000 yards ... and [conveyed] the information concerning the course, range and bearing of the incoming target to the APF set.'[140] Yet, despite a good start, the set was soon plagued by major technical and administrative problems. In 1940 nobody knew 'how to design a waveguide with a low-loss rotating joint, so as to make it possible to rotate only the antennas.' Another problem was the difficulty in obtaining sufficient quantities of magnetrons, cathode-ray tubes, and other vacuum tubes either from Canadian companies or from the United States.[141]

Despite the field tests of July 1941, the task of developing from a prototype into a mass-produced weapon was formidable, since 'over 300 engineering changes were necessary during the course of production.' Some of these were related to the emergence of new technologies, such as the need to integrate the Identification, Friend or Foe (IFF) system so that GL Mark IIIC–assisted anti-aircraft guns did not target friendly planes.[142] Another problem was the dispute between NRC physicists and Research Enterprises engineers over production schedules.[143] But the most bitter conflicts occurred at the management level, where Colonel F.C. Wallace and Research Enterprises manager R.A. Hackbusch engaged in a power struggle that was not resolved until the fall of 1943.[144] Although Wallace eventually prevailed, the delay in production seriously threatened the future of this weapon. In January 1944 British authorities abruptly cancelled their $26 million order for 400 GL IIIC sets, deciding, instead, to acquire the highly touted American SCR-584 gun-laying system.[145] It was a heavy blow to Radio Branch scientists, who saw their chance of placing a major Canadian weapon in the battlefields of Europe thwarted.[146]

This dilemma of 'too little, too late' also affected other NRC/REL radar projects. However, in some cases, the research went on to have valuable postwar use, with a special antenna, designed by W.H. Watson of McGill, that gave 'coverage up to 70 degrees in elevation,' and was relatively immune from jamming by any potential enemy.[147]

Radar and the Royal Canadian Air Force

The Royal Canadian Air Force was a major Radio Branch customer for Air to Air sets (AI), Early Warning Systems (EWS), Air to Surface Vessel (ASV) equipment, and Long Range Navigation Systems (loran). During the Battle of Britain, providing AI radar sets for the RCAF was the NRC's highest research priority, and the Council worked in close contact with Britain's Telecommunications Research Establishment and the U.S. Radiation Laboratory. By 1941, however, RCAF crews were increasingly being employed in the campaign against German U-boats operating in the western Atlantic/Gulf of St. Lawrence region, and they badly needed advanced microwave radar equipment to carry out their tasks effectively.

As one study has noted, most radar-fitted destroyers could search 75 square miles of ocean per hour searching for enemy submarines, while a similarly fitted airplane could cover 1000 square miles per hour with meter wave radar and 3000 square miles with microwave equipment.[148] Unfortunately, the RCAF's Eastern Air Command, which had responsibility for combating the coastal U-boat menace, could not obtain sufficient numbers of long-range British Liberator bombers or 10 cm ASV radar sets until the spring of 1943. This lack of microwave equipment was particularly serious since by this stage of the war most German submarines were equipped with Metox, which allowed them to detect the long-wave radar beams of Allied aircraft well before an attack could take place.[149] Additional problems for the RCAF were the dearth of ASV specialists in the RCAF and the fact that 'no one in senior command had any first-hand experience of anti-submarine warfare.'[150]

The Royal Canadian Air Force was better served by EW radar technology. By 1945 two EW systems were operative: the most extensive, consisting of twenty-two radar stations, protected the eastern approaches to Canada; another eight stations were located on the west coast. Both systems were closely integrated with the U.S. military and the Royal Canadian Navy.[151] But the Air Force was also interested in preparing itself for the postwar period by acquiring the sophisticated Microwave Early Warning Radar (MEW) system from the United States.[152] This meant, of course, using the NRC's liaison network. In January 1943 Colonel Wallace of the Radio Branch made arrangements with DuBridge of the Radiation Laboratory for Radio Branch scientists to consult with Rad Lab experts about microwave EW systems. It was a long-term arrangement: for the next two-and-a-half years D.W.R. McKinley and his associates made regular pilgrimages to MIT,[153] while top-secret components and progress reports flowed northward.[154] Rad Lab officials did not appear too concerned about safeguarding their valuable components. In August 1944, for example, Durnford Smith, who had joined the Radio Branch from McGill, was allowed to take a top-secret 4j34 magnetron back to Ottawa with him after his MIT visit, with no questions asked![155]

Another important Rad Lab project, which took form during the latter stages of the war, was the development of more effective loran equipment. The essence of this navigational system was explained in an April 1944 NDRC report:

LORAN is a system of position finding, on the sea or in the air, but the reception of radio signals from transmitting stations of known position ... the system aims to furnish reliable positions to navigators at greater distances from the transmitting than is possible by other methods of radio navigation ... Loran observations are possible at all times and in all weathers ... Loran transmitting stations emit steady successions of pulses (short, sharp signals) which travel outward in all directions at the speed of light. The stations operate in pairs, and a navigator must receive pulses from at least two pairs in order to find his position.[156]

Given its importance for naval and aerial warfare, British and Canadian authorities were anxious to acquire this technology. However, this initially proved to be somewhat difficult because of the U.S. Navy's protective stance, a position that contravened the exchange guidelines that the Office of Scientific Research and Development had worked out with its allies. In an attempt to rectify this problem, Vannevar Bush sent a letter to Rear Admiral J.A. Furer in February 1943 complaining about the proprietary attitude of the Office of Naval Research and its failure to provide the British with loran information.[157] It was another six months, after a formal complaint from British authorities and an official report from the OSRD, before the technology was released to the Royal Air Force.[158] Loran, along with another U.S. navigation aid, H2X, subsequently became a key component in the intense Allied bombing of German cities.[159]

The Royal Canadian Navy and the RCAF were also interested in loran. In October 1943, for example, F.H. Sanders of the Radio Branch requested the opportunity to meet with members of the Rad Lab navigation research group.[160] At the same time, the RCAF was considering loran as part of its postwar home defence system with links to an elaborate land-line communication system and a complex air command fighter control network. In 1944, Matt Nightingale, an electrical engineer with the RCAF's department of electronics and development, was one of those responsible for installing these complex defensive systems.[161]

The NRC and the Canadian Military, 1942–1945: A Critique

Canada profited from its access to U.S. radar secrets, for both its own and Britain's radar needs. The Tizard mission laid the groundwork for this cooperation, and its work was reinforced by the liaison activities of R.H. Fowler, E.G. Bowen, and Tizard, who had personal and professional associations with the NDRC scientific mandarins. The Tizard–Bush–Mackenzie connection was particularly important in developing an effective radar exchange system. This was not an easy undertaking, since it was necessary to overcome the reluctance of the British, U.S., and Canadian armed forces to relinquish control over such an important weapon system.

Mackenzie enjoyed early success in dealing with the Armed Forces, but as the war dragged on he increasingly had to overcome the resistance of some members of the Canadian military establishment who resented his influence. Some of these officers had visions of following the British example and establishing their own fully equipped research operations. This move toward Service-directed defence science was evident in March 1942, when the Army created its own Technical Development Board (TDB) with a mandate to produce new weapons and 'to coordinate all Army development projects except in such fields as chemical warfare and radar where standing boards and committees were already functioning.'[162] Both Mackenzie and Maass welcomed the TDB as a useful partner in developing new weapons. In contrast, the July 1942 decision by the RCN to establish its own Directorate of Technical Research came after a series of disagreements with the NRC over the quality of new weapons.[163] According to historian David Zimmerman, this RCN–NRC confrontation was more than a difference of opinion; it indicated a 'mismanagement of Canadian science' that was 'damaging to the nation's war effort':

The difficulties stemmed from ... the different institutional goals of the navy and the council. Mackenzie saw as his first priority the fulfilment of the council's pre-war dream of becoming an internationally recognized research centre. He did not see the future of the NRC as resting with military research, and did not give satisfactory attention to working out the organizational details that would have led to the more effective utilization of scientific resources by the RCN. The navy also did not provide the leadership necessary to pursue these goals.[164]

Given their heavy losses due to inadequate technology, the frustrations of the RCN are understandable. It should be kept in mind, however, that in 1939 the NRC was forced to establish an entirely new defence science system virtually overnight. Moreover, the urgency of the military situation demanded quick results, especially as the number of Canadian ships sunk by German submarines continued to grow. In addition, the British were not always helpful in supplying technical data and scientific manpower, despite the fact that many of the NRC's projects were in response to British demands.[165] Furthermore, none of Canada's universities was equipped to carry out the types of projects being routinely undertaken at the research laboratories of American universities such as Johns Hopkins, MIT, or Columbia. By the same token, none of Canada's industrial corporations could provide the kind of research and development that the OSRD received from Bell, Du Pont, or General Electric. In December 1943, Mackenzie explained these realities to a group of prominent Montreal corporate executives: 'When the war broke out we were the only people in Canada that could be contacted on secret research work ... repeatedly we had requests from England to have our industrial companies carry on certain development work but the answer always had to be there are no private research laboratories in Canada.'[166]

Conclusion

Radar was one of the wonder weapons of the Second World War. As the American historian Daniel Kevles has noted, 'To most American physicists before 1943 ... physics in World War II meant the Rad Lab ... The atomic bomb only ended the war. Radar won it.'[167] But in 1940 it was British not American radar scientists who were the most advanced, at least until the exchange system initiated by the Tizard and Conant missions. These complex negotiations also had important ramifications for Canadian radar scieniststs, especially after September 1940, when the National Research Council became a useful participant in the emerging Anglo-American radar partnership, even though it was highly dependent on the Rad Lab for advanced information and components. Still, Canadian university and Radio Branch scientists became involved in a number of significant radar projects on behalf of both Britain and the Canadian Armed Forces. Although the British connections remained important, by the latter stages of the war NRC officials were aware that the United States was now the dominant force in radar research, development, production, and operational use. This shift from British to American technological superiority would also occur in the case of most other major weapons systems.

4

Weapons Systems: Proximity Fuses and RDX

The work of the council is increasing day by day and the pressure at times has been great. The week that Banting died it looked as if my staff were vanishing ... Maass, who has been carrying a terrific load and has been of the greatest assistance to me, is on the verge of a nervous collapse and I cannot get him away for a rest; Flood, whom I think is today the greatest expert on chemical warfare on the American continent, has been working very hard with a resultant breakdown and we will have to send him away for two weeks. However, we will have to carry on.

C.J. Mackenzie to General A.G.L. McNaughton,
5 March 1941; cited in Thistle, ed., *The Mackenzie–McNaughton Wartime Letters*, 69

The year 1942 was a difficult one for Canada's military planners and political leaders, since most of the Allied campaigns against German and Japanese forces went badly. In addition, the military débâcles at Hong Kong and Dieppe had a negative impact on the Canadian war effort, reinforcing an image of government ineptitude. To make things worse, the Liberal Cabinet was seriously divided over the issue of imposing conscription for overseas service.

On the positive side, Canada's wartime economy was booming. The Department of Munitions and Supply, under C.D. Howe's direction, was overseeing the production of vast quantities of munitions, small arms, and artillery pieces, as well as sophisticated aircraft like the Mosquito and Lancaster, and a variety of naval corvettes, frigates, minesweepers, and destroyers. While one-third of Canadian war production went to the Canadian Armed Forces, over half was destined for the United Kingdom and other Commonwealth countries.[1]

During the previous three years, the National Research Council had worked closely with Munitions and Supply and the Armed Forces in developing various new weapons and improving old ones. This activity required a large component of scientific talent, and by 1942 the NRC had expanded its staff to about 1400 scientists, technicians, and support staff, and plans were being made to expand NRC activities further. In October the NRC's acting president C.J. Mackenzie asked the Treasury Board to recognize the Council as an official war unit, in addition to its status as official research station of the Army, Navy, and Air Force. This designation, he argued, would give the NRC equal priority 'in the purchase of equipment and supplies necessary for its war work,' with DND and M&S.[2] In preparing his case, Mackenzie had asked each NRC division to provide a summary of its war-related research. The following profile emerged:[3]

1. *General Physics*. This section is working exclusively on war problems generally for the Navy ... the work involves designing of chronographs, automatic gun sights, predictors, anti-submarine equipment etc.
2. *Radio Section*. This section is devoted entirely to war work. The peace time staff of three or four has grown to nearly three hundred, and operations consist of designing and construction of equipment for the military Services ...
3. *Chemistry*. About 95% of the work ... is connected with war projects and tests. The peacetime staff has approximately doubled.[4]
4. *Mechanical Engineering*. Of the eight sections of the aeronautical laboratories six are working exclusively on war projects ... The total laboratory staff numbers 114.[5]
5. *Applied Biology*. There are 50 members of the staff ... most devote their full time to war activities ... in the fields of fermentations, food storage and transport and plant science.

Between 1939 and 1942 each of these NRC sections was involved in the development of a number of important weapons. In the case of mechanical engineering, emphasis was placed on aircraft components, problems of de-icing aircraft, and improved techniques in aerial photography. The division also contributed to the construction of the Weasel, an imaginative snow vehicle with specially designed rubber tracks that later went into mass production in the United States and was deployed in the European and Pacific theatres of war.[6] The Physics Division was equally busy, perfecting a wide variety of radar equipment, developing countermeasures against German magnetic and acoustic mines, and expanding its asdic/sonar research. NRC physicists also participated in developing two of the war's 'wonder weapons': the proximity fuse and the atomic bomb.[7]

Despite claims that the Second World War was a physicists' war, NRC chemists also made major contributions to the allied war effort through their involvement with explosives such as RDX and DINA. Canada also became a major player in the field of chemical warfare after the establishment of the Suffield Experimental Station in Ralston, Alberta, in the summer of 1941. This shared facility, in which Canada provided the site and the staff, and Britain the advanced scientific information and weapon contracts, provided a model for other joint wartime projects.

Throughout 1942 and 1943 the National Research Council maintained a frantic pace in trying to meet the demands of the British government, the Department of Munitions and Supply, and the Canadian Armed Forces.[8] Although Mackenzie sometimes had difficulty dealing with strong-willed university scientists, his most frustrating experiences were in coping with the myriad of visiting British experts, whose missions often seemed 'without any anchor,' and who often failed to appreciate how Canada's war effort differed from their own.[9] Mackenzie was also displeased with the pattern of British procurement in North America, which he felt unfairly discriminated against Canadian companies and research institutions by favouring U.S.-based operations.[10]

Proximity Fuse Research and Development, 1940–1943

While proximity fuses assumed many functions during the war years, they were initially used to improve the efficiency of anti-aircraft artillery. During the Battle of Britain, operational research scientists such as Patrick Blackett had made great progress in improving the effectiveness of gun-laying radar sets by coordinating the actions of the different batteries and analysing the ratio of success, or 'rounds per bird.'[11] But while these tactical measures refuted the prewar dictum that 'the bomber will always get through,' the kill rate of anti-aircraft fire remained low, since direct hits on a fast-moving bomber were extremely difficult.

Proximity fuses represented an enormous technological improvement over conventional artillery fuses because they did not require a direct hit: both the radio and the electrical fuse would cause a shell to explode automatically when its trajectory passed within a lethal distance of the aircraft.[12] In 1940 the challenge was to create small but complex fuses that would fit into the head of an artillery shell and be strong enough to withstand enormous centrifugal forces when fired. British scientists addressed these challenges during the summer of 1940, when the limitations of conventional anti-aircraft fire threatened national survival during the German bombing blitz of 1940–1. At this stage, most researchers concentrated on the photoelectric and audio prototypes, although a group working on the radio-controlled fuse would soon achieve impressive laboratory and firing tests. Indeed, by the end of the year, the advantages of the radio system appeared overwhelming: it was more reliable, more accurate, and not as vulnerable to enemy countermeasures.[13]

At this juncture, two British organizations were involved with proximity fuse development. One was the Ministry of Aircraft Production (MAP), especially at its operation at Farnborough and Boscombe Downs; the other was the Air Defence Research and Development Establishment (ADRDE) of the Ministry of Supply, which carried out its research mandate under the direction of John Cockcroft. In February 1941, Henry Tizard, who carefully monitored this work, provided an encouraging assessment of British proximity fuse progress:

There are three lines of work, the proximity fuze to operate against aircraft, a fuze to burst bombs near the ground, and a similar equipment to indicate the height of armament above the ground. All these lines look promising, but too small a staff is engaged in developing them ... Hill says that the P.E. fuze can be safely recommended for use. It will not be 100% efficient but would be free from any danger due to pre-ignition.[14]

Meanwhile, American scientists were carrying out their own experiments. In August 1940, for example, a meeting of the NDRC and the Naval Research Laboratory (NRL) discussed the relative advantages of 'an anti-aircraft fuse operated by its proximity to an airplane ... to be actuated by one of the following means: acoustic, magnetic, optical, or radio.' Further cooperative work was inspired by the arrival of the Tizard mission.[15] On 18 September a select group of American scientists and military officers sat spellbound while John Cockcroft and R.H. Fowler explained the various ways proximity fuses could be used:

A. Two types of standard bombs, 250 and 500 pounds ... with proximity fuzes for use against hostile aircraft by dropping from friendly aircraft ...

B. Rockets; The rockets under immediate discussion were 4 inches in diameter, $2\frac{1}{2}$ metres in length ... provided with a photoelectric fuse ...

The acoustical fuses ... are provided with two paraboloid trumpets which focus the sound over a proper cone from the target ... made sensitive to sound of frequency above 5000 per second. They operate at about 150 feet from the enemy airplane.[16]

During the next four months, proximity fuse research proceeded rapidly on both sides of the Atlantic.[17] By May 1941, however, British scientists decided to concentrate all their efforts on the radio system, a priority that affected not only British proximity fuse work, but also that of the United States, since the visiting National Defense Research Committee delegation of L.R. Hafstad and C.C. Lauritsen were involved in this important decision.[18] Drawing on the recommendations of his NDRC colleagues, Richard Tolman, head of T Division, sent to Vannevar Bush a report reviewing the major contours of British research and outlining the progress of Merle Tuve's proximity fuse team at the U.S. Department of Terrestrial Magnetism.[19]

Tuve, the foremost American proximity fuse expert, had quickly recognized the enormous potential of radio proximity fuses, and during the next four years he would drive and inspire his NDRC colleagues into producing one of the wonder weapons of the Second World War.[20] At the early developmental stage, however, NDRC scientists depended on the expertise of British experts at MAP and ADRDE. The Americans also appreciated the work of the small proximity fuse research group at the University of Toronto, under physicist Arnold Pitt.[21] The initial achievements of the Toronto group were described by C.J. Mackenzie in a 5 March 1941 letter to General McNaughton:

We have been working on the development of a very rugged radio tube which can be inserted in a shell and take 36,000 g and stand also the necessary centrifugal forces. Such a tube has been developed and is being put into production, and is awaiting only the development of the radio devices to complete the proximity fuse. Cockcroft ... knows all about this development.[22]

The Conant mission to Britain in February 1941, however, had negotiated an agreement whereby the United States would have primacy in long-range complex joint projects such as the proximity fuse. This was an enormous opportunity for the NDRC since it would inherit much British proximity fuse expertise. The flip side, of course, was that Tuve would be responsible for providing fuses for both American and British military needs.[23] And the orders soon began pouring in, especially after Pearl Harbor.[24] In February 1942, for example, Tuve advised James Conant, chairman of the NDRC, that since the U.S. Navy required radio proximity fuses for anti-aircraft shells, drastic measures would be required 'to speed the rugged tubes and batteries into production.' This would mean, he added, 'immediate funds for added staff, space and facilities for our laboratory and field groups ... funds for factory reproduction ... to give us a supply of these units for field tests and development changes.' All of these requests were soon met, including a field test site.[25]

Now came the real test. Although the fuse had been miniaturized and made resistant to the enormous shock of firing, scientists still had to determine the wavelengths on which different radio fuses would operate.[26] Another difficulty was the high proportion of duds and premature explosions. There was also the difficulty of analysing performance data from a fuse in flight, which was further complicated by such factors as the type of artillery piece, the propellant charge, the target, and even the weather. According to one participant, 'it was a game of guessing what *might* be failing, what *might* fix it, and then assembling the changing units and firing them to see if one's guess had been correct.'[27] However, as field tests became more sophisticated, scientists began firing shells toward mock copies of Japanese and German planes 'to match fuze and fragmentation patterns to optimize burst efficiency.'[28]

By the spring of 1942, Tuve was able to convince Vannevar Bush that the majority of proximity fuse research and development should be carried out at Johns Hopkins University, a location that would facilitate 'joint activities ... by the NDRC, the Navy, the Army, the British, and the various manufacturers working on the radio proximity fuse.'[29] In addition, plans were made for an elaborate division of labour for each of the major functions: mechanical design, improvement of sensitivity and burst patterns, refinement of components, and development of anti-jamming fuses and defensive jamming techniques. This work was carried out by six operating groups, each concentrating on different aspects of manufacturing, assembly, and experimental field tests.[30] Tuve's operation was put to the test on 13 April 1942, when Section T gave an important demonstration of the radio-equipped five-inch shell at Parris Island, South Carolina, to a group of U.S. Navy observers. The shells fired from the 5"/38 AA guns of the battleship *Cleveland* were remarkably accurate in shooting down three radio-controlled drones. An elated Vannevar Bush, who observed the trials, sent a telegraph to Conant: 'Three runs, three hits, no errors.'[31]

Another challenge was to create new models that would combine both oscillator and amplifier in a single assembly and that could be fitted into the shell's nose cone. This innovation was achieved with the design of the Mark 45, the model for all future types of proximity fuse projectiles since it could 'accommodate different projectile spins.' As a result, arrangements were made for NDRC teams to produce different Mark 45 clones for the U.S. Army's 90 mm AA guns and the British Navy's 3.7 and 4.5 AA guns. This latter undertaking was the most formidable because of the problems of adapting the 2-inch radio fuse into British shells, which were meant for a 1½-inch fuse. Moreover, success in one gun did not guarantee success in another. After considerable consultation with Pitt's Toronto-based group, and with production facilities waiting, Tuve took a gamble and proceeded with the 4.5-inch 'without waiting for an acceptable reserve battery or small safety devices.'[32]

The difficulties of equipping the British Armed Forces with operational fuses were not only technological in nature. Politics on both sides, was also a factor. For example, the Ministry of Supply was reluctant to abandon the idea of a unique British fuse which could be used in conjunction with the Anglo-Canadian GL Mark III radar sets. A more serious problem was the reluctance of the U.S. Joint Chiefs of Staff to provide any operational fuses until all American military requirements were fulfilled. This situation was somewhat resolved in November 1943, when John Cockcroft travelled to Washington to carry out high-level negotiations. As a result of his mission, it was agreed that the British would acquire the superior U.S. gun-laying radar system (SCR 584) as well as 649,000 proximity fuses. This equipment arrived in Britain just in time to defend London from the German V-1 rocket attacks that began in June 1944.[33]

Proximity Fuse Work at the University of Toronto, 1940–1943

In the winter of 1940, a team of physicists at the University of Toronto began working on a Canadian proximity fuse, a project that soon attracted the attention of British experts. Links with British resources were expanded in February 1941, when D.W.R. McKinley of the NRC Radio Branch provided Arnold Pitt of the University of Toronto with a progress report of research he saw at England's Pye Laboratory:

You will note that a reference was made to 550 mc/s as being the resonant frequency of the shell and that work was abandoned because of insufficient power. Verbal reports about the end of December indicated that work was being done on this frequency once more because of newer valves capable of producing the required power. With the advent of high power valves of the 10 c.m., the P.F. boys anticipated going to town on a small dipole attached to the base of the shell which would give a much better radiation pattern than any of the longer wave systems.[34]

By the spring of 1941 Pitt's team was heavily involved in developing fuses for the British Army's 3.7-inch and 4.5-inch shells. This endeavour received high praise from William Lawrence Bragg, the Air Ministry's scientific liaison officer in Ottawa who in August 1941 told Mackenzie that he considered Pitt's research work so important that it should receive all available funding and material support. To make his point, Bragg cited Tuve's strong endorsement of the Toronto operation, and underlined the advantages of having 'an independent group thinking about details of mechanical and electrical design ... a form of insurance against missing some good feature of design.'[35] Although Mackenzie initially had reservations about the military potential of radio fuses, he soon changed his mind. If Pitt's team could make a significant contribution to the Allied war effort, as well as giving the NRC an entry into this new weapon system, it should be fully funded.[36] This message was conveyed to the Army Chiefs of Staff, who immediately agreed to finance the project because of 'the necessity of keeping Canada up to date on all future developments.'[37]

Once they had long-term support, the Toronto group began to concentrate on the technical aspects of this weapon. One of its projects was the development of a self-destruction switch 'based on the decrease of rate of rotation of the shell with time of flight,' a problem that had 'baffled' both American and British scientists. The team also conducted field experiments using different types of proximity AA shells. These highly technical operations attracted the attention of that old artillery officer, General Andrew McNaughton, who volunteered his advice while in Canada for a brief visit during the early months of 1942. In February, Pitt submitted the following progress report to the NRC:

For altitudes up to 25,000 the proximity fuse may have distinct advantages over other types of fuses. This applies particularly for A.A. on ships against the necessarily low altitude bomber attack.

For altitudes above 25,000 we would agree that co-incidence fuses would be the best assuming that radio co-incidence equipment can be made to follow accurately at this height ... The comparative value of the pulse and proximity fuses depends upon the position of the bursts relative to the target.[38]

By this stage in the war, the potential of proximity fuses had attracted new supporters. The Canadian Army, for instance, was interested in using fused artillery attacks on enemy ground forces. Otto Maass, head of the Directorate of Chemical Warfare, saw the possibility of equipping chemical bombs with 'a radio type fuse to produce an air-burst at about 100 ft from the ground.'[39] But the highest priority for Pitt's group remained the development of an effective fuse for British 3.7-inch shells, a task that required continual consultations with Tuve's operation at Johns Hopkins.[40] NDRC scientists regarded the Toronto group as useful partners, an outlook that was enhanced after the Canadians developed a special spin-type battery that provided 'indefinite shelf life compared to a life of approximately six months for the dry battery.'[41] This important innovation was confirmed by field tests carried out at Camp Borden and the Hamilton Bay firing range during the summer of 1942.

Pitt's team had now expanded to include a group of research students from the University of Toronto.[42] One of these was C.C. (Kelly) Gotlieb, who was recruited into the project by Eli Burton, the dean of physics-related war research at the university.[43] Gotlieb's major task was to carry out field experiments for using the radio method of studying the yaw, or 'angular motion of a shell in flight.' Tests were carried out at the Hamilton firing range. According to Gotlieb, these experiments were often quite exciting since shells were normally fired straight up from the sandbag bunker: 'You would hear for a couple of minutes the scream of the shell getting louder and louder and about one in ten would hit the dugout with a big thump.' Once the shell had landed, the components of the fuse were checked to determine how well it had withstood the shock, and the degree of 'spin and spin decay.' The results were then sent to Johns Hopkins.[44]

Since Pitt's team was entrusted with the task of adapting U.S. fuses for British shells, Gotlieb was sent to Britain in June 1943 with test results produced by both Toronto and Johns Hopkins. This mission was deemed so important that he flew VIP in a Lancaster bomber.[45]

Operational Use of Proximity Fuses, 1943–1945

Although the United States Army and Navy had placed special orders for a wide array of proximity fuses during 1942, operational use of these weapons remained firmly under the control of the U.S. Joint Chiefs of Staff. However, the JCS was inclined to be cautious in allowing the weapon to be used. In August 1943, for example, it warned 'that the compromise of the influence fuze at this time will result in its early adoption by the enemy for anti-aircraft fire and plane to plane bombing against our bomber flights, as well as permitting development of counter-measures against our anti-aircraft fire, and that this will do our offensive effort harm that will more than out weight any advantage we will gain in bombing operations in which there is a risk of compromise.' The JCS did, however, make a grudging exception for use by the U.S. Navy 'for certain important amphibious operations sufficiently to permit full use for defence against air attack in that area,' but only over open water.[46]

This policy did not please the British Chiefs of Staff, who now appreciated the vast potential of using proximity fuses 'for anti-personnel shells and bombs.' Nevertheless, they deferred to their more powerful ally.[47] And so the matter rested until February 1944, when Vannevar Bush, chairman of the Joint Committee on New Weapons (JCNW), attempted to break the-log jam by circulating a report showing that the advantages of using proximity fuses far outweighed the disadvantages. While impressed with Bush's arguments, the JCS preferred the cautious approach. So the fuses remained under wraps.[48]

There were, however, two notable exceptions to this restriction. One was related to the U.S. Navy's use of the advanced Mark 45 fuse for 90mm anti-aircraft guns against Japanese kamikaze attacks.[49] But the most spectacular use of this weapon occurred in June 1944, when the Joint Chiefs of Staff decided to allow their use against German V-1 rockets on the assurance that they were '2 to 3 times as efficient as ordinary anti-aircraft fuses ... [and] the risks entailed are not disproportionate to the advantages to be gained.'[50] Subsequent reports of this second Nazi blitz verified this claim. The use of the Mark 45 fuse, with the assistance of the SCR-584 gun-laying radar and its accompanying M-9 electrical predictor system, 'destroyed 17% of the flying bombs in the first week, 24% in the second, 27% in the third, 40% in the fourth, 55% in the fifth and 74% in the sixth week.'[51]

One month earlier, arguments had been made for widespread use of anti-aircraft fuses during the forthcoming Normandy landings. What made this deployment risky was that for the first time, proximity fuses would be used over enemy territory. The concern was that if Operation Overlord (the invasion of Normandy) did not bring about the rapid defeat of Germany, enemy scientists might be able to draw on American technology and develop their own proximity fuses and thereby decimate British and American bombing squadrons.[52] Such concerns greatly troubled the Joint Chiefs, but in the end they agreed to a compromise: proximity fuses would be used only in defence of the two artificial harbours and the beachheads, thereby helping to ensure the success of the D-Day landings.[53]

By mid-October the gloves were off. Even the JCS were prepared to sanction widespread use of proximity fuses as a way of 'materially hastening [the] defeat of Germany.' Their decision was made easier after assurances from Vannevar Bush that a ravaged Nazi Germany could not possibly develop operational fused weapons before its defeat. The battlefield impact of this weapon was felt almost immediately. In December 1944, the American Army deployed artillery proximity fuses against German tanks and infantry during the Battle of the Bulge, with devastating results.[54] Plans were also made to equip the RAF and the U.S. Army Air Force with fused 500-pound bombs for use against German cities since it was shown that such weapons would have 'a superiority factor of approximately four to one for air-burst bombs over surface-burst bombs.' By May 1945, proximity fuse weapons had become an essential part of U.S., British, and Canadian tactical arsenals, especially with the introduction of new types of fuses that were resistant to jamming and equipped with IFF identification systems.[55]

Would Germany or Japan have developed proximity fuses if the war had continued?[56] Yes, according to Reich Marshal Hermann Goering, who claimed during a 10 May 1945 interview with U.S. Air Force officials that 'in three or four months' German proximity fuses 'would have been in production.'[57] On the other hand, investigations by Allied ALSOS teams, assigned to interrogate leading enemy scientists, found no evidence that German production of proximity fuses was imminent.[58]

The Chemists' War: Explosives and Propellants

We have recently organized the explosives research on a comprehensive basis. We have a main explosives committee and three subcommittees covering the work at Valcartier ... another operating the Explosives Testing Laboratory on the Montreal Road site and a third subcommittee which directs extra-mural work. This latter subcommittee under Professor Wright, University of Toronto, has

now some fifty projects under way at the different universities, and I think the work being done in Canada on explosives will prove to be a bright spot of our many achievements.
C.J. Mackenzie to General McNaughton, 2 January 1943; cited in Thistle, ed., *The Mackenzie–McNaughton Wartime Letters*, 123

British scientists took the lead in developing improvements over World War One types of explosives. In 1939 trinitrotoluene (TNT) was the major explosive used in British and U.S. shells and bombs; ammonium picrate, picric acid, and nitrostarch explosives were also used.[59] Propellants, which differed from explosives in their rate of combustion, came in many forms; cordite, or smokeless black powder, was the most common.[60] By 1943, however, an assortment of new weapons had emerged: RDX explosives, aluminized explosives, plastic explosives, cordite N flashless propellant, incendiary and phosphorous bombs, and rockets.[61]

Despite their early lead, British authorities were willing to inform their American and Canadian allies about their various projects and to seek their assistance in developing new explosives and propellants. In July 1941 the NDRC sent a mission led by G.B. Kistiakowsky, head of its explosives program, with a list of inquiries about British techniques in dealing with the impact sensitivities of explosives; detonation and shock waves; the mechanics of fragmentation; and the use of phlegmatizers in RDX production 'to achieve greater power and safety.'[62] Some of the NDRC's other concerns related to tactical measures: how to control sensitivity on impact in order to achieve maximum projectile penetration, and how to create mixtures for large bombs 'providing relatively long-sustained blast waves.'[63] Once this information was forthcoming, NDRC chemists made great progress in developing new weapons, and by 1943 they had surpassed British efforts in the incendiary and propellant fields.[64]

Canadian scientists had been involved in explosives research and development even before the outbreak of war. Most of this activity was coordinated by the National Research Council and by the Explosives and Chemical Branch of the Department of Munitions and Supply (DM&S), which was administered by James Richardson Donald, a prominent Montreal chemical engineer and close friend of C.D. Howe's. Donald proved to be an able administrator who effectively used his contacts with major industrial companies and with university chemists, particularly those at McGill. Indeed, his chief lieutenant, J.R. (Jimmie) Ross, was given a wartime appointment to McGill's chemistry department so that he could work closely with Otto Maass and associates.[65] Another valuable contact was Arthur B. Purvis, president of Canadian Industries Limited (CIL), who had been appointed head of the British Purchasing Commission (BPC) in Washington.[66]

In 1940 Donald and Purvis shared a common concern: how to improve the quality and quantity of explosives and propellants for the British and Canadian war effort. It was a daunting task. 'We were ... faced,' Donald later wrote, 'with building large explosives plants, together with ancillary facilities of sulphuric acid, plus power, water, effluents ... [and] a wide range of chemical supplies for smoke bombs, activated charcoal for gas masks, pyrotechnics, solvents – all requiring new manufacturing facilities.'[67]

But under Donald's direction, the pieces began to fall in place. In February 1940, arrangements were made for the production of 12,000 tons of TNT and 15,000 tons of cordite at the new CIL plant at Nobel, Ontario, a task entrusted to Defence Industries Limited (DIL), a wartime subsidiary of CIL. However, by June 1940, British demands had stretched Canadian production facilities to the limit, and a new explosives plant was built at de Salabery (now Valleyfield), Quebec, with British technical assistance. But production still lagged behind demand. As a result, the Department of Munitions and Supply increasingly looked to the United States for badly needed equipment and chemicals, a process that was greatly facilitated by the March 1941 Hyde Park Agreement. Even more important was the subsequent creation of the Joint War Production Committee of Canada and the United States and its more specialized Technical Subcommittee on Chemicals and Explosives.[68] According to Donald, Canada's representative on the Subcommittee, the explosives exchange system was a great success:

Canadian–U.S. cooperation in the wartime chemicals and explosives programmes ... had its beginnings in the friendships and associations of pre-war days, in the common interests of technical and scientific societies, in business affiliations, and in the existing integration of the Canadian and American chemical industry ... From the first, this group of men, bound together by a common purpose, with common training and interests, and with a high regard for each other which has strengthened from year to year, have met and solved common problems in a spirit of informality, goodwill, and mutual respect.[69]

Equally important were the arrangements with British authorities both through the British Purchasing Commission, and the Inspection

Board of the United Kingdom and Canada, which was created in January 1941 to ensure that Canadian-produced ammunition met British standards. In order to meet these obligations, the NRC and the Department of Munitions and Supply arranged for the establishment of an artillery and small-arms testing facility at Valcartier, Quebec, especially to measure the effectiveness of explosives and propellants. The coordination of Canadian and British research and development was further improved in 1942 when the NRC Associate Committee on Explosives and Ballistics was created; during the next three years it worked closely with the Canadian Army's Directorate of Mechanization and Artillery and with officials of the British Ministry of Supply.[70]

The joint development of the explosive RDX by Canadian, British, and American scientists provides an excellent example of Alliance cooperation. Although RDX was originally discovered during the First World War, strategic use of this powerful explosive (known also as cyclonite and hexogen) had been hampered by its unstable qualities. Although some progress had been made during the interwar years by American and British chemists, mass production techniques remained wasteful and expensive. For example, the process used at the British Woolwich Explosives Station only 'converted about half the carbon in the hexamine to RDX [while] ... the other half went to low molecular weight bodies which were highly unstable ... and had a well deserved reputation for periodic violent decompositions.'[71] Yet, despite flaws in this RDX process, by 1940 British Explosives Research Laboratories (Woolwich) did not feel it could change production methods because of the enormous demands for high explosives and the absence of a viable alternative.[72]

Enter Otto Maass. While on a self-appointed mission to England in June 1940, Maass became fascinated by RDX's potential as a 'super-explosive.' His initial attempts to convince Woolwich that Canadian scientists should develop new forms of RDX received little encouragement. Maass was, however, undeterred. On his return to McGill, he mobilized his colleagues James Ross, Raymond Boyer, Charles Winkler, and Robert Nicholls into a special RDX research team. He also convinced George Wright of the University of Toronto to carry out his own RDX experiments.[73] By early 1941 the McGill group had developed a stabilizing process (the Ross-Scheissler process) that promised to revolutionize RDX production.[74]

While many scientists were involved in discovering this new formula, special credit should go to Ross's graduate student Robert Scheissler, who carried out the actual experiments. The essence of the discovery was that 'RDX could be synthesized in about 50% yield directly from the reaction of formaldehyde (in the form of its solid polymer, paraformaldehyde) with ammonium nitrate in acetic anhydride at 65 C.'[75] For a brief time the Ross-Scheissler process was viewed as the only method of making mass production of RDX viable. But this situation would soon change. In March 1941, Werner Bachmann, an organic chemist at the University of Michigan, devised his own RDX system 'which combined features of the nitric acid (Woolwich) and Ross-Scheissler processes,' and obtained high yields of RDX 'by reacting hexamine, ammonium nitrate, nitric acid and acetic anhydride.'[76]

The relative merits of these two techniques became the focus of much research activity in Canada during the spring of 1941, especially in George Wright's University of Toronto laboratory.[77] Wright, a superb organic chemist, has been described as a slave-driver who 'drove his team ruthlessly.'[78] A vivid account of Wright's RDX experiments is provided by Doug Downing, one of his wartime research assistants:

In the summer of 1941 a small unit was set up to carry out the nitration-neutralization-extraction sequence continuously. We called it a pilot plant but it was really a continuous laboratory scale set up. The elevator shaft of the old Chemistry Building was commandeered for a vacuum still which prepared nearly 100% RDX ... The elevator shaft was shut down in August 1941 and attention turned to ... the Bachmann process ... it was found that RDX could be prepared in excellent yields by feeding three liquid streams into a stirred reactor: (1) acetic anhydride, (2) hexamine in acetic acid, and (3) ammonium nitrate in nitric acid. This process, which could be called the liquid feed modification of the Bachmann process, was recognized as an improvement over the original form and was used in all the North American wartime plants.[79]

Driven by his desire to get the job done, Wright did not always observe normal safety procedures. In August 1941, for example, J.R. Donald, while on a site visit, was horrified to see substantial quantities of RDX 'stored in an old elevator shaft,' and warned Wright about this dangerous situation: 'As I pointed out, he and his whole group would probably have been blown to pieces and I would surely have been blown out of a job. While I think Wright felt I took an extreme view, the RDX was promptly removed much to my relief.'[80]

Donald's greatest concern, however, was not monitoring safety measures in war laboratories, but trying to coordinate allied RDX policies.[81]

Weapons Systems: Proximity Fuses and RDX   111

It was a delicate business, since neither the British nor American military hierarchies had committed themselves to using this new explosive. Without this endorsement, C.D. Howe was adamant 'that Canada should not force itself into the RDX situation.'[82] In July 1941, the RDX campaign received an important shot in the arm when British Air Marshal A.T. (Bomber) Harris told the Combined Chiefs of Staff that the RAF desperately needed bigger and better explosives; and this meant expanding RDX production facilities:

He [Harris] points out ... that the estimated output of these four [U.S.] plants is only 240 tons per week, of which he understands we are to receive half. Like Oliver Twist he therefore comes back for more, and requests that we make a further appeal for the provision of very greatly increased quantities. He states that the use of RDX in place of the Amatol filling increases the effectiveness of each bomb by some 40%, and he regards that result as practically equivalent to a similar increase in the size of our Bomber Force ... our need is demonstrated by the fact that we require 5,000 tons of high explosives per week for bomb filling in 1942, rising to 8,000 tons per week in 1943. Ideally, of course, all this should be RDX.[83]

Another British advocate for quick action was R.P. Linstead, an organic chemist and consultant with the British Central Scientific Office in Washington.[84] He recommended that Canada immediately build an RDX pilot plant, using the Ross-Scheissler process in anticipation of 'large scale operation.'[85]

While Donald and Howe were pleased with the Canadian RDX laboratory experiments, they remained wary about committing financial and industrial resources to creating such an expensive manufacturing facility, especially since DIL experts estimated that a plant producing 300 tons of cyclonite per month 'would cost about $6,500,000.'[86] Another deterrent was the fact that RDX production required vast amounts of synthetic methanol, which would involve building a Canadian plant at a cost of between $750,000 and $1,000,000.[87] Howe was not prepared to make such a commitment, proposing instead that the United States assume sole responsibility for developing a mass production RDX facility that would meet the military needs of the American, British, and Canadian armed forces.[88]

Howe's idea was premature. Before negotiations for new RDX plants could be finalized, a decision about the best production process had to be determined. This decision depended on the conclusions of George

Wright's University of Toronto team, who were investigating the relative merits of the Ross-Scheissler and Bachmann processes. In August 1941 they came to a decision: Bachmann![89] This message was personally delivered by Wright when he attended a high-level meeting of NDRC explosives experts in Washington, chaired by George Kistiakowsky of Harvard. After considerable discussion about recent research activities, it was agreed that RDX produced by the Bachmann process, now called RDX Type B, would dramatically improve the explosive damage of bombs, shells, rockets, torpedoes, and underwater depth charges.[90]

Across the Atlantic, Woolwich officials had reached the same conclusion, although their ability to assess the Ross-Scheissler process had been hampered by the fact that the Canadian RDX sample was lost in transit somewhere between 'Scotland and R.D. Woolwich.'[91] As a result, the visiting Canadian scientists were forced to produce a thirty-pound sample in the Woolwich laboratories, a task they completed in a few weeks.[92]

Once they were convinced that RDX could be economically mass produced, British authorities announced that they were prepared to purchase large quantities of the explosive from either the United States or Canada. This decision, however, placed the Department of Munitions and Supply in a difficult position. While C.D. Howe welcomed the opportunity to sell Canadian-produced RDX, he was concerned about the costs of production. As a compromise solution, Howe agreed to manufacture RDX Type B at the Shawinigan Falls pilot plant,[93] with an initial output of 100 tons a month.[94] He insisted, however, that this project was only being approved on the 'understanding that this will complete our explosives programme for Canada, and that no new plant or plant extension will be undertaken.'[95]

Plans to coordinate Canadian and U.S. production had already begun.[96] In September 1941 George Wright and McGill's Raymond Boyer, prominent members of the Canadian RDX Committee, attended a meeting of the American RDX Committee in Atlantic City, together with research teams from the University of Michigan, Cornell University, and the University of Pennsylvania.[97] A successful RDX partnership was soon forged.[98] In October, with the active assistance of Vannevar Bush and C.J. Mackenzie, a joint Canadian–American RDX Committee was formed. It consisted of Wright, Ross, and Boyer from Canada; W.E. Bachmann, R.C. Elderfield, Ralph Connor, Roger Adams, and F.C. Whitemore from the United States; and R.P. Linstead representing British interests.[99] There were also extensive consultations between RDX

PROFESSORS AND GRADUATE STUDENTS OF PHYSICS DEPT. UNIVERSITY OF TORONTO, 1922

PROFESSORS
1. BURTON
2. McTAGGART
3. CHANT
4. SATTERLY

GRADUATE STUDENTS
5. SHAVER
6. BATES
7. CURRIE
8. IRETON
9. LANG
10. YOUNG
11. SHENSTONE
12. McQUARRIE
13. BEALS
14. SMITH
15. AINSLIE
16. McLAY
17. SHRUM

University of Toronto physicists, 1920s.

Frederick Banting meets Ivan Pavlov, the famous Russian physiologist, while attending the 1935 International Congress of Physiology in Moscow.

British atomic pioneers of the 1930s. *From left*: Ernest Walton, Lord Ernest Rutherford, John Cockcroft.

Cutting the ribbon: the 1940 commemorative opening of the NRC aerodynamics administration building by the Honourable W.D. Euler (*left*) and Air Vice-Marshal E.V. Stedman (*centre*).

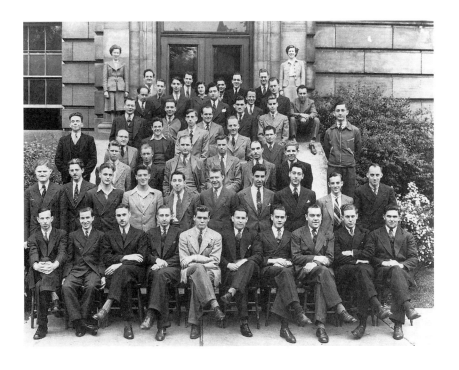

Off to war in 1942: radar technicians from the University of Toronto.

Canada's leading defence scientist: National Research Council president C.J. Mackenzie.

Frederick Banting prepares for his scientific mission to England, 14 February 1941.

Otto Maass of McGill University, chief of Canada's chemical and biological warfare operations during the Second World War.

Professor E.G.D. Murray of McGill, coordinator of Canada's biological warfare program.

General A.G.L. McNaughton and British prime minister Winston Churchill study a military map at Canadian Military Headquarters in London, March 1941.

Meeting of the Canadian Military Mission in Washington, General Maurice Pope presiding, January 1944.

American defence science mandarins. *From left*: Ernest Lawrence, Arthur Compton, Vannevar Bush, James Conant, Karl Compton, Alfred Loomis.

Red Army attaché and spy-master, Colonel Nikolai Zabotin.

Omond Solandt (*left*) greets Sir Henry Tizard during Commonwealth defence science meetings in Ottawa during the fall of 1947.

RCAF Air Marshal L.S. Breadner (*right*) and Captain V.S. Sokolov of the Soviet Embassy 'experiencing no linguistic difficulty' at a February 1943 reception.

Professor Raymond Boyer at his arraignment in a Montreal criminal court on charges of espionage for the Soviet Union, 15 March 1946.

Approaching the loading dock at Grosse Île, Quebec, site of Canadian biological warfare research, 1942–5.

Commonwealth defence scientists, led by Sir Henry Tizard, tour the Suffield weapons-testing station during the fall of 1947.

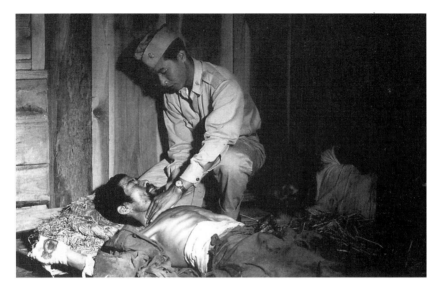

U.S. medical officer examining Chinese civilian for evidence of infection from biological warfare agents, 1944.

Unused poison gas: workers unloading drums of mustard gas stored at Stormont Chemical Plant, Cornwall, Ontario, January 1946.

Flame-throwing Canadian Army universal carrier 'consumes everything in its path,' July 1945.

This Black Brant high-altitude missile stands at the entrance of the defence research establishment at Valcartier, Quebec.

Towards the atomic bomb: scientists at the Chicago Metallurgical Laboratory achieve the first man-made nuclear chain reaction, 2 December 1942.

A giant ball of fire: atomic bomb test in Nevada, 1953.

scientists and representatives of the Canadian and American armed forces. In the United States these talks were coordinated by Colonel J.P. Harris, of the U.S. Army Ordnance Board, who was in charge of the American explosives program.[100] Because of the complexity and duration of these negotiations, Canadian chemists were required to make frequent visits to the Explosives Research Laboratory (ERL) at Bruceton, Pennsylvania. According to Andy Gordon, a prominent member of Toronto's RDX team, it was an exciting scientific and social occasion:

In the afternoon we went to Bruceton; they have made a good deal more progress than I had expected. They showed us their ballistic mortar, their fragment gun, their detonation velocity machine, their apparatus for measuring impulse and pressure for rocket propellants, and their impact sensitivity apparatus ... We had a very cheerful dinner together and consumed beer afterwards; in fact I had a thoroughly pleasant time even if I gathered a few more grey hairs from riding in U.S. government cars which never seem to travel at less than 65 mph.[101]

During 1942, Canadian and American RDX research and development operated in tandem. Although C.J. Mackenzie and James Donald were officially in charge of the Canadian operation, they gave George Wright responsibility for distributing extramural grants to scientists working at McGill, Toronto, and Queen's. By April 1943 sixty projects were being carried out in Canadian universities on problems 'related to the production of explosives now in use, or in the development of new explosives.[102]

While Mackenzie recognized that Wright was a scientific 'genius,' he became increasingly critical of Wright's 'eccentricities' and his autocratic manner. By November 1942 the situation came to a head. 'Wright thinks that he is running the British explosives show,' Mackenzie wrote, 'and that no one else has any interest in the plans. He [Wright] is demanding all sorts of impossible things ... and is giving orders to everyone.' In his efforts to clip Wright's wings and to centralize explosives research, Mackenzie decided to create the NRC Associate Committee on Explosives (ACE), with administrative authority in Ottawa not Toronto.[103] In mid-November the Committee held its first meeting, which focused on plans for building a new pilot plant at the Valcartier explosives and propellants complex.[104] In November 1942 Mackenzie, concerned about how the Valcartier operation was being managed, met with Brigadier Therriault, superintendent of the Dominion Arsenals. Mackenzie was offended by Therriault's administrative style, which he described as 'a combination [of] patriarch, dictator and father.' But what most concerned him were the dangerous working conditions at the plant, where 'an explosion [had occurred] about every two or three months.'[105] Despite these misgivings, Mackenzie felt that the situation was under control, a viewpoint he expressed in a January 1943 letter to General McNaughton:

> I have the feeling that with the explosives pilot plant, the filling plant, and the ballistics proving ground at Valcartier, and the gun plant at Sorel, there might develop a valuable centre of scientific interest in that part of Quebec which could be carried on after the war ... We have also tried to organize all of the ballistics work which is going on in Canada in the different departments ... As you will realize, the ballistics field is large and intricate, and the contacts with the services are so intimate that it is difficult to draw a broader line between development, research and production. However, we hope that the co-operative group which we have formed will serve a useful purpose.[106]

Mackenzie was also pleased that the Associate Committee had been able to provide effective leadership and direction for Canada's various explosive and propellant research and development programs, which, by 1943, included the Allied War Supplies Corporation, the Chemicals and Explosives Division of DMS, and the five technical agencies of DND (Directorate of Staff Duties [weapons], Directorate of Engineering Development [Army], Directorate of Artillery, Directorate of Inspection, and the Army Technical Development Board).[107] Plans were also made, despite Wright's strenuous objections, to have all RDX reports sent directly to Adrien Cambron, secretary of ACE, on the grounds that it was unsafe to have sensitive reports accumulating 'in outside laboratories.'[108]

For security reasons, discussions at the Canadian–American RDX Committee were kept confidential.[109] Although meeting agendas were determined by the respective chairs (Ralph Connor and George Wright), most of the administrative work of the Canadian Committee was carried out by Raymond Boyer of McGill. Not only was Boyer an outstanding chemist, but his reports were characterized by astute analysis and a clear writing style.[110] Between meetings, which alternated between Canadian and American universities, RDX chemists of the two countries engaged in extensive correspondence. Robert Nicholls of McGill, for instance, corresponded with Alfred Blomquist and Walter McCrone of

Cornell about the pilot plant project at Shawinigan Falls, especially about 'the sensitivity of mixtures of RDX and HMX.'[111] Both men were concerned that an explosion at the plant might occur. For McCrone, such an accident would have disastrous consequences for the RDX Type B campaign, since its critics in the U.S. Army would be able to say 'We don't want any of that stuff. We'll stick to TNT.'[112]

It was a timely warning because final arrangements for RDX production were being worked out at the highest military and scientific levels. In Canada, the National Research Council, the Department of Munitions and Supply, and the Armed Forces had already committed themselves to the Bachmann continuous process. So too had the British.[113] This was evident when J.R. Donald and Otto Maass visited England during the summer of 1942. The British Army, they noted, was well advanced in the development 'of defensive ammunition against armoured columns in tank warfare, involving high velocity ammunition, use of RDX fillings and rocket projectiles.' They also observed that the Royal Navy was using RDX in its Torpex explosives, 'for torpedo warheads, hedgehogs, and other anti-submarine weapons,' as was the RAF in its 2000- and 4000-pound bombs.[114]

The most difficult challenge, however, was to convince the United States military that RDX Type B explosives were superior to existing munitions. This campaign gained momentum throughout the fall of 1942, when Canadian, British, and American scientists marshalled a variety of arguments in favour of change. One of their most convincing arguments was irrefutable evidence that the RDX/TNT mixture was 40 per cent more powerful than TNT. This meant that allied bombing raids could be more devastating, and scarce allied chemical resources utilized more effectively. 'The question arises,' Donald stated, 'as to whether it is easier to increase bomber production by 40% or to build further explosive capacity.' For Donald, the answer seemed obvious:

It is roughly estimated that if the present U.S. programme for RDX were doubled sufficient RDX/TNT mixture would be available for the most effective weapons. It is a reasonable assumption that the capital cost ... would be a great deal less than the capital cost of creating plants to increase bomber production by 40%. The raw materials required for RDX by the anhydride process are relatively easy to obtain ... Expansion of the plane programme would require expansion of plants for the production of engines and plane accessories and would necessitate increased quantities of aluminum, rubber, and other essential materials.[115]

Another convincing argument was that the Axis powers were 'known to be using RDX in increasing quantities.'[116] But the clinching factor was the report from the Tennessee Eastman (Kodak) pilot plant in Kingsport, Tennessee, that the Bachmann continuous process was viable for mass production, and that a full-scale plant could produce 170 tons of RDX a day by July 1943.[117] In February 1943 the U.S. Army Ordnance Board made it official: RDX Type B would be mass produced for use by the American Armed Forces.[118] In six months, the gigantic Tennessee Eastman complex at Kingsport reached full production. On 29 September the Canadian–American RDX Committee celebrated this event by touring the Kingsport plant, with its 'gleaming steel and glass.'[119] It was one of their last meetings: these scientists would soon move on to other military projects.[120]

Several technical and legal issues associated with this joint project had to be resolved, however. Vannevar Bush, anxious to avoid patent problems, distributed applications for letters patent to the leading American and Canadian RDX scientists. Surprisingly, given his obstreperous nature, George Wright quickly agreed 'to assign to the United States Government his United States patent rights to his discoveries.'[121] Protecting RDX secrets was another concern shared by both Bush and Mackenzie. In January 1943, for example, OSRD officials advised W.L. Webster, secretary of the British Central Scientific Committee, that because the term RDX was widely known, 'it might be preferable not to advertise the movement of such shipments.'[122] By common agreement, the code word 'golf' was used instead of RDX in cross-border communication, a situation that produced such interesting messages as 'Golf in Montreal on February 15' and 'Ship 1000 lb. golf balls.'[123] It was also agreed not to share information about the RDX Type B production process with the Soviet Union.[124]

Safeguarding the RDX secret was not the only concern for McGill and Toronto scientists. There was also the ongoing danger of working with highly volatile chemical substances in laboratories which often lacked effective ventilation and fire control systems. In addition, there was the intense pressure to complete their assignments rapidly, and without error. George Wright's team of researchers from the University of Toronto, for instance, remember their wartime work on RDX with a combination of pride and amazement: pride that they accomplished so much; amazement that they allowed Wright to work them so hard. In nostalgic moments, they also remember the famous 1942 'revolt of the molecules':

Weapons Systems: Proximity Fuses and RDX 117

Within a few hours of each other on the same winter afternoon in 1941–42, three hot sulphuric acid melting point baths blew up at U of T, McGill, and a university in the United States, all of them working on the RDX program. The Toronto event only left a surprised graduate student with his glasses cracked and minor burn spots, but one of the others put a student temporarily in the hospital. The very next day the newspapers reported explosion of a milk test device using sulphuric acid. A science buff in the labs announced that what was taking place was The Revolt of the Molecules, with sulphuric acid leading the way. Would chlorine, or maybe caustic soda, be next? Fortunately the molecules calmed down and the revolt did not spread.[125]

Canadian Explosives Research, 1943–1945

After RDX became operational, Canadian chemists were diverted to other wartime projects.[126] George Wright's services were especially in demand. In October 1942, in response to a request of the NRC, Wright prepared a survey of the twenty-three major extramural explosives projects being conducted at Canadian universities. Of these, the most important related to RDX, the use of NENO as a shell detonator, the development of high-pressure propellants, and the preparation of explosive polymers, which were deemed superior to the existing nitrocellulose form of explosives.'[127] Wright was particularly enthusiastic about his own work on NENO, a substance he claimed could be used in plastic explosives more effectively than RDX/TNT: it could be produced more cheaply and quickly than RDX; it had more power; it was easier to use in shell fillings; and it was durable. Since Wright always assumed that his logic would be automatically accepted by others, he wasted no time in demanding the immediate construction of a NENO pilot plant, in anticipation of substantial orders from the British and the Americans.[128]

Wright was also working on the military use of dinitrodiethanol nitramine, code named DINA, either as a propellant or an explosive.[129] In his campaign for DINA, Wright found a ready ally in General McNaughton, who had long championed the need for more efficient propellants. In May 1942, for example, the General had written C.J. Mackenzie about this issue:

In order to increase the muzzle velocities of our new guns without having to provide excessively large chamber capacities, I would like to raise the permissible chamber pressure from about 28 tons per square inch to, say, 50. This means, I think, a new kind of propellant, for existing types do not give the loading density needed, or they tend to detonate at the higher pressure. So what we need to do is to find some new chemical compound for use as a propellant; it should be unaffected by moisture and should have the smallest possible temperature coefficient. If the search for such a propellant is successful, it will open an easier road to a practical high velocity gun than the conical bores we have been considering.[130]

Six months later, the Army Technical Development Board, with General McNaughton's steady prodding, called on the Associate Committee on Explosives to prepare a new propellant 'which would produce pressures up to fifty tons per square inch without detonation.'[131] Although DINA appeared to meet these and other specifications, Wright wanted the approval of the American scientific elite before he proceeded further. For that reason, in December 1942, he accepted an invitation from George Kistiakowsky, head of the NDRC's explosives division, to carry out tests on DINA's 'rate of detonation' at the Bruceton, Pennsylvania, test laboratory.[132] With Kistiakowsky's personal endorsement,[133] Wright managed to convince ACE to carry out the production of ten tons of DINA at the DIL pilot plant at Valleyfield.[134] Plans were also made to erect a pilot High Explosives and Propellants Laboratory at Valcartier, next to the Inspection Board's Explosives Proof Establishment.[135]

Wright's vigorous campaign to promote both NENO and DINA generated opposition within ACE. But his most formidable critics were British chemists at Woolwich, who questioned whether either substance met their rigorous standards. This scepticism was reinforced after they actually analysed samples of NENO and discovered that it had stability problems. At this juncture, they requested that Wright come to England and defend his project. This invitation, in turn, produced another bitter confrontation between Wright and C.J. Mackenzie. In part, it was a clash of personalities: Wright was brash, temperamental, and dictatorial, and was a scientific chauvinist; Mackenzie was a calm, methodical administrator, with an engineer's disdain for erratic scientists. Matters came to a head when Wright, with the support of U of T president H.J. Cody, refused to accept the NRC's request that he go to England and consult with Woolwich authorities. Instead, Wright demanded that C.D. Howe give firm assurances that 'the experimental manufacture of this explosive would proceed provided the British experts so recommend.[136] Mackenzie was outraged by Wright's attempt to dictate policy:

This suggestion, of course, is silly and unpracticable and something that no Minister would do or could do. The trouble with Wright is that he is a first class research man, active and energetic, [but] completely inexperienced in the world outside of a laboratory and is not content to stay in his laboratory. He has done the laboratory work on his new explosive NENO and he wants to have field tests on it. He wants to get $50,000 to buy material to have tested in a proving ground. Obviously when the matter gets outside of the laboratory it is not a research man's job but Wright insists ... [on moving] into this field. On the other hand the fact that he does not jump at the chance to go to England but rather pulls back on it means that he is not willing to go through with the thing one hundred percent ... he should be willing to go to England, Timbuktu or wherever it is necessary to go.[137]

While Wright eventually did go to Woolwich, the results were disappointing: British authorities rejected both NENO and DINA because of their instability. Wright, however, was able to enjoy a partial triumph when DINA was eventually used by both the Canadian and United States navies.[138] In 1945 the Associate Committee on Explosives began making plans to stockpile DINA for postwar use.[139]

In June 1945, Raymond Boyer, former secretary of the Canadian–American RDX Committee, proposed a survey of research at university chemistry laboratories that could 'contribute knowledge to the work of the Explosives Experimental Establishment at Valcartier.'[140] With the enthusiastic approval of C.J. Mackenzie, Boyer began his cross-Canada tour. First on his list were the smaller Ontario universities – Queen's, Western, and McMaster, followed by facilities at the universities of Manitoba, Saskatchewan, Alberta, and British Columbia. Although Boyer's investigation did not include Toronto, McGill, or the Maritime universities, his final report was considered useful by NRC officials. Deemed of particular relevance was his observation that most academic chemists he encountered complained about problems of coordination during the war that 'had resulted not only in inefficiency and duplication of effort, but also led to discouragement and, in some cases, even to hostility toward government-sponsored programs.'[141] According to Boyer, these problems could be overcome if a more collective approach were adopted with 'members of science departments (Chemists, Physicists, Mathematicians and Engineers) ... pooling specialized knowledge and ideas for the solution of problems instead of merely satisfying his scientific curiosity in his own fashion.'[142]

The NRC Associate Committee on Explosives had little time to consider Boyer's report, because in August 1945 responsibility for explosives and propellants shifted to the Canadian Armament Research and Development Establishment (CARDE), an interservice organization operating out of Valcartier. It combined the resources of the NRC's explosives experimental establishment, the Inspection Board's artillery proof establishment, the ballistics laboratory, the Army's field testing wing, and a new innovative group 'charged with the responsibility for the design and development of projectiles and weapons up to the production and engineering tests of prototypes.'[143] The NRC and university scientists did, however, maintain a role in policy matters and in research through a new Associate Committee on Explosives and Ballistics, which was created in May 1946.[144] Significantly, the Committee's first meeting, chaired by Omond Solandt, the newly appointed Director of Defence Research, included most of the chemists who had been active in explosives research during the war.[145]

Conclusion

The RDX saga demonstrates how the war changed scientific career patterns in Canada. For example, in 1939 George Wright was a relatively obscure chemist at the University of Toronto; in 1941 he became an RDX research 'star' and in October 1942 was appointed chairman of the new NRC Research Sub-Committee on Explosives.[146] As an important scientific-military administrator, Wright was expected to carry out a variety of functions, the most demanding of which was to coordinate Canadian research in the fields of explosives and propellants with work being done in the United States.[147]

Wright and the other Canadian chemists made a significant contribution in the development of RDX and DINA, both of which were important allied weapons. While Canadian scientists were ostensibly working to fulfil British demands, their most significant relationship was with American RDX scientists. The positive legacy of this partnership was described in the NDRC's official history of U.S. explosives research:

The activity of the RDX Committee represents an example of perfect international cooperation in research ... The experimental programs were thoroughly integrated ... Samples were freely exchanged, and a laboratory having special analytical functions often used these for the entire group of workers. The complete unselfishness and devotion to the program of the various laboratories led to a relationship in which there was no distinction between the nationality of the workers, and did much to make the entire development proceed rapidly.[148]

Parallels can be found in the experiences of Canadian scientists working on the proximity fuses. For three years, Arnold Pitt and his Toronto team tried to develop components for the proximity fuse, carrying out seemingly endless tests to prepare fused shells for operational use. Significantly, responsibility for the mass production of this new weapon was the exclusive preserve of the United States, despite the fact that British scientists had pioneered its early development. As usual, Canada was caught in the middle of this power struggle between Britain and the United States, although the achievements of Pitt's group did enhance their status with the Americans, as useful junior partners.

It could be argued that Canadian involvement in explosives and proximity fuse research illustrates what Canadian foreign policy specialists call the functional principal, namely that the level of a country's contribution determines its international status.[149] Although it is difficult to rate the performance of Canadian defence scientists in these fields, at least the NRC was involved in the critical research and development stage of each of these weapons systems. This sense of partnership would be even more pronounced in the wartime collaboration between Canada, Great Britain, and the United States in developing chemical and biological weapons.

5

Chemical Warfare Planning, 1939–1945

> It is most satisfactory to note that all goes well with the development of Suffield. I am certain that chemical warfare will be used by the enemy on a large scale, when the particular circumstances suit, and for this reason we must be on the alert against all new forms of gas and we must put ourselves in a position to retaliate with even greater effect.
>
> General McNaughton to C.J. Mackenzie, 6 August 1942;
> cited in Thistle, ed., *The Mackenzie–McNaughton Wartime Letters*, 113–14

In the 1915 Battle of Ypres Canadian troops were among the first to face the poison gas of the Germans. This experience was instrumental in Canada's decision to support the 1925 Geneva Protocol ban on the use of chemical and bacteriological weapons. What Canada ratified, however, was essentially a 'no-first-use' treaty in which every nation retained the right to use chemical weapons in a defensive or retaliatory situation.

The late 1930s witnessed considerable public debate in Britain and in Canada about the threat of poison gas. This was reflected in the publication of such books as *Breathe Freely! The Truth about Poison Gas* by James Kendall, *A.R.P.* by J.B.S. Haldane, and *Commonsense and A.R.P.: A Practical Guide for Householders and Business Managers* by Major General C.H. Foulkes. The latter study was the most authoritative account, since Foulkes had been in charge of Britain's chemical warfare operation during the First World War. He had also studied the use of terror weapons by fascist governments, for example, the Italian gas attacks against Ethiopia in 1935 and the Nationalists' bombing of Spanish cities during the Spanish Civil War. Foulkes did not believe that poison gas would be used, 'not because of its supposed inhumanity, but because it would not be as effective, weight for weight, as high explosives.' He did concede, however, that chemical weapons could be used, with deadly effect, as a follow-up weapon, contaminating cities and battlefields.[1]

Concerned that the Germans and Italians might resort to chemical warfare, the British Chiefs of Staff in 1937 belatedly authorized the development of offensive chemical warfare munitions and delivery systems. This included gas-filled artillery shells and aircraft sprays designed to make chemical weapon attacks more effective. A wide range of defensive measures was adopted as well, notably the issuing of over 30 million gas masks and the establishment of civil defence anti-gas measures. During the early months of the Second World War, Britain also arranged to employ the scientific, industrial, and military resources of its Canadian ally just in case Germany reneged on its September 1939 pledge to abide by the Geneva Convention and not initiate chemical warfare.

Remarkably, poison gas was never deployed strategically during the Second World War. That is not to say, however, that either side believed that its adversary would desist from using chemical weapons if an advantageous situation emerged or if a deadly new gas munition were suddenly discovered. Indeed, on at least seven occasions during the war, a chemical warfare exchange appeared imminent. Yet each crisis passed, with neither side resorting to gas.

Historians offer three major explanations for this reluctance to initiate a poison gas attack. The first is that international prohibitions and massive public revulsion towards the use of poison gas prevented the British Commonwealth and the United States from using this weapon. The second viewpoint holds that there was no decisive advantage to be gained from using chemical warfare since the other side could retaliate. The third, and most convincing, argument points out that chemical weapons did not fit into the military culture of any of the belligerents; that it was inconsistent with the prevailing 'way of war' for the British, German, and American armed forces.[2]

For most of the Second World War, chemical munitions were improved versions of the major gases of the First World War. These included choking gases, such as chlorine, phosgene, and chloropicrin, as well as blood gases, such as hydrogen cyanide and cyanogen chloride. In a separate category were the vesicant or blistering weapons: mustard gas (HS + HT), and its derivatives – sulphide mustard, nitrogen mustard, and mustard-lewisite.[3]

All of these agents were analysed, produced, stockpiled, and tested by Canadian chemists during the war years. Most of this research was coordinated by the Canadian Army's Directorate of Chemical Warfare and Smoke (DCWS) in its own laboratories and in university laboratories through NRC-administered grant programs. Created in 1941, the Suffield Experimental Station in Ralston, Alberta, became the nerve centre of Canadian and British field tests involving a variety of gas munitions and delivery systems, most notably mortars, artillery shells, bombs, and sprays. Smoke weapons and flame-throwers were also tested, as were biological weapons.[4]

The Banting–Rabinowitch Mission, 1939–1940

In 1938 Frederick Banting attempted to convince Canadian and British military authorities to prepare for chemical and biological warfare. Although his campaign failed, Banting's interest in these weapons remained intense. With the outbreak of war in September 1939, he immediately volunteered to lead a mission to England in order to determine how Canadian scientific resources could be best mobilized in support of British chemical and biological warfare planning. Banting was accompanied by Israel Rabinowitch, 'one of Canada's most outstanding toxicologists,' who had convinced General McNaughton that he was uniquely qualified to coordinate British and Canadian chemical warfare research and development.[5] Although their backgrounds were different,[6] both men shared a concern that Germany would deploy not only the 'popular' war gases of the First World War, but also deadly new gases derived from 'other known industrial poisons.' They also agreed that if the Nazis believed they could gain a strategic advantage by using chemical weapons, Germany would not adhere to the Geneva Protocol of 1925. Bantings' and Rabinowitch's call for chemical warfare preparedness was readily accepted by General McNaughton, who encouraged the two scientists to familiarize themselves with 'knowledge of the recent advances in this field in Great Britain' especially 'in the latest methods of diagnosis and treatment of war gas poisoning.'[7]

On 25 November Banting and Rabinowitch reached London. After a briefing by High Commissioner Vincent Massey, they were turned over to RAF Air Marshal A.V. Richardson for their tour of British defence science facilities. But they had not reckoned on British military red tape, which was even more elaborate in the case of colonial intruders. Fortunately, Sir Edward Mellanby, Secretary of the British Medical Research Council (MRC), intervened and gave the visiting Canadians 'free access' to all MRC documents and files pertaining to tetanus and gas gangrene research, and arranged meetings with key British immunology and disease control researchers. One reason for Mellanby's generosity was his appreciation that research being carried out in Banting's laboratory could assist British defensive measures against a possible enemy chemical weapons attack. He was particularly impressed by the salve developed by Edward Hall and Colin Lucas that relieved the pain and decreased the inflammation of mustard gas poisoning.[8]

During the next three weeks Banting and Rabinowitch gathered information about the British Army gas school at Porton and about research being carried out at various British universities. Demonstrations of various forms of gas munitions gave the two Canadians the opportunity to assess the effectiveness of decontamination techniques and protective equipment against choking and blister gases. By the end of his visit, Banting was convinced that close relations between Canadian scientists and Porton were essential, and advised C.J. Mackenzie that Canadian scientists from several disciplines should be immediately recruited for overseas service. Each field, Banting argued, had its own specific contribution for a successful chemical warfare program:

> Engineering chemists are interested in the production of toxic gases and the means for their distribution, and also in the means of defence against toxic gases. In physics the greater part of the research has to do with the colloidal side of the gas dispersion, vapour tension and physical properties of the gases. Meteorology is important because gas warfare depends so much upon humidity, temperature, winds and weather. The duties of the military engineer appear to be from the tactical point of view, but he must have an understanding and working knowledge of the uses and properties of the gases. This also applies to the aviation engineer, who has in addition the special problem at present of designing the apparatus for the spraying of gas from aeroplane.[9]

While Mackenzie was initially attracted by the Banting–Rabinowitch proposal, he subsequently changed his mind, largely because of Otto Maass's advice that there were more British chemists 'free to undertake [chemical warfare] work than there are special problems.'[10]

Preparing for Chemical Warfare, 1940–1941

Was chemical warfare a serious threat in 1940? Within the Canadian and British military hierarchies, opinion was sharply divided. Those who dismissed the threat argued that poison gas was not as effective as high explosives, and that countermeasures could neutralize most known chemical weapons. This was not, however, the view of Israel Rabinowitch, who on 1 March sent a strongly worded letter to National Defence headquarters, warning that large-scale chemical warfare would 'break out ... probably within three months,' since, in his opinion, the use of gas-filled artillery shells and aircraft sprays was consistent with proven German blitzkrieg tactics. Rabinowitch was particularly concerned that Canadian troops in Britain were vulnerable since they 'did not even know simple defensive procedures such as how to test the respirators or how to operate the gas detectors.'[11] Would it not be possible, he asked, for the Defence Department to establish an Army gas training school, with highly qualified instructors, state-of-the-art equipment, and a constant flow of information about British operational research?

Discussion about Anglo-Canadian chemical warfare defensive preparedness was temporarily stilled by the realization that the dazzling victories of the Wehrmacht in western Europe had been achieved without chemical weapons.[12] By June 1940 the central issue was not whether the Germans would use gas against British military and civilian targets, but whether the British military should use gas to defend Britain. This debate reached the highest political level on 15 June, when Sir John Dill, Chief of the Imperial Chief of Staff, asked the War Cabinet to consider the strategic advantages of such a policy:

> Enemy forces crowded on the beaches, with the confusion inevitable on first landing, would present a splendid target. Gas spray by aircraft under such conditions would be likely to have a more widespread and wholesale effect than high explosives. It can moreover be applied very rapidly, and so is particularly suitable in an operation where we may get very little warning ... Besides gas spray, contamination of beaches, obstacles and defiles by liquid mustard would have a great delaying effect. The use of gas in general would have the effect of slowing up operations, and we believe that speed must be of the essence of any successful invasion of this country.

On the other hand, Dill warned that the immediate strategic advantages of using chemical weapons might be offset by certain negative factors: by resorting to chemical weapons, Britain would violate its 'no-first-use' pledge, which was an integral part of the 1925 Geneva Protocol; and such a policy also ran the risk of seriously alienating American public opinion and sympathy. It was also sure to bring immediate German retaliation against military and civilian targets. But what other options, Dill asked, did Britain have 'at a time when our National existence is at stake, when we are threatened by an implacable enemy who himself recognises no rules save those of expediency'?[13]

While other members of the British Chiefs of Staff shared Dill's concern about Britain's military vulnerability, they were not willing to embrace a first-use policy, arguing that Britain would be 'throwing away the incalculable moral advantage of keeping our pledges ... for a minor tactical surprise.'[14] Prime Minister Churchill was not swayed by these moral arguments. On 30 June he instructed General H.L. Ismay, his chief of staff, to prepare a report on available stocks of mustard gas for artillery and air spray use, arguing that if invasion occurred there would be 'no better points for application of mustard than on those beaches and lodgements.'[15]

During this period of gas preparedness, General McNaughton kept troops of the First and Second Divisions on battle readiness, determined that 'Canadians will not be caught as they were at Second Ypres [1915].' He also arranged to establish a mobile laboratory to determine 'the type and volume of gas used by enemy and the countermeasures required.'[16] These defensive measures were coordinated by Major Israel Rabinowitch, who had joined the Canadian Army as McNaughton's special chemical warfare adviser and personal physician.[17]

Canada's role in Britain's wartime chemical warfare planning was also affected by the deliberations of the August 1940 Tizard mission to North America. When Tizard first reached Ottawa, he held prolonged discussions on chemical weapons with Otto Maass, who had been given the job of streamlining Canadian chemical warfare operations and establishing effective links with the U.S. Army Chemical Warfare Service (CWS). These conversations continued in Washington when Maass, now acting with the Tizard mission, met with representatives of the CWS in an attempt to arrange free exchange of information between Porton and Edgewood, Maryland, the focus of U.S. chemical warfare research and development.

Although Porton scientists believed that their chemical weapons projects were more advanced than those of the United States, they gave Tizard a list of scientific and logistical questions to ask the Americans: What type of mustard gas was used in U.S. artillery shells? Were they armed with delay or air burst fuses? How effective were U.S. air spraying equipment and techniques? Do they use proximity fuses for shells or bombs? Have any chemical anti-tank weapons (gas or flames) been considered or developed? Can any credence be placed on rumours of German 'nerve' gases?

The Americans' responses contained few surprises. As the Porton group had predicted, the United States had not progressed far beyond its First World War chemical warfare technology and techniques. There were, however, three notable exceptions: high-altitude spraying of persistent and nonpersistent gases; the operational use of flame-throwers mounted on tanks 'for cleaning up scattered resistance in narrow trenches and pill boxes'; and the use of white phosphorus in artillery smoke shells because of its 'combined incendiary, personnel, and smoke effect.' Tizard's group were also told that the United States had not developed biological weapons, although some CWS experiments suggested the 'possible use of bacteria in spores and affected insects.'[18]

The possibility that the Germans had developed some kind of nerve gas was another subject of mutual interest to British and American authorities. However, both sides concluded that there was no validity to the rumours British intelligence had received about a deadly new poison gas.[19] What allied scientists didn't realize was that four years earlier Gerhard Schrader, a chemist working for Germany's most powerful chemical company, had discovered the world's first nerve gas – tabun. Tabun was a colourless, odourless gas that brought rapid death through convulsions and asphyxiation. Fortunately for the allies, Germany did not make plans for its production until January 1940, when the Army Division of the Wehrmacht ordered a tabun plant to be built in the Silesian village of Dyhernfurth. Even then, the agent remained nonoperational because of production problems related to tabun's extreme toxicity and due to Hitler's prohibition of the use of chemical weapons.[20]

In the fall of 1940, therefore, there appeared to be no reason to be alarmed about a new gas, and British, American, and Canadian scientists arranged their cooperative ventures on the assumption that if chemical warfare did occur it would be carried out with the familiar gases of the First World War. To make their system of research and development more effective, however, all three sides agreed in October 1940 to exchange chemical warfare information through the medium of Otto Maass, who became the representative of both the Canadian and British gas programs.

Maass was soon busy touring U.S. facilities at the Edgewood Arsenal, where he and Colonel E.W. Flood, a former NRC chemist, were allowed to examine the lewisite and mustard gas plants, the gas mask testing laboratory, as well as the white phosphorus shell-filling plants. On their way back to Ottawa, the two Canadian scientists stopped off at MIT. There they met with key members of the newly created NDRC Committee on Explosives and Chemical Warfare, who told them about chemical warfare research being carried out by American university chemists.[21] Maass was particularly pleased when his old friend James Conant, head of the NDRC's chemical warfare division, promised full cooperation with Canada 'in order that problems may be dealt with more efficiently.' This set the stage for the tripartite agreement of 5 November, which Maass and Porton's E.L. Davies negotiated in Washington with high-ranking officials of the CWS and NDRC.[22] Under this arrangement, all three countries agreed to share information about new chemical weapons and defensive measures.

While American scientists were primarily interested in British technology and operational experience, they were also curious about what progress Canada had made in this field.[23] Indeed, following the Washington agreement, W.K. Lewis of the NDRC visited National Research Council chemical warfare laboratories in Ottawa. He discovered that most of the NRC's work was still focused on the defensive aspects of gas warfare through the NRC Committee on Container Proofing and Research.[24] This committee was sufficiently prescient to recognize that if chemical warfare did break out, 'the capacity of British plants to produce war gases will be exceeded by chemical plants now available to Germany and Italy by a very considerable margin.'[25] In order to cope with this deficiency, Canada would have to manufacture large quantities of the two major war gases: phosgene and HT mustard gas.[26]

Another ambitious Canadian project was the development of toxic smoke from cadmium mixtures through the combined efforts of chemists and physiologists at Toronto and McGill universities and the NRC.[27] The most spectacular breakthrough however, was, the November 1941 discovery by McGill chemists of Compound Z, a hexafluoride 'impurity' that had three times the toxicity of phosgene and could 'penetrate any service or civilian gas mask in the British Empire, United States or Russia.'[28] During the next three years Canadian and British scientists tried to turn Compound Z into an operational weapon, a project that was seriously hampered by its enormous production costs, almost ten times higher than for HT mustard gas.[29]

The Suffield Experimental Station and Canadian Chemical Warfare Research

While Porton officials welcomed these ambitious research undertakings, what they wanted most in 1940 was a large-scale experimental testing facility. The campaign to obtain such a site in Canada began on 25 September when Major General H.D.G. Crerar, Chief of the General Staff (CGS), received the following telegram from General McNaughton:

Developments in chemical warfare require experimental area much larger than can be found in U.K. Until last May use was made of French area of North Africa but this is no longer available. Ministry of Supply would like to transfer this urgent work to Canada provided suitable area approximately 20 miles by 20 miles can be found ... They have suggested finances on 50–50 basis. If this proposal acceptable to Government of Canada in principle British Ministry of Supply experts will proceed to Canada at an early date to study available sites and submit definite proposals for consideration.

In case Crerar did not appreciate the seriousness of the situation, McNaughton reminded his colleague that British intelligence showed conclusively that the Germans were just waiting for the opportune moment 'to use chemical warfare on large scale.'[30] Crerar got the message, and urged his boss, Minister of Defence J.L. Ralston, to accept McNaughton's proposal.

The next step was for Otto Maass and Porton's E.L. Davies to agree on a proper site: either the Maple Creek area of southern Saskatchewan, or the Suffield region of southern Alberta.[31] In early December, Suffield was declared the winner, because of its size (1825 square miles), its greater isolation, and its cheaper cost – one-tenth that of its rival.[32] Work soon began on the physical infrastructure, including residences, machine shops, and laboratories to accommodate the teams of chemists, physiologists, physicists, and meteorologists who were soon to arrive. A few of the key administrative positions were initially held by British scientists, notably Porton's E.L. Davies, who was appointed superintendent of the station. But the vast majority of the station's nineteen scientists, twenty-three technical assistants, fifty laboratory assistants, and thirty support staff were Canadians. In addition, the station was staffed with a military component of fifty-four Army and forty-eight RCAF personnel.[33]

What was now required was a formal chemical warfare agreement between Canada and Britain.[34] On 17 December, negotiations began in London between generals McNaughton and Crerar, Minister of Defence Ralston, High Commissioner Massey, and Major Rabinowitch, representing Canada, and Lord Weir, Director General of Chemical Defence Research, R. Kingan of the Ministry of Supply, and his deputy, J. Davidson Pratt, for Britain. The initial deliberations got off to a bad start when Lord Weir declared that the British government 'would be very pleased' to assist Canada's chemical warfare plans 'in any way that it could.' After an awkward pause, General McNaughton bluntly explained the Canadian position: 'chemical warfare was a [British] War Cabinet problem,' and while Canada was prepared to help, the fact remained that Ottawa was responding to Britain's needs, not vice versa. Once this reality was acknowledged, negotiations proceeded smoothly.[35]

Under the final agreement of 1 March 1941, it was stipulated that the Suffield Experimental Station (SES) 'would consist of two branches, one for research and development, and the other for field trials comparable with the actual use of gas in warfare.' The costs of land, buildings, equipment, materials, and maintenance were to be shared equally,[36] despite last-minute British attempts to have Canada shoulder more of the burden.[37] The Alberta government, for its part, agreed to lease the 700,000-acre site to Ottawa for ninety-nine years at a cost of one dollar per year.[38] Responsibility for administering the station would rest with the Department of National Defence, a commitment the Army Chiefs of Staff reluctantly agreed to assume. This represented a triumph for Mackenzie and Maass, who managed to convince Ralston that the National Research Council had neither the financial resources nor the authority required for the job.[39] The NRC, however, was willing to administer DND-funded chemical warfare extramural research projects because of its excellent contacts with the country's university chemists.[40]

Canada's chemical warfare operation was greatly enhanced by a number of administrative changes made during the spring and summer of 1941. In April, provision was made for a separate NRC chemical warfare facility in Ottawa that would keep close contact with Porton and maintain a library for accumulating British chemical warfare reports. In May the CW Inter-Service Board was created, consisting of representatives of the three services, the NRC, and the British government.[41] In August the status of the Directorate of Chemical Warfare (DCW) was also clarified: it would function as an independent subdivision of the Directorate of Technical Research under the jurisdiction of the Master General of Ordnance (Army). Most important, Otto Maass was named head of the Directorate of Chemical Warfare and chairman of the CW-Inter-Service Board.[42]

Meanwhile, the Suffield Experimental Station was taking form with the construction of testing and toxic storage facilities, and accommodation for over 200 employees. Arrangements had been made with the Canadian Pacific Railway to build a spur line to the station and to supply water from its facilities in the town of Suffield.[43] The Alberta government constructed an all-season connecting road to the station and established telephone service.[44] Protecting the SES from enemy agents, or curious locals, was another concern. After much consultation, it was decided to supplement local RCMP detachments and military police with civilian range riders who would ensure that the extensive barbwire fences remained intact. It was also recognized that in such an isolated community certain social amenities would have to be provided for the large and diverse SES staff. Plans were therefore made for recreational facilities 'to keep the morale ... at a high pitch.'[45]

Maass was able to attract capable and respected scientists to this unique scientific establishment by offering higher salaries than those paid by Canadian universities, a bonus Maass justified because of the urgency of the work and the dangerous working conditions.[46] Among the original scientific recruits were two scientific 'stars': University of Toronto chemist H.M. Barrett, who was appointed superintendent of research,[47] and Major J.C. Paterson of the Royal Canadian Army Medical Corps, who was named head of the physiological and pathological section.[48] Other candidates for Suffield were interviewed by a special NRC selection board, with Maass making the final decision, usually after consultations with prominent university chemists such as George Wright and 'Andy' Gordon of Toronto, Christian Sivertz of Western, and T. Thorvaldson of Saskatchewan.[49]

Another of Maass's functions was the distribution of extramural grants to university chemists, engineers, and medical researchers involved with chemical warfare. He was particularly insistent that other universities adopt the McGill model and allow promising graduate students to focus on problems of chemical warfare while completing their dissertations, a strategy that also kept these young researchers out of the clutches of National Selective Service.[50] In financing these grants, Maass initially drew funds from the War Technical and Scientific Development Committee's 'Santa Claus' fund. In 1940, for example, the WTSDC allocated over $119,000 to thirty-four chemical researchers for both explosives and gas projects.[51] Since Maass was involved with both fields he was ideally situated to supervise such cooperative ventures in both fields.[52]

While these internal administrative matters were important, Maass's most challenging undertaking was to coordinate Canadian-American-British chemical warfare planning. And there was much to be done. One of the most serious problems was dealing with the complaints of the U.S. Army Chemical Warfare Service that it had 'obtained very poor cooperation from the Canadians.' Under Maass's watchful eye, these problems were soon rectified: there was an immediate improvement in the volume of chemical warfare reports and equipment moving across the border, and a representative of the CWS was appointed to Suffield. Another of Maass's initiatives was the dispatch of Directorate of chemical warfare scientists to American chemical warfare laboratories and pilot plants to observe how these complex and dangerous operations were conducted.[53] One early result of this Canada–U.S. collaboration was the establishment of the HT mustard gas plant at Cornwall, Ontario, which greatly profited from American equipment and technical assistance. Once the plant became operative in 1942, it was able to provide the CWS with over 2,600,000 pounds of HT mustard gas and 766,000 pounds of phosgene.[54]

It was at the Suffield Experimental Station, however, that Canadian-British-American cooperation was most pronounced. Here scientists and military personnel from the three countries carried out numerous experiments and field tests. One project was combining 'true gases and aerosols or particulates,' since, in order to use toxic aerosols effectively, it was crucial to know 'the most effective size of particle for penetrating the enemy's respirators ... the amount of dosage which would be required to be lethal ... and the technique for laying down such an effective dosage.'[55] Some of these weapons also combined gas and explosives. On 15 May 1941, for example, it was reported that an explosion of cadmium oxide and RDX had produced 'fumes ... as toxic as phosgene.'[56] Another type of test involved the use of gas-filled artillery shells and airplane sprays in battlefield conditions. C.J. Mackenzie observed such a demonstration when he made his first visit to Suffield in March 1942:

For the first time, as far as the united nations [sic] are concerned at any rate, a field test was made using over 100 tons of phosgene in one place ... The Americans cooperated, gave gas, and sent a large delegation up. There were officers from the Air Force, Navy and Army from the Atlantic and Pacific coasts there and altogether it was a remarkable show. Probably the most expensive experiment ever performed in Canada – probably $200,000 spent in something like a couple of hours but the results are well worth it.[57]

Maintaining good relations with the Army brass was another essential part of Maass's job. And he did it well. Throughout the latter part of 1941, for instance, he worked closely with Colonel G.P. Morrison, Director of Technical Research, in the development of improved gas masks and protective clothing. It was estimated that, by 1942, a total of 440,000 masks would be needed to equip the Canadian Army and two million respirators would be required for the newly established Civil Defence Branch of the Department of Pensions and National Health.[58] Most of this material was eventually produced at the two government-owned assembly plants, in conjunction with the Inspection Board of the United Kingdom and Canada, which had been established in 1941 to maintain quality control over Canadian war materiel produced for the British Armed Forces.[59] In addition, in December 1941, it was decided to equip Canadian troops overseas with chemically impregnated clothing that would protect its wearer from mustard gas, an undertaking that became more crucial as the possibility of chemical warfare became more acute.[60]

The Military Threat of Chemical Warfare, 1942

On 29 December 1941 the Directorate of Chemical Warfare issued a warning that since 'the possibility of the outbreak of C.W. in the Pacific sphere of operations would seem very likely ... there is every chance that it would spread to other zones.'[61] What made the situation menacing were reports that Japan had already used these weapons against Chinese troops and civilians.[62] On 2 February 1942, Chief of Staff Lieutenant-General Kenneth Stuart advised the Defence minister of five ominous trends:

(a) The confirmed use of vesicant gases on a substantial scale by the Japanese in the China theatre.
(b) The apparent failure of the Japanese Government to make a satisfactory reply to Great Britain's request of 11 December 1941 that neither power use lethal gases in the present war.[63]
(c) The strong prima facie evidence of increased German concentration of vesicant gas weapons on the invasion coast of Europe, coupled with a recent

movement of their main chemical warfare plants farther away from the range of bombers operating from the United Kingdom.
(d) The entrance of the United States into the war on 9th December, and their non-adherence to the Geneva Anti-Gas Protocol of 1925.
(e) The reported intention of the United States to take the initiative in the use of vesicant gases against any attempted invasion of this Continent.[64]

Throughout the early months of 1942 the Canadian Chiefs of Staff and the Cabinet War Committee reassessed the country's chemical warfare policy. On the logistical side, it was decided that the Directorate of Chemical Warfare and Smoke, as it was called after 1942, would coordinate the mass production of chemical warfare agents, with emphasis on mustard gas, so that Canada's Armed Forces would have sufficient gas munitions to launch a retaliatory attack in the European or Pacific theatres of war.[65] Plans were also made to equip all Canadian troops, both at home and abroad, with quality protective equipment and gas masks.[66] But the most controversial issue was whether Canadian Forces should use chemical weapons against a possible Japanese assault on the coast of British Columbia.

Officials had several reasons to fear a Japanese invasion. First, the Japanese had achieved spectacular military success during the four months after Pearl Harbor in both southeastern and southwestern Asia. Second, the March 1942 Japanese occupation of the Aleutian Islands provided the necessary staging area for an amphibious landing of Alaska or northern British Columbia. Third, the presence of almost 22,000 people of Japanese ancestry along the coastal areas of British Columbia created great concern about 'fifth-column' espionage and sabotage. Popular anti-Japanese feeling was intensified after reports that Canadian troops captured at Hong Kong were being brutally treated in Japanese prisoner of war camps.[67] On 24 February 1942 the King government ordered all persons of Japanese ancestry to be removed from the coastal protective zones and placed in camps scattered across the country for the duration of the war.[68]

Meanwhile, military preparations for the defence of British Columbia were being carried out at a feverish pace. And strategic planning involved chemical weapons. On 7 February Lieutenant-General Stuart had issued a statement that 'in the event of [Japanese] attempted landings, particularly on the Pacific Coast ... the operational value of mustard spray is rated highly, both in attack on the enemy during landings, and in denying alternative beaches to him if he adopts infiltration tactics.'[69] This recommendation soon became a strong policy initiative. On 22 March the CW Inter-Service Board called for dramatic changes in Canadian chemical warfare policy:

The offensive aspects of Chemical Warfare should be strongly emphasized in all plans for the defence of Canada and contiguous territories. Canadian policy should envisage the initiative in offensive C.W. action against any enemy move towards a possible attack in force against this Continent, and that such initiative will be taken whenever it is considered by the appropriate Service authorities, that the strategic advantage lies with our forces in so doing, subject only to prior consultation with the War Council of the United Nations, and especially with the United States.[70]

While the Chiefs of General Staff debated the merits of this recommendation, there was growing evidence that the German Army was preparing to use chemical weapons as a knock-out blow against the Soviet Union's beleaguered Red Army. Although the Soviet Union was only a recent ally of Britain's, the British government decided to intervene forcefully. On 10 May 1942 Winston Churchill issued his famous warning: 'We shall treat the unprovoked use of poison against our Russian ally exactly as if it were used on ourselves, and if we are satisfied that this new outrage has been committed by Hitler we will use our great and growing air superiority in the west to carry gas warfare on the largest possible scale far and wide upon the towns and cities of Germany.' On 5 June U.S. President Roosevelt gave a similar warning to Japanese authorities: 'I desire to make it unmistakably clear that if Japan persists in this inhuman form of warfare against China or against any other of the United Nations, such action will be regarded by this government as though taken against the United States, and retaliation in kind and in full measure will be meted out.'[71]

Did this mean that neither the United States nor Britain would consider a first-use poison gas attack? This was not clear, since Churchill had been prepared to use gas in July 1940 and would again consider this option in 1944. Moreover, immediately after Pearl Harbor, U.S. Secretary of War Stimson indicated that he would consider 'using vesicant gas against any enemy attempting an invasion of this continent.'[72] Under these conditions, it is not surprising that the Canadian General Staff should consider using mustard and phosgene to repel a Japanese landing force. On 29 June Major General J.C. Murchie sent a memorandum to Defence Minister J.L. Ralston requesting that chemical weapons be stockpiled and held in readiness. Effective operational use of gas, he argued, would require at least forty-eight hours' warning so that the Air Force, Army, and Navy would be in a position to attack enemy forces with mustard gas just when they would be 'seeking to consolidate beach-head positions and before [their] main force has been tactically deployed.'[73] Murchie's appeal fell on deaf ears. In July 1942 the Cabinet War Committee rejected the use of poison gas on a first-use basis, even to defend Canada – a policy made easier when the threat of Japanese invasion had diminished.[74]

The War Committee did, however, attempt to bring Canadian retaliatory policy in line with that of the United States and Great Britain. It also accepted Murchie's recommendations that all three services use Suffield's offensive chemical warfare training facilities,[75] and that all future Canadian chemical weapons and apparatus designed for operational use 'follow or be adapted to U.S.A. designs.'[76] Maass had already taken steps to ensure that the DCW&S was well situated to gain access to American chemical resources: both through the Joint Technical Sub-Committee on Chemicals and Explosives, and by convincing General William Porter, head of the U.S. Army Chemical Warfare Service, 'to support Canadian demands for procurement of materials in the U.S.'[77] This mutually beneficial exchange system – a deadly form of North American free trade – soon took shape. During 1942 the CWS sent large quantities of HS mustard gas and phosgene to Suffield, along with a number of 4'2" C mortars and flame-throwers, and assisted the Cornwall HT mustard gas operation.[78]

New and Improved Poison Gases, 1942–1945

Throughout 1942 and 1943, Otto Maass urged the Army Chiefs of Staff to consider 'having as wide a variety of chemical weapons as is consistent with industrial production capability.' This meant, he explained, that where there were 'two or three gases essentially similar in technical use and not differing markedly in casualty producing power,' the costs of production would be the determining factor whether his Directorate would undertake either production or weaponization.[79] While Maass did not specifically refer to Compound Z, which McGill chemists had discovered in 1941, it was obvious that its prohibitive costs of production still prevented operational use. A similar problem existed in the manufacture of agent W, a new war gas reportedly 'from 50 to 500 times as toxic as phosgene.' However, Maass was impressed with the agent's potential and allocated $50,000 for a pilot project to manufacture a small portion of the substance.[80]

But the most potentially lethal new gas was an American discovery. In July 1942 James Conant, chairman of the NDRC division on chemical warfare, informed his British counterpart, J. Davidson Pratt, that tests on KB-16 had demonstrated a 'high degree of toxicity,' and that concentrations of 0.03 milligrams per litre would incapacitate victims and produce death within ten days. While Conant was prepared to keep Pratt informed, he suggested that the production of this deadly agent remain exclusively under U.S. control, since the British had only limited quantities of ethylene oxide to make the agent and since it was important to keep all information about KB-16 secret.[81] Conant's real motives were revealed when he discussed the issue with General Porter, head of CWS. Wasn't it unfortunate, he said, that 'it was impossible to cut off all communication with the Canadians and British on this subject,' in part because they had already received a sufficient number of reports on the capabilities of KB-16, and were 'actively at work on the experimental phases of the subject themselves.'[82] Conant's frustration in having to participate in a joint venture was intensified when British physiological experiments indicated that KB-16 was considerably less toxic than the NDRC claimed and he was forced to send a small sample of the substance to England for further testing.[83]

At this stage, however, it was the Canadians not the British whom Conant found most difficult to control. On 12 August he was notified by Colonel E.A. Flood that DCW&S scientists intended 'to carry out large scale field trials with this material,' which would be produced by a civilian chemical company in Montreal. Conant was furious. Not only would the Canadians become involved in the production phase of the new agent, but the KB-16 secret might be jeopardized. Conant's first response in dealing with the DCW&S challenge was to forward a memorandum to Ottawa emphasizing the compound's dangerous instability.[84] Maass, however, was not buying this negative line; indeed, his reply stressed KB-16's potential as a multipurpose gas weapon:

My feeling is that anything that is as good as TL-186 [the Canadian code name for KB-16] appears to be, most certainly could be used very effectively in a number of weapons ... The material could be used effectively from light case bombs that would break on impact without explosive charges. Providing that the material could be kept for a reasonable length of time in suitably lacquered shells, spray tanks, etc., I see no reason why it could not be used in base ejection shell of the piston type where the content is not appreciably exposed to the explosive blast. The material could be used in aircraft-spray tanks, mines and a variety of other weapons.[85]

Maass also indicated that DCW&S was prepared to carry out necessary research 'toward finding a stabilizer and corrosion inhibitor,' and would be prepared to send one of its scientists to the University of Chicago in order to assess the KB-16 work being carried out there.[86] This request was denied. While Conant could not prevent a separate Canadian KB-16 project, he still had the means to ensure that it was not successful. On 3 October he sent the following message to Colonel Flood:

Of course we cannot have a valid objection to your own organization undertaking the manufacture of TL-186, but all of us here think it unwise to do so at this time. We are putting all possible efforts into our work here and are pushing the investigations with great speed. The hazards and dangers involved are serious. Our English colleagues ... have work under way, and full information is being supplied ... I understand Mr. Pratt will probably visit Canada during his coming trip to North America, and decisions on the whole problem can then be reached with his knowledge and consent. In view of these remarks and the added limitations of scarcity of scientific personnel in Canada, I believe this might well be one of those projects for which you might depend on us.[87]

The KB-16 saga had many important ramifications. One of these was that ultimately Maass and Flood were forced to accept Conant's terms and withdraw from the project. Another issue was Conant's determination to maintain a U.S. monopoly on KB-16 on the grounds that the United States' superior scientific and technological capabilities gave it special rights in developing the most advanced weapons systems. This reasoning had already prevailed in the case of air-to-air radar and the proximity fuse, and it would soon dominate in the development of nuclear weapons. Indeed, in December 1942, Conant specifically cited the KB-16 example as one of his reasons for advocating a U.S. monopoly on the atomic bomb.

Preparing for Gas Warfare, 1943-1945

In both Canada and the United States, jurisdiction over chemical warfare was shared by military and civilian agencies. At times this was an uneasy partnership because of the military's inherent suspicion of outsiders. In addition, there were formidable difficulties in trying to develop a nonoperational weapon. As James Conant explained to Vannevar Bush, 'since chemical warfare is not as yet employed in this war, it is not possible to speak ... definitely in tactical and strategic terms.'[88]

The ambivalent status of chemical weapons also made Otto Maass's job more difficult, although he busied himself making the Directorate of Chemical Warfare and Smoke as efficient as possible while awaiting the inevitable outbreak of gas warfare. On 9 February 1943, he unveiled his new organizational plan for the 900 scientific, technical, and support personnel, along with the host of university scientists[89] who were now connected with the various DCW&S establishments. Under the new scheme, the work of the DCW&S would be divided into four sections: service, technical, financial/administrative, and extramural, each headed by a lieutenant-colonel.[90] The DCW&S would also have sole responsibility for administering the biological warfare research being carried out at the War Disease Control Station at Grosse Île, Quebec.

Although the Directorate was an adjunct of the Army, Maass was generally pleased with the cooperation he received from the Royal Canadian Air Force and the Royal Canadian Navy.[91] Such cooperation was mostly due to the effective operation of the CW Inter-Service Board, which included high-ranking representatives of all three services.[92] Operational planning, however, was most extensive between the DCW&S and the RCAF. In November 1942, for example, Air Marshal L.S. Breadner, Chief of the Air Staff, submitted a detailed memorandum outlining how his service could best provide 'support of the army in chemical warfare.' The RCAF was particularly concerned about 'what areas of the East and West Coast [were] likely to require the use of these weapons,' 'what type of gas and smoke weapons [would] be required,' and 'the relation ... of the various gas weapons and high explosive bombs.'[93]

In his efforts to ensure that the DCW&S was kept abreast of British chemical warfare developments, Maass utilized the services of Suffield's superintendent E.L. Davies, who, despite his shift from Porton, retained close personal ties with his former colleagues. In addition, Maass carried out his own personal brand of scientific diplomacy. In December 1942 he invited J. Davidson Pratt, Controller of Chemical Defence Research (Ministry of Supply), to inspect the Directorate's facilities. Pratt's subsequent report praised the Suffield and Ottawa laboratories for 'the high quality of the work being done.'[94]

Maass had long been aware that getting along with Britain's chemical warfare hierarchy was important but that having the support of the Americans was critical. Throughout the war years he had cultivated the support of General Porter and James Conant, who, he correctly assumed, not only appreciated the potential of Canadian chemical warfare resources and facilities, particularly Suffield, but also recognized that the DCW&S was a useful bridge to Porton.[95]

However, a number of aspects of the Anglo-American relationship threatened Canadian interests. One of the most serious of these involved the guidelines for retaliatory attack, which were amended by the Combined Chiefs of Staff (CCOS) on 31 December 1942. The new arrangement specified that 'gas warfare will be undertaken by both the United States and the British Commonwealth forces on the order of the Combined Chiefs of Staff after approval by U.S. and U.K. Governments; or independently in retaliation on the decision of the Government concerned.'[96] Surprisingly, the Canadian response to this high-handed decree was slow in coming. In September 1943 Major General Maurice Pope, head of the Canadian Joint Staff Mission, formally criticized the CCOS's changes:

Canada as a separate country having signed and ratified an international treaty prohibiting the use of gas as a method of warfare finds herself unable to express here adherence to this statement of policy as presently defined. The United Kingdom Government is not in a position to take a decision in a matter of this kind which remains within the competence of the Canadian Government and obviously the U.S. Government would not propose to do so.

Pope's brief called for an amendment to the CCOS directive so that 'Commonwealth Governments concerned' must also give their consent before the Combined Chiefs of Staff launch a retaliatory attack.[97] This challenge to an Anglo-American bilateral gas warfare policy brought an immediate response from the U.S. Joint Chiefs of Staff, who argued that Pope's brief was unacceptable, since 'there would seldom be time to consult all the Governments concerned when a decision as to the use of gas had to be made.' After extended consultation in Washington between generals George Marshall, Sir John Dill, and Maurice Pope, a compromise was reached. The Americans dropped their objections on the understanding that it was 'morally certain that the Commonwealths concerned would most certainly see eye to eye with the U.S. and British Governments as to the necessity in any given set of circumstances of resorting to gas warfare and would answer with the utmost promptitude any request put to them for their concurrence.'[98]

Once their national sensibilities had been assuaged, the Canadian government resumed its accommodating stance.[99] This cooperative spirit was evident when the Chiefs of Staff instructed the DCW&S to prepare Canadian troops for an effective retaliatory capability, and the Suffield Experimental Station was asked to accelerate development of chemical weapons, which were to be ready 'for operational use before 1 Jan. 1945.'[100] Plans were also made to coordinate Canadian policy on poison gas production and storage with that of the United States, especially through the recently created Canada–U.S. Chemical Warfare Advisory Committee (CWAC).

Otto Maass had long wanted an organization like CWAC, but he first had to convince General Porter and the Canadian General Staff that the committee would 'only deal with the allocation of research projects,' and would in no way disrupt command structures or threaten national initiatives.[101] On 22 September 1943 the committee met for the first time with General William Porter in the chair and a high-powered Canadian delegation led by Maass, Davies, and Flood. Most of the meeting was spent appointing technical subcommittees, which were responsible for assessing some twenty projects deemed essential to the allied war effort. In addition to its analysis of various forms of gas warfare, the committee assessed the effectiveness of smoke weapons and flame-throwers, which had become increasingly important for American and Canadian troops in the battlefields of North Africa, Sicily, and Italy, and whose use would greatly expand during the 1944–5 battles in western Europe and in the Pacific campaign.[102]

Smoke weapons came in different forms: smoke was used for defensive screening; coloured smoke facilitated joint air and ground attacks; and toxic smokes were deployed to incapacitate enemy troops.[103] However, by 1943, DCW&S officials were concerned that despite their research activities German research in these fields was more advanced.[104] As a result, the Canadian Chiefs of Staff, without waiting for British approval, ordered a major series of smoke field trials at Suffield for all three Services. The exercise involved twenty-three different types of weapons, including smoke generators, mortars, and assorted smoke shells and grenades.[105]

Flame-throwers were increasingly used by the Canadian military as the war progressed. Initially, Canada depended on British equipment and supplies, which, by 1943, included a flame-thrower mounted on the

Mark IV (Churchill) Tank for use against enemy pillboxes, a smaller device that was attached to a Bren gun carrier, and a pack type of flame-thrower, or 'lifebuoy,' that was carried and fired by one specially trained infantryman.[106] The coordination of British flame-thrower research and development was the responsibility of the Petroleum Warfare Development Department (PWDD), and its New York–based liaison office, headed by Sir William Wiseman. Maass was not pleased with this arrangement, and much preferred dealing with either the U.K. Military Mission or the British Supply Council because these bodies were familiar with the Directorate's activities and priorities. Maass also developed an intense personal dislike for Wiseman, especially after the latter insisted that all Canada–U.S.–U.K. flame-thrower exchanges go through his New York office.[107] Maass's response to this directive was blunt and direct: the DCW&S would deal only with 'authorized' chemical warfare officials. Such authorities did not include Wiseman.[108] Maass's firm stance reflected his growing frustration in dealing with British administrators, and his belief that most of the innovative flame-thrower technology was now coming from U.S. rather than British sources. Maass's assessment was correct: by the end of 1943 American napalm weapons far surpassed their British rivals in range, reliability, and killing power.[109]

During the last two years of the war, Suffield scientists made their own contribution to the development of flame-thrower technology through such devices as the long-range carrier Barracuda model, and the gun design Rattlesnake II. By July 1943 some 1300 Ronson flame-throwers were being manufactured in Canada, of which 300 were designated for the Canadian Army.[110] Plans had also been made to produce the Barracuda and the pack unit Crocodile at the DCW&S plant in Belleville, Ontario.[111] This work was accelerated after the November 1943 creation of the U.S.–Canada Sub-Committee on Flamethrowers.[112]

Canadian, American, and British chemical warfare experts were also preoccupied with the possibility of using new types of mustard gas in the battlefields of the Pacific theatre.[113] This subject was raised in a 17 September 1943 memorandum by James Conant, who argued that specially prepared HT mustard gas could have a devastating impact on Japanese troops entrenched in tropical island bunkers. Conant urged Vannevar Bush to create 'a joint program involving the Canadians and the British since both of these countries have experienced meteorologists with knowledge of the use of gases in warfare.'[114] Although a fully integrated joint program did not emerge, Suffield scientists, because of their active work with HT mustard gas, did carry out a variety of experiments with the U.S. Chemical Warfare Service, and participated in the April 1944 field tests in the jungles of San José Island, Panama.[115] However, with the Normandy landings imminent, all the resources of DCW&S were mobilized for the reconquest of western Europe.

Allied Planning for the Use of Chemical Weapons, 1944–1945

Using chemical weapons in retaliation against a German or Japanese attack had been official allied policy since the spring of 1942. But by the fall of 1943 British and American military planners began to consider the possibilities of using chemical weapons to regain territory from the enemy by bombing their cities. This change in policy was evident in the spring of 1944, when Major General G. Brunskill, Director of the Special Weapons Branch of the British War Office, carried out a comprehensive survey of the activities of the Directorate of Chemical Warfare and Smoke (DCW&S) and of the role Suffield was playing 'as a proving ground for both British and United States weapons.'[116] In his meetings with Canadian officials, Brunskill was outspoken in his criticism of the retaliation-only policy, arguing that it 'placed the initiative with our enemies.'[117] An even stronger case for the offensive use of chemical weapons had already been raised by Brigadier General Alden H. Waitt of the U.S. Chemical Warfare Service. On 13 January 1944 he told the Canadian CW Inter-Service Board that,

> properly used, gas can shorten the war. By the proper use, I mean 'overwhelming attack with gas' with non-persistent as well as persistent use on a very large scale ... The U.S. figures 40 to 80 tons of mustard per square mile and 150 to 200 tons of phosgene per square mile. I think phosgene will kill more people per ton than high explosives. It can only be done by the use of large enough quantities and I do not believe in harassing the enemy. I want their resistance eliminated completely ... We believe strongly in an overwhelming attack with non-persistent gas, the use of the 500 lb and 1000 lb bomb. On the British side it is the 500 lb bomb and 250 lb bomb. Our filling will be phosgene, hydrogen cyanide and cyanogen chloride. British filling will be phosgene.

Waitt reassured his Ottawa audience that the existing U.S. policy made it unlikely that the CWS would be allowed to unilaterally launch a chemical warfare attack, despite its strategic advantages. But what was most important, he stressed, was that Canada and the United States continue their close cooperation:

I believe we are going to use about 3,000 tons of purified mustard for 4.2 shell and about 9,000 tons for spray ... We consider the Canadian HT production ... as contributing to the common pool of HT munitions ... When this European phase of the war is over there may be a new chapter in U.S.–Canadian C.W. co-ordination on the operational side in the Pacific theatres.[118]

For both moral and strategic reasons, however, neither President Roosevelt nor the U.S. Joint Chiefs of Staff were prepared to grant Waitt and his CWS colleagues permission to deploy gas in any theatre of war. The British Chiefs of Staff (BCOS) shared this viewpoint, although they were once again being pressured by Churchill to consider possible operational use because of intelligence reports that the Germans might use gas 'to oppose OVERLORD.'[119] To ensure that Germany got the message, Churchill proposed that he and Roosevelt reissue their 1942 warning 'that if any gas or toxic substance is used upon us or any of our Allies, we shall immediately use the full delivery power of our Strategic Air Forces to drench the German cities and towns where any war industry exists.'[120] On 28 April the British Chiefs of Staff responded to Churchill's outburst. On the one hand, they readily acknowledged that German leaders understood 'that a successful OVERLORD will spell their doom'; and that the Wehrmacht might 'be faced with a grave temptation to use gas or any other weapon which they might consider would help them to defeat this operation.'[121] On the other hand, the BCOS continued to assume that the German High Command were sufficiently prescient to recognize 'that the tactical advantage they might gain would ... be quickly reversed by Allied retaliation before it could prove decisive.' Fear of massive reprisals, aided by Allied air superiority, they argued, would be enough to deter German leaders, especially since the Allied invasion force would have sufficient defensive equipment to withstand a mustard, phosgene, or lewisite attack.[122] Although Churchill was sceptical of this analysis, he eventually gave way.

But how well prepared and protected were British and Commonwealth invasion troops against a German poison gas attack? According to the British Chiefs of Staff, the troops had ample chemical munitions for retaliatory purposes: 344,000 65-pound bombs (mustard), 41,000 500-pound bombs (phosgene), 42,100 250-pound bombs (phosgene), 4,240 500-pound spray tanks (mustard), 1080 250-pound spray tanks (mustard), 152,600 30-pound bombs (mustard), 14,600 250-pound bombs (mustard), 2,500 500-pound airburst bombs (mustard); 28,125 400-pound spray tanks (mustard).[123] However, since the BCOS was convinced that Germany would not use gas, they decided that the initial assault wave of British and Commonwealth troops would not be equipped with a full array of anti-gas equipment. This decision was not well received by the DCW&S, which insisted that the protective ointment BAL, developed at Suffield for use against lewisite and other arsenicals, be issued to Canadian troops.[124] Although British experts claimed that DCW&S's obsession with BAL had more to do with national pride than with scientific evidence, the Canadians won their case.

What allied military planners did not realize was that Germany would soon have devastating nerve gases, such as tabun, sarin, and soman, at their disposal.[125] In 1945, the plant at Dyhernfurth was producing 12,000 tons of tabun, most of which was filled into weapons: '10,000 tons in bombs, 2,000 tons in shells.'[126] Although effective delivery systems lagged behind munitions production, there was evidence to indicate that if the war had continued, the Luftwaffe would have had sufficient quantities of the 85-kg tabun bombs, armed 'with a very large explosive charge,' capable of creating a 'lethal area of about 3,000 to 5,000 sq. m. [metres].'[127]

Allied intelligence, however, was oblivious to the existence of these weapons. As a result, British and U.S. military planners assumed that because of allied air superiority they would have the upper hand in any poison gas exchange with Germany. This was one reason why the chemical warfare 'hawks' stepped up their campaign for a first-use policy, arguing that deploying chemical weapons was no more inhumane than fire-bombing German and Japanese cities. By the summer of 1944 support for this point of view came from another influential source: Winston Churchill.

On 6 July 1944, outraged by the ever-increasing German V-1 rocket attacks on London, Churchill renewed his demands that the British Chiefs of Staff consider the possibility of launching a massive chemical warfare attack against German cities: 'If the bombardment of London really became a serious nuisance and great rockets with far-reaching and devastating effect fall ... I should be prepared to do *anything* [Churchill's emphasis] ... We could drench the cities of the Ruhr and many other cities in Germany in such a way that most of the population would be requiring constant medical attention ... I do not see why we should always have all the disadvantages of being the gentlemen, while they have all the advantages of being the cad.' Although the British Chiefs of Staff remained convinced that a chemical warfare exchange would delay, not accelerate, victory over Germany, they agreed to commission another special study. This review, however, was quite different from the April–May 1944 report, since it considered the merits of using both chemical and biological weapons.[128] In the end, however, these feasibility studies were shelved, and the BCOS held to its position of using poison gas only in retaliation against a German chemical weapon attack:

(i) By its very nature, gas was a defensive weapon; the Allies were on the offensive on all fronts and the use of gas would therefore be less likely to assist them than it would the enemy; (ii) The area which we had conquered in Normandy was still restricted, and the supplies of food and ammunition for our armies were of necessity stocked in large dumps, which would be very vulnerable to attack of a persistent gas [mustard] ... (v) There was a distinct danger of reprisals by the enemy against captured airmen and other prisoners of war.[129]

While British and U.S. military leaders agreed that chemical weapons should not be used against Germany, the United States held a different attitude toward Japan. In part, this stemmed from the Americans' intense hatred of their Japanese enemy, and was reinforced by the 'war without mercy' being waged in the Pacific theatre. Predictably, the U.S. Army Chemical Warfare Service took the lead in calling for a more aggressive poison gas policy. By the spring of 1945, their arguments were gaining new adherents, particularly after the death of President Roosevelt in April and the high U.S. casualty rates during the invasion of Iwo Jima and Okinawa. In May 1945 General Porter and his CWS colleagues finally received the green light from Chief of Staff General George Marshall, who proposed that the strategic use of poison gas against Japanese island strongholds now be considered. He explained the reasons for this dramatic change in policy to Secretary of War H.L. Stimson on 29 May 1945:

It did not need to be our newest and most potent – just drench them and sicken them so that the fight would be taken out of them – saturate an area, possibly with mustard and just stand off ... The character of the weapon was no less humane than phosphorous [sic] and flame throwers and need not be used against dense populations or civilians – merely against these last pockets of resistance which had to be wiped out but had no other military significance.[130]

During the next two months, the Chemical Warfare Service carried out special field tests to determine which chemical munitions should be deployed and which delivery system would be most effective. Meanwhile, the Joint Chiefs of Staff had been considering a special study of the Operations Divisions that recommended the extensive of gas for the projected November 1945 invasion of Kyushu (Operation Olympia) and the March 1946 landing on Honshu (Operation Coronet). 'Gas is,' the report stated, 'the one single weapon hitherto unused which we can have readily available which assuredly can greatly decrease the costs in American lives and should materially shorten the war.'[131] However, it was the atomic bomb attacks on Hiroshima and Nagasaki in August that eventually brought about Japan's surrender.

American and British planning for offensive chemical warfare had a major impact on the operation of the Directorate of Chemical Warfare and Smoke. In March 1944, for instance, DCW&S was advised that before initiating gas warfare, U.S. and British authorities would 'exchange notice of the proposed release ... of information on new chemical agents (including enemy chemical agents) for training or public discussion.' Canadian chemical warfare officials were also aware of U.S. offensive research through liaison bodies such as the Toxicity Committee and the joint C.W. Advisory Sub-Committee.[132] But it was not until the 6 July 1944 meeting of the CW Inter-Service Board that the strategic deployment of chemical weapons in the Pacific theatre was fully discussed.[133] There was, however, still confusion over how Britain would respond to major changes in U.S. chemical warfare policies. This situation was clarified somewhat on 8 January 1945, when Major General G. Brunskill and Lieutenant-Colonel C.H. Wansborough-Jones gave DCW&S officials an extensive account of how the War Office was considering deploying smoke, flame, and gas weapons in Burma:

As regards gas weapons, Major General Brunskill said that he was impressed with the tactical value of persistent gas used in a semi-persistent way against the Japanese. He mentioned that the sites ... were usually of a type which could be attacked better by gas than by HE weapons. He believed, however, that there had not been a situation of sufficient size or urgency to justify a recommendation for the initiation of CW even if the Government had not given a pledge not to start this method of warfare.[134]

Brunskill also endorsed the expanded operational use of powerful, lightweight flame-throwers. Canadian officials pointed out the advantages of their own Rattlesnake Mark II flame-thrower over existing British models because of its range (over 300 feet), its ignition system, and its reliability. The meeting concluded with an appeal from Maass for a continuation of Anglo-Canadian chemical warfare cooperation after the war. He was understandably gratified when both Brunskill and Pratt 'voiced deep appreciation of the work carried out by the Experimental Station and expressed strongly their personal desire for its continuation as a post-war technical station.'[135]

Conclusion

In September 1939, Canadian scientific mandarins such as Frederick Banting and Otto Maass were determined that Canada assume a major role in allied chemical warfare research, development, production, and testing. Neither believed that poison gas would remain on the military shelf, and both agreed that it was inconceivable that 'a major weapon [gas] employed in one conflict was not carried forward to be used in a subsequent conflict.'[136] Between 1940 and 1945, on at least seven occasions, the use of poison gas was considered by one or more of the warring nations. Military planners on both sides assumed 'that gas, once used, would become accepted as a legitimate weapon and so employed in all theatres of war.'[137] This expectation of poison gas deployment was reflected in the enormous stockpiling of chemical weapons: by the end of the war, the United States had amassed 87,000 tons of toxic chemicals, and Germany 70,000 tons.[138]

While there was virtually no battlefield use of chemical weapons by the great powers during the Second World War, that did not mean that these weapons were not used in other circumstances. Italy used gas against Ethiopia in 1935, and there were isolated Japanese attacks on Chinese troops and civilians before Pearl Harbor. Nor should it be forgotten that the Germans used zyklon B to kill millions of people in their concentration camps. Significantly, this massive use of poison gas against helpless civilians did not produce an allied warning similar in tone and substance to the Roosevelt–Churchill ultimatums of 1942. Despite attempts by the Hebrew Committee of National Liberation in September 1944 to convince the U.S. Joint Chiefs of Staff to threaten Nazi leaders that unless this 'practice ... ceases forthwith, retaliation in kind will be immediately ordered against Germany,' no effective retaliatory policy emerged.[139]

Generally speaking, the chemical warfare alliance between Canada, Britain, and the United States was a success. Of particular note were the close relations between the Directorate of Chemical Warfare and Smoke and the two major American agencies: the U.S. Army Chemical Warfare Service, and the chemical warfare group within the National Defense Research Committee. In many ways, Otto Maass was the architect of this alliance. In 1947 he was awarded the United States Medal of Freedom for his contributions to the allied chemical warfare effort and received the personal congratulations of General William Porter, his wartime partner and friend, who praised Maass' contribution in making Canadian and American chemical and biological warfare activities 'so close that they have been practically one, and the results certainly have been good.'[140]

Another key to Maass's success was the respect and confidence he received from the Canadian Chiefs of Staff, who appreciated his nononsense approach and his ability to get things done. From the perspective of military leaders, the Directorate was one of Canada's most successful wartime operations. With few exceptions, Canadian scientists and military officials worked harmoniously on a complex variety of offensive and defensive weapons and devices.[141]

Although the DCW&S was dependent on British and American scientific information, equipment, and supplies, it had one unique asset: the Suffield Experimental Station. Here the British and Americans not only tested their own chemical and biological weapons, but they also cooperated to develop joint weapons systems with the assistance of the Suffield scientists. Canadian chemists at McGill, Toronto, Queen's, and Western investigated the potential of new war gases such as 'Z,' 'W,' and KL-16, as well as enhancing the effectiveness of mustard gas. Biological agents and munitions, even more frightening in their potential to kill, were another weapon system that preoccupied Suffield scientists during the war years.

6

Canadian Biological and Toxin Warfare Research: Development and Planning, 1939–1945

> The highly secret BW unit at Grosse Île is well under way. It is a co-operative enterprise between the United States and Canada and for administration is under Dr Maass as director of Chemical Warfare. This is another of the many projects which Sir Frederick Banting was keen about in the early days of the war. When I think of all the projects which Sir Frederick started and observe now how many of them have proven eminently successful, the stark tragedy of his early death bears heavily upon me.
>
> C.J. Mackenzie to General McNaughton, 2 January 1943; cited in Thistle, ed., *The Mackenzie–McNaughton Wartime Letters*, 123

Canada's chemical warfare and biological and toxin warfare (BTW) activities between 1939 and 1945 were guided by the National Research Council and the Army's Directorate of Chemical Warfare and Smoke (DCW&S). Both projects involved the participation of Canadian scientists in developing potentially deadly weapons of mass destruction. However, unlike most poison gas developments, biological warfare research was secret and hidden, its practitioners shadowy and self-conscious. Most were medical researchers who were committed to saving, not taking, lives, but whose weapons included some of the world's most lethal diseases: bubonic plague, typhus, typhoid, yellow fever, tularaemia, brucellosis, anthrax, and the deadly botulinum toxin. Unlike chemical weapons, biological weapons were primarily living pathogens whose battlefield use raised alarming questions: Could they be controlled? Would both sides be crippled in a biological warfare exchange?

A number of factors determined the extent of Canada's participation in biological warfare research, development, and planning in 1939: first, its involvement with the British military establishment through the

Committee of Imperial Defence and, scientifically, through the British Medical Research Council; second, the presence in Canada of a number of talented medical scientists, most notably Frederick Banting and J.G. Craigie of the University of Toronto, E.G.D. Murray and J.B. Collip of McGill University, and Guilford Reed at Queen's University; and third, the institutional links between the various university laboratories, and the two patrons of this research – the National Research Council (NRC) and the Department of National Defence (DND). Cooperation between these two government organizations had already been established during the late 1930s by General Andrew McNaughton, and continued during the war years thanks to the efforts of C.J. Mackenzie, Otto Maass, and Sir Frederick Banting.

These scientific administrators also took the lead in trying to convince, first, British authorities and then the Americans to adopt defensive measures against a biological warfare attack and to develop a retaliatory capability. After Banting's death in February 1941, E.G.D. Murray assumed the responsibility for coordinating Canada's biological warfare policies with its wartime allies. During the next four years, Murray presided over the organization of Canadian biological warfare scientists in the secret M-1000 Committee and was instrumental in creating the War Diseases Control Station at Grosse Île, Quebec. He also supervised the testing of biological warfare munitions at the Suffield Experimental Station and coordinated the vital liaison with British scientists at Porton Down and their American counterparts at Camp Detrick, Maryland. Although final authority for Canada's biological and toxin research remained with the Canadian Cabinet War Committee, Murray and his boss, Otto Maass, were given great latitude in making arrangements with U.S. and U.K. scientists for developing and testing various biological warfare agents and munitions. Canada was a junior partner in this Anglo-American biological warfare alliance, and had neither the capability nor the responsibility of deploying these weapons in a retaliatory attack. This heavy burden rested with Canada's allies – the governments of the United States and Great Britain.[1]

Planning for Biological Warfare, 1938–1940

While the 1925 Geneva Protocol had outlawed the use of both chemical and biological weapons,[2] there was little discussion in Canada about the latter weapon in the ten years after the agreement was signed. Even the 1934 publication of British journalist Wickham Steed's sensational disclosures about German biological warfare activity attracted little attention from the Canadian media.[3] However, in March 1935, the *Toronto Star* ran a series of revelations under the lurid title 'Disease Germs Going to Flood Cities When Next War Comes.'[4] Not until 1937, largely through the efforts of Frederick Banting, did Canadian defence scientists begin to consider seriously the threat of an enemy biological warfare attack.

Banting's concerns first found public expression on 11 September 1937, when he warned General McNaughton that, according to scientific and military evidence, Germany was prepared to initiate germ warfare. With McNaughton's encouragement, Banting prepared a memorandum detailing the various air-, water-, and insect-borne agents that could be used 'as a means of destroying an enemy.'[5] For Banting the threat was grave and immediate: 'An epidemic of disease would have a paralysing effect on both the military and civilian population ... By the invention of an automatically air-cooled shell, bacteria such as gas gangrene, tetanus and rabies could be added to the danger of the wounds inflicted, so that even a scratch would be deadly ... In the same way botulinus toxin could be introduced with deadly effect.' Banting was also concerned that since the filtered water supply of Canadian cities was stored in open reservoirs, 'if an enemy aeroplane were to drop a ton of live bacteria or water-borne diseases such as typhoid, cholera or dysentery, an epidemic would be a certainty.'[6] He did not believe that the 1925 Geneva Protocol against chemical and bacteriological weapons would necessarily deter either Germany or Italy in the eventuality of war, since 'scientists of these countries are carrying out research on practical utilization of these deadly weapons.'[7] Banting's prestige, combined with McNaughton's endorsement, meant that the memorandum was passed on to the Canadian Army Chiefs of Staff and the British War Office.[8]

Not everyone shared Banting's concerns. For example, the British geneticist J.B. Haldane, in an October 1937 public lecture entitled 'Science and Future Warfare,' declared that he was 'very sceptical' about the possibility of using either bacteria or viruses as weapons.[9] Similar reservations were expressed by N.E. Gibbons of the NRC, who had been given the task of analysing Banting's memorandum.[10]

Undeterred by these arguments, Banting resumed his campaign against biological weapons during the 1938 Munich crisis.[11] In a 16 September letter to General McNaughton, he emphasized the need for Canada to obtain 'adequate supplies of anti-serum and vaccines, particularly anti-tetanic serum, typhoid vaccine and supplies for active and passive immunity, both of soldiers and civilians.'[12] McNaughton was quick to act, since many of Banting's warnings had been confirmed by four secret documents on bacterial warfare that had been sent to Canada by the Committee of Imperial Defence and its newly created Biological Warfare Sub-Committee.[13]

Visiting Ottawa at this time was Sir Edward Mellanby, Chairman of the British Medical Research Council (BMRC). Mellanby took advantage of his meeting with the NRC Associate Committee on Medical Research to outline British biological warfare policy and to explain the role that the BMRC would assume in 'organizing medical facilities for protection against bacterial warfare ... [and] to build up a supply of tetanus and gas gangrene antitoxin.' He assured his audience, however, that there was good reason to believe that biological warfare would *not* occur.[14]

This was also the view of the Canadian Army Chiefs of Staff. On 24 October, Major General L.R. LaFlèche informed McNaughton that while the Defence department was concerned about the military aspects of biological warfare, it had decided that the problem was 'of national interest rather than one solely for defence.' LaFlèche had, therefore, referred the matter to the Department of Pensions and National Health, who in turn, were instructed to appoint a subcommittee to investigate the defensive aspects of biological warfare issues.[15] Further devolution of authority occurred when the Department of Agriculture was assigned the task of protecting Canadian livestock 'from potent animal diseases such as rinderpest, pleura-pneumonia and foot and mouth.'[16]

Banting's British Mission, 1939–1940

Although Banting was disappointed that his 1938 campaign for chemical and biological warfare preparedness had failed, his interests in these fields increased in the following year.[17] After the outbreak of war in September 1939, Banting announced that he was prepared to lead a Canadian scientific mission to England, along with I.R. Rabinowitch of McGill University.[18] While most of their time was spent touring British chemical warfare installations, Banting was primarily interested in biological warfare, and he was grateful when Sir Edward Mellanby arranged for him to meet many of the experts in this field.[19] His fears about a German biological warfare attack were reinforced after a discussion with Major General H.M. Perry, Royal Army Medical Corps Director of Pathology and the War Office's principal adviser on biological warfare. Perry told Banting about British intelligence reports that indicated that 'the Germans were experimenting with shells which carried bacteria.'[20]

At the official level, however, most high-ranking British officers and scientific advisers continued to minimize the dangers of biological weapons, a response that infuriated Banting. In December 1939, at McNaughton's request, he prepared a detailed memorandum of the biological warfare threat, stressing the importance of immediate defensive measures and the need to develop a retaliatory capability so that 'British Forces should be in a position to give back in a ten fold measure any attack that the Germans may attempt.' Banting also emphasized that effective biological warfare defence could only be achieved 'by means of a study of offensive warfare.' On 3 January 1940 he presented his nineteen-page 'Memorandum on the Present Situation Regarding Bacteria Warfare' to General McNaughton. He also explained its contents in a private meeting with McNaughton, Canadian High Commissioner Vincent Massey, and Britain's Lord Hankey, chairman of the Biological Warfare Sub-Committee.[21] Hankey responded favourably to Banting's report, and complimented him for being the first 'scientist of standing' to present a 'definite proposal.'[22] But Hankey's interest was more diplomatic than real, for neither he nor his committee were swayed by Banting's 'somewhat alarmist view' of the danger of a German biological warfare attack.[23]

When Banting returned to Ottawa in February 1940 he found little support for his report from either the Department of Defence or the National Research Council.[24] Although disappointed with this lack of institutional support, he continued to press his University of Toronto colleagues to investigate the capabilities of different agents and methods of dissemination. By June 1940, Banting was convinced that the dazzling German military successes in Holland and France had been partly achieved through the use of biological weapons, and he predicted that the Germans would undoubtedly use them again 'if England ... is invaded.'[25] On 24 June he wrote C.J. Mackenzie, stressing the importance of initiating large-scale experimental work on both the defensive and offensive dimensions of biological warfare. Mackenzie agreed. But if the program was to get off the ground, Banting would first have to convince Colonel J.L. Ralston, the newly appointed minister of Defence, that Canada should take the lead in developing biological warfare operational weapons, with or without British support.[26] Although Ralston was not prepared to go that far, he did give Banting a limited mandate to proceed with biological warfare experiments.[27] But it was obvious that Canada had neither the scientific nor technical resources to proceed unilaterally, even with Banting's dedicated program. In February 1941 the campaign was further weakened when Banting was killed in an air crash while en route to England to lobby for, among other things, accelerated research on biological warfare.[28]

Creating a Biological Warfare Policy, 1941–1942

Although Banting did not live to see Canada's full-scale biological warfare operation, he had established the essential components for its success in the last four years of his life. Under his guidance, a talented research team emerged at the University of Toronto, consisting of physiologist G. Edward Hall; bacteriologists P.H. Greey and Donald Fraser; and James Craigie of the world-famous Connaught Laboratories.[29] Banting also brought other Canadian scientists into the project. One of these was Charles Mitchell, animal pathologist with the Department of Agriculture and an authority on rinderpest.[30] Another recruit was the noted McGill microbiologist E.G.D. Murray, who had already been involved in biological research as a member of the Sub-Committee on Gas Gangrene.[31] He and Banting were regular correspondents, since both believed that an 'unscrupulous' Germany would undoubtedly use biological weapons, and that 'the only safe defensive position against any weapon is afforded by a thorough understanding which can only be gained by a complete preparation for the offensive use.'[32]

Banting also established links with American biological warfare scientists, who had taken great interest in British and Canadian work in this field since the Tizard mission.[33] In December 1940, for example, M.V. Valdee, Chief of the Division of Biological Control, National Institutes of Health, contacted Banting regarding the formation of a secret American committee 'to consider the possibilities of the use of bacteria and other similar substances in warfare.'[34] However, it was not until October 1941 that an effective U.S. biological warfare organization was appointed, code named the WBC, under the chairmanship of Edwin B. Fred of the University of Wisconsin, and with the guarded support of the Office of Scientific Research and Development and the Army's Chemical Warfare Service.[35] Four months later, the WBC submitted its first report, calling for a comprehensive study of biological warfare 'from every angle,' with special attention to defensive measures that would 'thereby reduce the likelihood of its use.'[36] The WBC also warned that the Japanese, long regarded as one of the world's leading exponents of biological warfare research, might deploy these weapons operationally in the Pacific theatre and possibly against the American mainland.[37]

Fear of Japan's ability and willingness to use biological weapons increased dramatically after Pearl Harbor. In late December 1941, Lord Hankey, chairman of Britain's Biological Warfare Sub-Committee, notified the War Cabinet that the Porton research team under Paul Fildes had been carrying out experiments with biological agents and toxins that could be deployed in a retaliatory attack, although their only operational weapon was 'the use of anthrax against cattle by means of infected cakes dropped from aircraft.'[38] Impressed by Hankey's brief, the War Cabinet and the Chiefs of Staff reluctantly agreed to support offensive biological warfare research and development, under stringent guidelines.[39]

Given their depleted scientific and industrial resources, it is not surprising that British officials should soon look to the National Research Council for assistance, especially since the Council had temporarily assumed responsibility for Canada's biological warfare program.[40] The first NRC initiative came in November 1941, when J.B. Collip, Banting's successor as chairman of the Associate Committee of Medicine, made arrangements for the creation of a secret biological warfare committee, code named M-1000.[41] Its first meeting took place at the NRC offices on 19 December with E.G.D. Murray in the chair.[42] The discussion was lively and the recommendations numerous. High on the list of concerns was the possibility of enemy agents using biological weapons against Canada's civilian population through acts of sabotage.[43] Another perceived problem was the vulnerability of Canada's eight million cattle to the rinderpest virus, which was known to spread 'rapidly, extensively and disastrously.' Since the disease was unknown in North America, and because foreign sources of supply were so remote, it was recommended that Canada develop its own vaccine, hopefully with U.S. assistance.[44] The most important dimension of the committee's deliberations was, however, its emphasis on large-scale biological research and experimentation. Each of the eleven participants was given a specific biological warfare problem to investigate, with preliminary results to be available at the next committee meeting.[45]

In addition to heading the M-1000 unit, Murray assumed responsibility for plague research and for obtaining information about U.S. defensive measures.[46] To this end he travelled to Washington in late December to meet with Edwin Fred and other members of the WBC

Committee and with liaison officers of U.S. Navy and Army Intelligence. Murray, pleased with Americans' interest in Canadian experiments,[47] was allowed to examine secret WBC files concerning ongoing research on typhus, malaria, yellow fever, plague, tularaemia, psittacosis, and botulinum toxin.[48] He also invited the WBC to send a representative to the forthcoming meeting of the M-1000 Committee in the hope of establishing a joint rinderpest research station.[49]

That meeting, held in Ottawa 28 January 1942, and attended by Canadian, American, and British scientists, had an imposing agenda.[50] The primary topic was which pathogenic or toxin agent the Germans and Japanese might deploy.[51] Guilford Reed believed that botulinum toxin would be used, because of its extreme toxicity and because large-scale production was 'very easy.' Other potential toxins included the viral diseases yellow fever and rift valley fever. The latter was deemed particularly dangerous, since the virus was 'hardy and the disease difficult to diagnose.'[52]

Another subject of debate was how best to organize biological warfare research. As chairman of M-1000, Murray related his intention of establishing long-term contractual arrangements with both Canadian universities and private companies. This, he cautioned, depended on whether the committee received funding from National Defence. Sir Henry Dale, visiting Ottawa on behalf of the British Medical Research Council, urged Murray to build his operation on the British model, which, he claimed, combined 'the decentralization of laboratories' with an effective 'register of pathologists, bacteriologists, building and equipment.' Dale also recommended that arrangements be made for the efficient transmission of biological warfare documents 'from the English Committee to this and the American W.B.C.'[53]

The most contentious issue was Murray's proposal for a joint Canada–U.S. rinderpest research station at Grosse Île.[54] For Murray, the advantages of the site were obvious and compelling. The island was isolated, yet only thirty-five miles from Quebec City. It was uninhabited, but had a physical infrastructure left over from its days as an immigrant quarantine station. And above all, it was available, and could be immediately brought under DND control. Significantly, Murray's arguments produced divergent responses from American and British representatives. The former were enthusiastic about the location and downplayed safety concerns. According to Captain C.S. Stephenson, of the U.S. Navy's Bureau of Preventive Medicine, the yellow fever virus had been 'produced in the middle of New York City'; he wondered 'if the virus of rinderpest was really any more difficult to handle than that of yellow fever.'[55] By contrast, Sir Henry Dale claimed that British biological warfare experts 'were not very much disturbed about the possible use of rinderpest in England.'[56]

Murray's most formidable task, however, was maintaining the wavering support of the Department of National Defence for biological warfare research. What purpose, the generals asked, would be served for Canada to pursue this form of warfare when neither Britain nor the United States would 'engage offensively in Bacteriological Warfare.'[57] Faced with this internal opposition, Murray turned to Washington for support. In June 1942, he asked WBC officials to endorse the Grosse Île rinderpest project, since it 'would carry great weight in deciding the question.'[58] Murray's case was greatly enhanced in July, when U.S. Secretary of War Henry Stimson officially approved the undertaking.[59]

By this stage Murray had already submitted his proposal for the establishment of the War Disease Control Station to the Canadian Cabinet War Committee.[60] Under the final agreement of August 1942, Grosse Île was placed under the jurisdiction of the Joint U.S.–Canadian Commission, which consisted of eight members chosen by Otto Maass and General William Porter, head of the U.S. Chemical Warfare Service (CWS).[61] Financial support for the project, code named GIR, was divided between the Canadian government ($300,000 and the site) and the U.S. Army, which paid 75 per cent of all costs associated with new buildings, equipment, and general operating expenses.[62] Significantly, despite its joint character, most of the decisions affecting the Grosse Île War Disease Control Station came from Ottawa, specifically from E.G.D. Murray and Otto Maass.[63]

Allied Biological Warfare Research and Development, 1942–1944

During the summer of 1942 major organizational changes in American defence science greatly affected Canadian biological warfare planning.[64] Of particular importance was the appointment of pharmaceutical mogul George Merck as chairman of the newly formed War Research Services (WRS), a secret civilian organization that was given responsibility for coordinating biological warfare developments in the United States and maintaining close contact with British and Canadian research teams.[65] Indeed, one of Merck's first initiatives was to send two WRS scientists overseas to study the operational potential of British anthrax munitions, particularly since anthrax was known to be a hardy and virulent agent whose 'dissemination was not an impossible problem.'[66] In October, after meeting with E.D.G. Murray in Washington, Merck also made arrangements for a special visit to Quebec City and Ottawa.[67] Although he was primarily anxious to monitor the progress of the rinderpest (GIR) project, he also wanted to discuss anthrax and botulinum toxin with officials of the National Research Council and the Directorate of Chemical Warfare and Smoke, who had now assumed official control over Canada's biological warfare operation.[68]

Canadian expertise in these two biological munitions had been enhanced by the September 1942 visit of Porton's Paul Fildes to Ottawa. In his presentation, Fildes was particularly enthusiastic about the battlefield potential of anthrax spores as a result of recent experiments and tests at Porton. His report described how Porton had developed a cloud chamber 'for sampling concentrations,'[69] how anthrax meal cake could be used against the cattle of enemy countries, and how the four-pound Chemical Warfare bomb 'had been modified to carry a biological filling.' Fildes's most important revelation, however, pertained to field experiments that had been carried out during the summer of 1942 on Gruinard Island in northern Scotland demonstrating that an anthrax bomb explosion could create a spore cloud 'effective against a man one mile distant.'[70]

Fildes's briefing was accompanied by a request for Canadian scientists to help Porton develop sufficient quantities of anthrax munitions (code named N) for a retaliatory response against a possible German or Japanese germ warfare attack. An agreement was quickly signed.[71] On 25 November E.G.D. Murray and Guilford Reed made an emergency trip to Grosse Île in order to devise the 'most satisfactory plan for reconstruction of the [anthrax] building.' Murray was also able to convince Maass to provide $40,000 to cover the purchase of scientific equipment, 'which must be bought immediately in order that the "N" project can be put into production quickly.'[72] Murray outlined the terms of reference of the anthrax (GIN) project in a 8 December letter to Maass:

It is entirely Canadian; It is designed to provide 'N' for the British authorities immediately and in suitable quantity. If practice comes up to our calculations we hope to produce 300 lbs of 'N' spores per week; this would provide for 1500 of the 30 lb. bombs per week. Should the U.S. wish to join us in the 'GIN' Project, CI [C-I Committee] would not raise any objections, but there are no indications that they are prepared to do so immediately.[73]

Murray's pessimism about U.S. intentions was misplaced. The WRS was keenly interested in the GIN project, and in the spring of 1943 became a full partner. Under this second joint agreement, War Research Services assumed three-quarters of GIN's operating costs and provided additional technical support and equipment.[74]

During its early stages, however, the GIN project depended primarily on British not American assistance, since Porton scientists were carrying out the most advanced experiments and possessed the most sophisticated equipment. And Fildes's team was eager to help. Throughout 1943 they forwarded a wide variety of technical reports to the C-1 Committee of the Directorate of Chemical Warfare and Smoke, which was now responsible for coordinating Canada's biological warfare efforts, with most of the material going to Guilford Reed, who was conducting experiments to increase the virulence of anthrax in his Queen's University laboratory.[75] Equally important was the arrival of the anthrax charging machine, which Fildes sent to Grosse Île during the summer of 1943.[76] In exchange for this assistance, the British assumed that Murray's team would be able to produce enough anthrax for the projected anthrax bomb tests at Suffield.[77] It was an obligation the DCW&S took very seriously.[78] On 23 July Murray informed Maass that the GIN (anthrax) project had reached the stage 'where weapons development and Field Trials are essential.'[79]

The problem was that anthrax production at Grosse Île was seriously behind schedule.

Biological Warfare Research and Development at Grosse Île, 1942–1944

During its first six months, the War Disease Control Station at Grosse Île experienced many difficulties. First, construction of the GIN and GIR laboratories was delayed.[80] Second, Murray's request for an anthrax research team of fifteen scientists, including three qualified bacteriologists, had not materialized.[81] Third, staff morale was low due to the station's isolated location and the failure of local military officials to provide the necessary supplies, including sufficient numbers of laboratory animals. Yet, despite these difficulties, by July 1943 considerable progress had been made, especially by the rinderpest vaccine team led by Lieutenant-Commander Richard Shope of the U.S. Navy.[82] Unfortunately, just as the rinderpest project was gaining momentum, Shope was informed by his U.S. Navy superiors that he would be transferred to another American biological warfare project, where, it was reasoned, his experience at Grosse Île would be most useful.[83]

The health risks associated with working on the anthrax project were an ongoing concern. In October 1943, for example, Murray reported a problem with leaking gaskets in the three autoclaves, which, he claimed, were only kept in operation because of the pressure 'to produce sufficient "N" for the experiments at Suffield this fall.' Nine months later, the possibility of anthrax infection became a reality when some of the staff were hospitalized in Quebec City 'with an unexplained fever.'[84] Fearing a serious outbreak, Murray ordered the immediate requisition of special protective clothing, masks, and special equipment to supply 'fresh filtered air.' For a time these measures were effective, but serious health hazards continued to disrupt the GIN operation.[85]

By August 1944, however, there seemed little point in undertaking costly repairs to the existing facilities, since Washington had given notice that the Canada–U.S. anthrax contract would be cancelled at the end of the month. This meant that all Canadian GIN personnel would be transferred to Suffield, along with most of the equipment,[86] except for the anthrax charging machine, which was sent to Camp Detrick, along with a 100 cc sample of 'high quality' anthrax.[87]

Once the anthrax laboratories had been shut down, scientists assoicated with the Grosse Île project (GIR) had the island virtually to themselves, at least until June 1945, when they too were disbanded. By this stage Canadian and American scientists at the station had produced over one million doses of the vaccine for military purposes. A full assessment of their scientific contribution was provided by the WBC's E.B. Fred in a May 1946 memorandum to a special committee of the U.S. Department of National Defence:

The mission ... was of a two-fold character. In the first place it called for the prompt development of ways and means for the safe production and stocking of a type of rinderpest vaccine which ... some years previously had been successfully used in eradicating the disease from the Philippine Islands ... it was likewise highly important to institute research studies to determine whether an effective vaccine could be produced without having to utilize cattle as the source of virus ... After a number of attempts, success was attained in establishing the virus in developing chick embryos ... Aside from the insurance provided by the rinderpest investigations ... the findings and knowledge gained from this work will be of great post-war value.[88]

Biological Warfare Testing at Suffield, 1944–1945

While Grosse Île gradually returned to its prewar isolation, quite the opposite was happening at Alberta's Suffield Experimental Station, where biological and chemical field tests became more numerous and sophisticated. Although these tests greatly enhanced the reputation of Canada's biological warfare program, they also raised serious health concerns. In December 1943, Murray insisted that Suffield scientists observe great care in carrying out anthrax spore experiments lest contamination 'be spread by the large number of gophers and antelope there.'[89] He was also adamant that field tests follow the example of the U.S. Chemical Warfare Service's Granite Peak test facilities and use only anthrax simulants.[90]

In 1944 the Directorate of Chemical Warfare and Smoke turned its attention to an even more potent munition: botulinum toxin, 'the most powerful poison in nature.'[91] Its potential for use as a weapon was greatly enhanced after Canadian and American scientists succeeded in producing a particularly lethal strain, labelled Type A, as well as the somewhat less potent Type B.[92] It did not appear that the Allies were the only ones investigating botulinum toxin's offensive capabilities. In December 1943 Murray and Reed were told about British and U.S. intelligence reports suggesting that the German High Command were preparing to use 'bot tox' (its laboratory name) against Allied invasion troops. This alarming news produced two responses from the DCW&S: the implementation of a crash program to produce large quantities of the Type A toxoid in Reed's laboratory for possible immunization purposes and an acceleration of plans to develop botulinum toxin (code named X) weapons for retaliatory purposes.

During the next eight months Suffield scientists made great progress in developing a variety of 'bot tox' weapons.[43] This was evident in August 1944, when Murray advised Suffield's superintendent E.L. Davies that if the DCWS could 'bring off satisfactory trials and prove the ESS 4 lb [bomb] can be a satisfactory weapon, we shall have X nearly as far advanced as the U.K. have N.' He also indicated that Detrick had 'no [X] weapon near production.'[94] Murray's scientific boosterism was given further credence by Suffield's next series of experiments, which used a newly designed weapon – a complex 500-pound cluster bomb composed of 30,000 small darts contaminated with anthrax. On 25 November, 1944 Davies relayed a field report about the testing of this weapon to Paul Fildes in Porton: 'Declustering in all cases occurred at about 3000 ft ... In every case about 10,000 square yards were contaminated to a density of from 1 to 5 darts per square yard ... This data indicates darts can be aimed with considerable accuracy ... penetration through battledress ... into flesh of unsheared sheep ... of 6 inches or more.'[95]

During the early part of 1945 'bot tox' cluster bombs were made even more effective by means of a dart coating machine, which increased the contamination densities. These improvements in killing power were described by H.M. Barrett, Suffield's scientific director, in January 1945. The impact of 'bot tox' cluster bombs, he reported, was much 'greater than other [high explosives] attacks on slit trenches and is more efficient than fragmentation bombs for men out in the open.'[96] Even previously sceptical British experts had a 'new appreciation' of the potential of 'bot tox' cluster bombs.[97] This was evident in April 1945 when the British War Office provided $50,000 to further the postwar development of 'bot box' cluster bombs.[98]

Throughout the latter stages of the war, Suffield and Camp Detrick scientists were also busy testing two other anti-personnel biological weapons: brucellosis (code named US) and tularaemia (code named UL). The former was regarded as the most promising because of its infectivity, its stability, and its potential to incapacitate rather than kill. One British expert ranked it as '100 times more potent than N or X.'[99] By November 1944 the Detrick pilot plant had produced sufficient quantities of brucellosis for trials, some of which were carried out at Suffield in December 1944 and January 1945. A Detrick technical report described the results: 'Two types of bombs, the Mark I and the shot-gun shell, were fired statically during the Canadian trials and the resulting clouds sampled by means of impingers and animals stationed on area of 50 to 100 yards from the site of the bomb-burst ... From 10 to 35 percent of the organisms were recovered from the clouds, more than sufficient to cause animal infection.'[100]

Before brucellosis could be used as a battlefield weapon, researchers needed to develop both a toxoid and a medical treatment for its effects, for the 'protection of friendly troops.'[101] Since none existed, every brucellosis test was fraught with danger; indeed, seventeen Detrick scientists became infected in late 1944. In February 1945 Captain Leroy Fothergill of the U.S. Chemical Warfare Service advised Maass of these problems, and asked whether Detrick should send one of its medical safety officers to Suffield 'for the purpose of treating the personnel at the station and giving instruction for such further treatment as may be necessary.' Maass readily accepted the offer.[102]

Many of these same problems affected the study of other bacterial agents at Detrick, such as bubonic plague, glanders and melioidosis, and the psittacosis group of chlamydia. In addition were the fungal agents such as coccidia, and the deadly paralytic toxin produced by shellfish. More 'user' friendly were the plant agents developed by the CWS with British and Canadian assistance: brown spot of rice, rice blast, and chemical plant growth regulators primarily designed as sprays for low-altitude attacks on Japanese crops.[103] But as the European war drew to a close, the only biological weapon ready for use was anthrax.[104]

Biological Warfare Diplomacy and the D-Day Controversy

During 1944, military officials in Canada, Britain, and the United States became increasingly interested in the viability of biological weapons.[105] In part this was related to concerns that the Germans might use biological weapons to repel the Allied invasion at Normandy. Even more important were the rapid technological improvements in biological munitions that Suffield and Detrick scientists had achieved.

These developments brought about important administrative changes in Britain and the United States. In August 1944, the British Chiefs of Staff decided to transform the Porton operation into a military agency under the administrative control of the Inter-Services Sub-Committee for Biological Warfare (ISSCBW), chaired by Air Marshal Sir Norman Bottomley.[106] Although it officially stressed the need 'to develop defensive measures against those agents in which ... the Axis powers might be interested,' the ISSCBW was primarily interested in gaining access to all aspects of the rapidly expanding American program.[107] Accordingly, two ISSCBW representatives were assigned to Washington: Colonel H. Paget, on general policy and operational matters, and Lord T.C. Stamp, on scientific and technical matters.[108]

A similar process of centralization was under way in Washington with the formation of the United States Biological Warfare Committee (USBWC) in September 1944.[109] Its mandate emphasized internal planning and coordination. But the USBWC chairman, George Merck, also made a point of insisting that both British and Canadian representatives should participate in the committee's deliberations.[110] While this may have appeared to be an enlightened gesture, Merck had not taken British pride and imperial assumptions into account. Just before his committee's first meeting, ISSCBW representatives Lord Stamp and Colonel Paget asked that Otto Maass be excluded from the deliberations.[111] Maass's appointment, they claimed, would establish 'a dangerous precedent' for Commonwealth unity; and would complicate the work of the Combined Chiefs of Staff.[112] Neither Merck nor General William Porter, head of CWS, was impressed with arguments concerning Commonwealth status or, for that matter, with Porton's usefulness in developing operational biological munitions; by contrast, they increasingly appreciated Canada's contribution. On 2 December Merck advised U.S. Secretary of War Stimson that direct liaison between Canada and USBWC 'was important and desirable,' even if it meant offending British sensibilities. He gave three reasons for his decision: '(a) We are engaged in important joint projects with the Canadians; (b) Our countries are contiguous and our b.w. policies should therefore be closely coordinated, especially in defensive matters; (c) The Canadians have made important progress in this field, considerably more than the British in some respects, and direct exchange of ideas with them should be facilitated in every practicable way.'[113]

Canada's relationship with Britain in planning for Allied biological warfare continued to be a contentious issue throughout the remainder of the war.[114] E.G.D. Murray, chairman of the C-I Committee, had become increasingly frustrated with the tendency of ISSCBW representatives in Washington to act arbitrarily. In October 1944, for example, Lord Stamp, without informing the C-I Committee, made arrangements with Camp Detrick scientists to conduct brucellosis trials at Suffield.[115] He also disagreed with British proposals for a new classification system that Colonel Paget had forwarded to Ottawa that same month:[116]

There has been a tendency to treat all BW activity as TOP SECRET with resultant loss of efficiency in military prepared and ordered staff consideration. Since the defensive aspect dominates the situation as far as the United Nations are concerned, and involves no morally compromising consequence. It is suggested that down-grading to CONFIDENTIAL would be warranted. Research and development of offensive BW should be considered as TOP SECRET.[117]

Murray's most bitter confrontations, however, centred on Suffield's safety procedures, and whether properly qualified scientific personnel were monitoring laboratory and field tests. In November 1944, he reminded Suffield's superintendent E.L. Davies that before tests planned for 'your area E' were carried out, it would be necessary to assign an 'accredited bacteriologist ... whose advice must be taken on the bacteriological aspects of any such trial.'[118] Part of Murray's concern stemmed from his suspicion that Suffield had become unduly influenced by the British biological warfare agenda and its assessment of possible enemy use of these weapons.[119] In particular, he rejected Porton's assumption that neither Germany nor Japan had developed operational biological weapons.[120] This issue would seriously divide British and Canadian scientists on the eve of the Normandy invasion.

On 20 December 1943 the U.S. Joint Chiefs of Staff had received an alarming report from the Office of Strategic Studies (OSS) that the German military were preparing to use botulinum toxin (code named X) against Allied civilian and military targets. The brief outlined the scientific and tactical advantages the enemy would obtain by using this weapon:

a. This toxin is not contagious and its military use by the Germans, would, therefore, not backfire onto the continent.
b. The air-borne dust would have no odor or taste, and thus a population would have no reason to protect itself with gas masks. Symptoms do not develop until four or five hours after contact, when death invariably follows.
c. The cause of death is an embolism and would tend to bewilder medical opinion.
d. The toxin ... if mixed with high explosives might not be detected. [121]

The OSS brief referred to Canadian tests that had demonstrated how airborne botulinum powder was lethal even 'in extremely great dilution,' and how casualty rates could reach '70% mortality.'[122] On 2 February 1944 the JCS were further advised by Detrick officials that they had carried out extensive consultations with Porton about Germany's biological warfare potential and the most effective Allied response. In keeping with their earlier position, Porton scientists dismissed the threat as exaggerated, since there was 'no reliable evidence available' that the Germans were capable of a large-scale botulinum toxin attack. Porton's Paul Fildes did not believe that mass immunization would be effective 'because of uncertainty as to the precise agent the enemy may employ.' The only viable policy, he concluded, was retaliation in kind.

This approach was, of course, central to the British, American, and Canadian response to an enemy biological or chemical weapon attack, and all three countries frantically began to stockpile the necessary munitions in case this type of warfare erupted. In addition, the U.S. Joint Chiefs of Staff continued to plan for the mass immunization of American invasion troops.[123]

The Canadian Chiefs of Staff and the Canadian Cabinet War Committee were even more disposed than the Americans toward the defensive approach. On 17 December 1944 Master General of Ordnance J.V. Young notified Minister of Defence J.L. Ralston about the OSS report and the response of the U.S. Joint Chiefs of Staff. He also indicated that E.G.D. Murray and Guilford Reed had met with CWS officials and 'confirmed that U.S. policy was to provide immunization for American troops.'[124] A major factor in the Canadian deliberations were reports about Suffield experiments showing that botulinum toxin was now an effective weapon. Not only were there viable delivery systems, with 'bot tox'–coated darts,[125] but there was now a vaccine to protect Allied troops and civilians.[126]

In their brief to the Canadian Chiefs of Staff, Murray and Maass noted that Camp Detrick had been instructed to ship 1,138,000 units of Type A toxoid, sufficient for treating 300,000 American military personnel, with another 46,000 units of Type B toxoid held in readiness.[127] Over 60,000 units of the Type A toxoid, code named 'Esoid,' would be required for Canadian troops overseas.[128] As a result of Murray's and Maass's appeal, arrangements were made for the necessary Esoid supplies to be produced at Guilford Reed's laboratory.[129] In short order, 200 litres of A toxoid a week were being turned out, later expanded to 400 litres. By April 1944 Reed was in a position to ship 25,000 doses overseas, with another 150,000 available by mid-May.[130] To ensure that the inoculations were carried out properly, Murray requested that both he and Reed be sent overseas.[131]

It was not an easy undertaking. During their stay in England, Murray and Reed found themselves increasingly harassed in their attempts to carry out plans to immunize Canadian troops. The pressure became more intense on 18 May when Lieutenant-General Kenneth Stuart, Chief of Staff, Canadian Military Headquarters, London, was informed by the British War Office that neither British nor American troops would receive the toxoid, since new British intelligence provided irrefutable evidence that the German military did not have the capability of launching an effective botulinum toxin attack against Operation Overlord. Stuart was impressed by the evidence and by the determination of British and American military planners to abandon all inoculation plans. Could Canadian troops, he wondered, be exempted from this policy without seriously undermining morale among the assembled troops?[132]

Ottawa had a different response. Neither Chief of Staff J.C. Murchie nor the Canadian Cabinet War Committee was impressed with the British War Office's logic or with the quality of its evidence. In a sharp rejoinder, the committee reminded Stuart that his primary responsibility was 'the safety of Canadian troops,' and that it 'would be most reluctant ... to accept the view that Canada should take responsibility for foregoing potential protection of Canadian personnel because other Allied personnel and civilians are not afforded the same protection.' The committee also referred to a number of strategic and tactical considerations that British and American military officials seemed to have conveniently overlooked. One of the most important of these was the argument that the enemy might claim that 'bot tox' was a chemical 'not ... BW agent' and more likely to be used, given the precedent of the First World War when chemical weapons had been routinely displayed.[133]

Once the Normandy invasion had been successfully carried out, the debate over inoculation subsided. And since the troops had not encountered botulinum toxin, the British Chiefs of Staff could congratulate themselves that their analysis of German BW capabilities and intentions had been vindicated. But was it more luck than good planning? Clearly, British intelligence was not aware that within the Third Reich many powerful advocates, including SS leader Heinrich Himmler, were pressing for the deployment of biological weapons.[134] In the last analysis, the BCOS had tossed the dice and won. But was the gamble worth taking? Murray, Maass, and members of the Canadian War Committee did not think so.[135] Nor did most Camp Detrick scientists, an opinion summarized by CWS historian Rexmond Cochrane in his 1946 report: 'Although the material was available for inoculating the assault forces landing on the Normandy beaches in June, 1944, it was decided not to use it. Our troops went in wearing protective clothing and carrying their gas masks, ready for chemical attack, but not biological attack.'[136]

Once the western allies broke out of the Normandy beachhead, the next question was whether the Germans would use biological weapons in a desperate attempt to defend their homeland. While the British Chiefs of Staff remained sceptical of such a possibility, it did agree to appoint two Canadian liaison officers to the War Office in order to have access to intelligence reports about German biological warfare activity.[137] Murray, however, did not believe that the danger had passed. In January 1945 he warned the Canadian Chemical Warfare Inter-Service Board of the importance of continued vigilance since 'the enemy may have devised more effective bacterial warfare than we are aware of and these might contribute to a combination of weapons with terrible effect.'[138]

Allied Planning for the Offensive Use of Biological Warfare, 1944–1945

The German military were not the only group considering the use of biological weapons in 1944. British and American planners were also engaged in feasibility studies. One of the more celebrated and controversial incidents was the July 1944 report of the British Chiefs of Staff to Prime Minister Winston Churchill that outlined the relative merits of attacking German cities with chemical or biological weapons in retaliation for the V-1 rocket attacks on London. While the report assumed that phosgene and mustard gas would produce a casualty rate of between five and ten per cent, a large-scale anthrax bombing attack would be even more devastating:

'N' is the only Allied biological agent which could probably make a material change in the war situation before the end of 1945. There are indications which lack final scientific proof, that the 4-lb bomb charged with 'N' used on large scale from aircraft might have a major effect ... There seems little doubt the use of Biological warfare would cause heavy casualties, panic and confusion in the areas affected. It might lead to a breakdown in administration with a consequent decisive influence on the outcome of the war.[139]

Significantly, the report did not mention that the quarter-million four-pound anthrax bombs required for such an attack could not be delivered by the U.S.-based Virgo plant until after the war was over. But it did not really matter: the Chiefs of Staff report was only a feasibility study and was never incorporated into British military strategy.[140]

The same ambivalent situation prevailed in the United States, where the Chemical Warfare Service had responsibility for planning the possible use of biological/toxin munitions. High on its list of concerns was how the biological weapons research carried out at Detrick and the operational priorities of the American Armed Forces could be integrated more effectively.[141] Lieutenant-Colonel Oram Woolpert, commander at Detrick, addressed this problem in a memorandum of 25 October 1944. He was particularly concerned that while the U.S. strategy emphasized

Canadian Biological and Toxin Warfare Research, 1939–1945 171

a retaliatory response, 'no policies have been laid down in respect to the types of agents which will be accepted for this purpose.' So that Detrick scientists might have a better grasp of military priorities, Woolpert asked for an early decision on 'the selection of classes of agents which will be acceptable in respect to persistence, transmissilibility.' To ensure that the Joint Chiefs of Staff appreciated the concerns of Detrick scientists, he posed seven questions:

(a) Will agents that are highly persistent or highly transmissible be acceptable to the War Department should research and development be carried to the point where their use appears to be practicable?
(b) Should attention be directed primarily to agents which can be used tactically against combat formations, or rather for strategic effects against civilian centres?
(c) Is there a requirement for agents which will quickly produce large numbers of fatalities or for agents which will cause only temporary or prolonged disability?
(d) Will the using Services accept agents which, because of their instability, will require special handling in storage and transportation, particularly in respect to refrigeration?
(e) Will an agent be accepted even though adequate protection of our Forces is not yet assured, as in the case of 'N'?
(f) If an agent is to be used for retaliation, on what scale should preparations be cast?
(g) Should research installations confine their investigations strictly to authorized classes of agents, or explore in addition other possible agents, on the assumption that circumstances might later alter policies?[142]

One of these changes in circumstances was Germany's unconditional surrender in May 1945. This meant that the the American military shifted its attention almost exclusively to the Pacific theatre. As was discussed in Chapter 5, at this stage of the war the United States was planning to use chemical weapons if the war against Japan had continued. What is perhaps not as well known is that the United States was also preparing to use anti-plant fungal agents, chemical plant-growth inhibitors, and defoliants against Japanese crops.[143] In March 1945, the U.S. Joint Chiefs of Staff agreed that such tactics would not violate the Geneva Protocol, since 'the compounds used are comparatively harmless to human beings.'[144]

Japanese Biological Warfare Balloon Attacks

As the United States and its allies were preparing to invade Japan, new intelligence reports indicated that the Japanese army was planning to use biological weapons against its enemies.[145] These rumours gained credibility with the dramatic launching of hundreds of Japanese bomb balloons, which were carried by the prevailing winds to the west coast of North America. Between 4 November 1944, when the first sighting occurred, and 28 August 1945, over a thousand balloons, many of which were armed with incendiary or high-explosive bombs were identified by military intelligence, although there were only a few instances when explosions occurred.[146] American scientific leaders were most concerned, however, that the next payload would contain biological agents. What made the situation even more threatening was that these paper balloons, floating at altitudes of between 20,000 and 30,000 feet, could drift as far east as Michigan without being detected or intercepted. They could also carry 400 pounds of equipment – more than enough to disseminate agents such as plague, Japanese B encephalitis, anthrax, and rinderpest.[147]

Throughout this crisis, the Canadian Directorate of Chemical Warfare and Smoke was kept fully briefed about the new Japanese weapon by the U.S. Chemical Warfare Service and the Office of Scientific Research and Development.[148] Maass effectively used this information in his consultations with the Canadian Chiefs of Staff.[149] Indeed, he and General Murchie both agreed that immediate action was needed 'in regard to the defensive aspects of BW' and that the Inter-Service Board would appoint a special subcommittee to report on 'the threat of BW from ... [Japanese] balloons.'[150] After almost three months of investigation, the subcommittee submitted its report. It concluded that the balloons were difficult to intercept and could not be easily detected by radar. The report also stated that the balloons were 'well suited for the carrying of dangerous bacteria and the pay loads of the balloons already studied would be sufficient to cause serious wide spread damage to animals, crops or humans.' The report recommended that the Directorate of Chemical Warfare and Smoke coordinate the medical and veterinary resources of the Department of Agriculture, the Department of National Health and Welfare, and provincial boards of health 'in view of the danger both to the civilian population and agriculture as well as to the Armed Forces.' Given the seriousness of the situation, the Chiefs of Staff decided that the subcommittee should report directly to the Cabinet War

Committee so that 'effective defensive and offensive measures could be quickly implemented.'[151]

By July 1945 the Japanese balloon threat seemed less serious. In Washington, a relieved Vannevar Bush, President Roosevelt's primary scientific adviser, turned his biological warfare files over to the War Department, or more specifically to George Merck, now special adviser to Secretary of War Stimson. Merck, however, preferred to remain cautious, and he arranged for his staff to continue 'the detection, reporting and analytical work,' which, he indicated, was 'fully coordinated ... with those of the Canadian Government.'[152]

The full extent of Japanese biological warfare research, which included inhuman medical experiments, were only fully appreciated after August 1945, when the activities of the infamous Units 731 and 100 of Harbin, Manchuria, were discovered.[153] But did Japanese authorities ever intend to use balloons to transport biological warfare agents? Absolutely not, was the response of members of Unit 731 during postwar interrogations. On the other hand, American intelligence reports show that in 1943 a prominent Unit 731 scientist had visited the Fu-Go balloon project on Honshu.[154] Whether the Japanese would have resorted to biological weapons if the war had continued and their homeland had been invaded will remain a historical mystery. What is known is the extent of Japanese biological weapon development and testing, often against civilians and prisoners of war. Equally shocking is that most of those involved with these atrocities went unpunished during the American postwar occupation of Japan.[155]

The fact that these Japanese experiments were classified top secret by American military officials meant that few Allied biological warfare scientists had an opportunity to reflect on the differences between their wartime work and that carried out by the Japanese. If they had, most would have undoubtedly stressed that human subjects were never used in any of the laboratory or field tests, and that their countries were primarily involved in biological warfare research for defensive reasons. More specifically, in order to prevent either the Japanese or the Germans from using biological weapons, an effective retaliatory capacity was required.

Biological Warfare Planning in the Postwar Years

The question of Canada's continued involvement with biological/toxin warfare was a matter of much debate during the summer of 1945. In many ways the answer to this question was closely associated with developments in Britain, or, more specifically, with the special study being prepared for the British Chiefs of Staff.[156] In February 1945 the Joint Technical Warfare Committee, chaired by Henry Tizard, had begun its survey of Britain's postwar defence and weapon systems. One of its first moves was to ask the British Inter-Services Sub-Committee for Biological Warfare (ISSCBW) to provide an assessment of the postwar strategic role of biological/toxin weapons, particularly since leading members of the committee had recently visited the United States and discussed the strategic value of different munitions with the Army's Chemical Warfare Service.[157] The committee's report of March 1945 concluded that the United States's biological warfare capabilities went 'far beyond the scope of anything that this country can provide.' The ISSCBW believed that close cooperation with Canada's biological warfare program would allow Britain to become at least a junior partner of the United States in the postwar years.[158] The subject of a tripartite approach to biological weapons was further explored in June 1945, when Otto Maass and General Alden Waitt of CWS attended the ISSCBW meeting of 11 June 1945.[159] Although nothing was resolved at this stage, these consultations would gain momentum with the advent of the Cold War.[160]

Conclusion

By August 1945, the team of Maass, Murray, and Reed had advanced Canada's biological warfare operation to a level that even Frederick Banting would have found incredible.[161] They had organized Canadian bacteriologists, biochemists, and veterinarians into an effective organization that carried out research and development in a number of fields. On the defensive side, at Grosse Île, DCW&S scientists, together with their American counterparts, conducted successful experiments in the development of the rinderpest vaccine. In addition, a Canadian version of the botulinum toxin (A and B) was developed by Guilford Reed in his Queen's University laboratory to immunize Canadian troops overseas.

On the offensive side, GIN scientists at Grosse Île were able to prepare sufficient quantities of virulent anthrax spores for use as munitions in Suffield field tests and by Camp Detrick scientists in their development of anthrax as a weapon of mass destruction. Murray and Reed were also among the first Allied scientists to recognize that anthrax munitions had serious limitations and to encourage the development of other pathogenic agents and toxins. High on their list was botulinum toxin for use in four-pound bombs and in 500-pound cluster/dart bombs. They also supported the military use of brucellosis, an agent with high infectivity but low morbidity.

Most of this research was initially carried out at the request of the British Biological Warfare Sub-Committee and its successor, the Inter-Service Sub-Committee for Biological Warfare. By the fall of 1944, however, Maass and Murray were much more involved with the work at Camp Detrick than with that at Porton Down. They recognized, even if the British had not, that the only biological warfare Allied super-power was the United States. They also appreciated that George Merck and General William Porter treated them as scientific equals, not as colonial technical assistants.

Did the wartime work of Banting, Murray, and their associates violate contemporary scientific and medical values? Did Canadian scientists regret their involvement in biological warfare research?[162] There are no simple answers to these questions. Clearly, Canadian biological warfare scientists were divided about the value of their war work; indeed, some shunned military research after 1945. But biological warfare research was viewed as a viable option for most scientists during the Second World War, when the country was forced to confront ruthless and powerful enemies.[163] Although the biological weapons tested at Suffield were never used, the possibility that Canada might require a retaliatory capacity remained acute during the postwar years.

7

Atomic Research: The Montreal Laboratory, 1942–1946

> More than four hundred scientists and engineers are now engaged on research and development of atomic energy in the project and in addition, there are more than five hundred mechanics, electricians and other workers of various kinds engaged in supplying the industrial and auxiliary services. The 10,000 acres of sparsely populated land expropriated by the Government as a site for the Project ... has become the centre of a great scientific activity in the most exciting and promising field of the moment, and Canada for the first time in its history has the privilege of being an effective pioneer in a great world development.
>
> C.J. Mackenzie's radio broadcast of 3 June 1947 on the status of Canada's nuclear operation at Chalk River[1]

Canada's involvement in the development of nuclear weapons was a further manifestation of its wartime alliance with Great Britain and the United States. It also reflected the enormous changes that occurred in the nuclear field in these years. Under the pressures of war, atomic research rapidly progressed from the laboratory fission experiments of Otto Hahn's German scientific team in December 1938 to the first atomic bomb tests in New Mexico on 16 July 1945. While innovative research was carried out by scientists in Great Britain, aided by the 'Free French' contingent under Hans von Halban, only the United States had the technological and industrial resources to actually manufacture the bomb. By war's end, the Manhattan Project would spend over $2 billion – and employ thousands of scientists at the University of Chicago Metallurgical Laboratory, the Oak Ridge and Hanford production facilities, and the weapons assembly laboratory at Los Alamos.[2] Canada became an aspiring member of the nuclear club with the establishment of the Anglo-Canadian Montreal Laboratory in 1942 and the Chalk River

Nuclear Laboratories in the autumn of 1945. Over the next thirty years, Canadian nuclear technology would continue to evolve with such technological achievements as the NRX (National Research X-perimental), which became operational in 1947, and the acclaimed CANDU (Canada-Deuterium-Uranium) system, which emerged in the 1960s.[3] Canada also participated in the tortuous international discussions about the postwar use of atomic energy and nuclear weapons.[4]

This chapter focuses on nuclear research during the war years, with particular emphasis on a number of important issues arising from the atomic research partnership between Canada, the United States, and Great Britain. First, how did scientists of the three countries, and especially the key scientific administrators, view cooperation and the exchange of information before the Quebec Conference of August 1943? Why did U.S. scientific and military administrators such as James Conant and General Leslie Groves continue to have misgivings about the Montreal Laboratory? In what ways did the Montreal Laboratory assist in the work of the Manhattan Project after the technical agreement of June 1945, which provided for a variety of cooperative ventures? And, finally, how did this agreement provide the basis for Canada's postwar nuclear development?

Early Development of the Atomic Bomb

The establishment of the Montreal Laboratory was directly related to British efforts during 1941 and 1942 to develop an atomic bomb. Despite the reservations of some prominent defence scientists, including Henry Tizard, the British government authorized the establishment of a special committee, code named Maud, to investigate the possibilities that the uranium 235 isotope could be used as a super-explosive. In August 1941, the British Maud Committee submitted its final report to the special subcommittee of the British War Cabinet. After eighteen months of deliberations, this elite group of British scientists arrived at three major conclusions about the viability of an atomic bomb: it could be built within five years; its explosive power would dwarf all existing chemical explosives; and massive casualties would be caused by the blast and by the resulting radiation.

We have now reached the conclusion that it will be possible to make an effective uranium bomb which, containing some 25 lb. of active material, would be equivalent as regards destructive effect to 1,800 tons of T.N.T. and would also release large quantities of radioactive substances which would make places near to where the bomb exploded dangerous to human life for a long period.[5]

The report also included recommendations concerning the technical problems of creating a uranium 235 bomb. The most important of these involved the immediate establishment of a uranium 235 gaseous diffusion separation plant that would be able to have 'material for the first bomb ... ready by the end of 1943,' unless there was 'a major difficulty of an entirely unforseen character.'[6]

All that was required was the endorsement of Britain's military establishment. This was achieved, in part, in November 1941, when the War Cabinet's Scientific Advisory Committee, chaired by Lord Hankey, strongly endorsed the conclusions of the Maud Report. The committee stated that the project would have to go through six stages before a uranium bomb could be produced: direct measurement of the fission cross-section of uranium 235; design of the fusing mechanism; completion of the design of the pilot gas diffusion separation plant; construction of the pilot plant; developmental work on chemical processes; and installation of a full-scale separation plant. The committee also emphasized the need to establish 'one pilot plant and full-scale separation plant ... in Canada.'

Meanwhile, Prime Minister Churchill had been advised by Lord Cherwell that while the atomic bomb was a long shot, 'it would be unforgivable if we let the Germans develop a process ahead of us by means of which they could defeat us in war or reverse the verdict after they had been defeated.'[7] Churchill was convinced, and quickly brought the Chiefs of Staff on side. Their only qualification, which the prime minister accepted, was that all work on the British atomic bomb remain totally secret, even from most members of the War Cabinet. After almost a year of deliberation, the nuclear project was ready to proceed under the auspices of a new organization, the Directorate of Tube Alloys, which was directly responsible to Lord President Sir John Anderson.

Anderson's selection as czar of Britain's atomic bomb venture was based on three factors: his political skills, his scientific background, and his links with the Department of Scientific and Industrial Research (DSIR), which provided the administrative support for the project.[8] Executive duties for the project were shared between Sir Edward Appleton, head of DSIR, and William Akers, Research Director of Imperial Chemical Industries, who was named chief executive officer of Tube

Alloys.[9] In addition, two consultative committees were appointed: one composed of five prominent politicians and administrators who would advise Anderson on broad questions of policy,[10] and a Technical Committee, consisting of four outstanding atomic scientists: James Chadwick, Rudolph Peierls, Hans von Halban, and Francis Simon.[11]

From its beginnings, the Tube Alloys Directorate was controversial. One problem was the government's failure to consult with those scientists, such as Mark Oliphant, John Cockcroft, and George Thomson, who had initially been involved with the project in 1940. Oliphant was particularly upset that 'the people put in charge of this work should be commercial representatives completely ignorant of the essential nuclear physics upon which the whole thing is based.'[12] Nor did matters improve during the next nine months. By August 1942, the British atomic effort, so boldly proclaimed in the Maud Report, was in serious trouble. Three major reasons for this situation were: the intense competition for vital technological and manpower resources in a country fighting a desperate war of survival; the lack of effective support during this time from either the War Cabinet or from Churchill; and, perhaps most serious of all, a naïve and condescending attitude toward the fledgling U.S. nuclear operation.[13]

Between 1940 and 1942, the British were not interested in an atomic partnership with the United States because they believed their project was superior and because of concerns about U.S. security standards.[14] Such an attitude was unfortunate because at this stage in the war the American authorities were eager to establish a collaborative arrangement. This was evident in August 1941, when Vannevar Bush, director of the U.S. Office of Scientific Research and Development (OSRD), proposed that the development of atomic bombs be a joint effort between Britain and the United States. President Roosevelt went one step further in an October 1941 letter to Churchill: 'It appears desirable that we should soon correspond or converse concerning the subject which is under study by your M.A.U.D. Committee and by Dr. Bush's organization in this country in order that any extended efforts may be coordinated or even jointly conducted.' Two months went by before Churchill gave FDR his answer: no thanks![15]

Bush and Roosevelt did not renew their offer. In part their response was out of pique, but a much more important factor was the enormous progress the American atomic project was to achieve by June 1942. On this date responsibility for the American atomic bomb shifted from the OSRD to the U.S. Army's Manhattan Engineer District (MED) under the direction of General Leslie R. Groves. Six months later, on 2 December 1942, American scientists achieved their first major goal: the chain-reacting pile at the Univesity of Chicago Metallurgical Laboratory became operational. This modern miracle was enough to convince President Roosevelt of the viability of the Manhattan Project; as a result, he authorized $400 million for uranium separation plants and a plutonium-producing pile. With a virtual blank cheque at his disposal, Groves proceeded with plans to construct gaseous diffusion and electromagnetic uranium separation plants at Oak Ridge, Tennessee, as well as a massive plutonium pile complex at Hanford, Washington.[16]

These tremendous scientific and technological achievements had a powerful influence on the attitude of U.S. scientific administrators toward their wartime allies. From their vantage point, there was no reason to include the British in their nuclear bomb project, aside from a few nuclear experts who could be seconded to the Manhattan Project for the duration of the war. This sense of confidence was reflected in Vannevar Bush's opposition to the May 1942 proposal that Hans von Halban and his slow-neutron team join the Chicago Metallurgical Laboratory: 'The fundamental difficulty is that of having a group of English scientists here without being essentially responsible to anyone in this country. In my mind, it would be very much better if they were to work in Canada where there would be no question of "extra-territoriality."'[17]

Creating the Anglo-Canadian Nuclear Laboratory, 1942

The establishment of the Montreal Laboratory was based almost exclusively on British priorities – or, more accurately, on their belated recognition that a wartime atomic partnership with the United States was impossible without having direct access to the resources of the Manhattan Project.[18] On 30 July 1942 Sir John Anderson informed Prime Minister Churchill that the British pioneering work on atomic bomb research was 'a dwindling asset and that, unless we capitalize on it quickly, we shall be outstepped. We now have a real contribution to make to a "merger." Soon we shall have little or none.'[19] With Churchill's endorsement, Anderson took the next step: to convince Mackenzie King's government that a joint Anglo-Canadian nuclear laboratory was essential for the war effort. On 6 August 1942 Anderson instructed British High Commissioner Malcolm Macdonald to pursue the matter immediately with Canadian officials, since the British atomic project was 'in danger, owing to our comparative lack of resources, of being outstripped by the

Americans who are working on four alternative methods.' In order to remain competitive, Anderson noted, it would be necessary to shift the British uranium 235 fast-neutron team to the United States, 'where it would be incorporated in the U.S. program on equal terms.' On the other hand, he argued, the Cambridge slow-neutron team of Hans von Halban should be moved to Canada not the United States, since their primary task was to concentrate on 'the production of the element [plutonium] by using "heavy water" or even ordinary water, if uranium enriched with U 235 can be obtained.'[20]

While Anderson acknowledged that Halban's team was involved with the longer-term project, he expressed confidence that it would 'in the end prove the more efficient method and the one which will eventually hold the field for the purpose of power production.' The key question was whether Canadian authorities would agree that the slow-neutron project was a vital wartime undertaking. Anderson, shrewdly assessing the scientific 'boosterism' of his Canadian allies, concluded that they would come on board, especially since the project meant 'the development of the use of a raw material indigenous to Canada.'[21]

The prospect of being 'hewers and drawers' of uranium was of minimal importance in convincing Canadian authorities to endorse Anderson's proposal. A more important factor in Canada's ready acceptance was that by August 1942 a series of collaborative arrangements between Britain and Canada were already in place, most notably in the areas of radar and sonar, explosives and propellants, and chemical and biological warfare. For Mackenzie and his NRC associates, therefore, the Montreal Laboratory was another important Canadian contribution to the British war effort.[22] Mackenzie quickly recognized the advantages of getting in 'on the ground floor of a great technological process' and made great efforts to convince C.D. Howe, Minister of Munitions and Supply, and Prime Minister King that this venture would have long-term benefits for Canadian science, technology, and industry.[23]

The October 1942 agreement between the two countries was based on a shared cost and responsibility formula. The two national research agencies, DSIR and NRC, would jointly supervise the operation, with High Commissioner Macdonald and C.D. Howe mutually responsible for general policy matters.[24] Administratively, William Akers, director of Tube Alloys, played a central role both in coordinating the activities of the Montreal Laboratory with those of the British fast-neutron team and in establishing a working relationship with the newly established Manhattan Project.[25] Laboratory research was to be coordinated by the

NRC, and by a scientific committee chaired by the Research Director, Austrian-born scientist, Hans von Halban.[26]

At first, the prospects of the Montreal Laboratory appeared favourable.[27] Despite initial concerns about site location, equipment, and supplies, the NRC handled the logistical requirements effectively. Also encouraging were reports that Vannevar Bush welcomed the arrival of the British nuclear team and that they could count on the OSRD 'for American supplies.'[28] Considerable headway was also made in recruiting high-quality atomic scientists both from abroad and within Canada. Two of the earliest recruits, Pierre Auger and Bertrand Goldschmidt, were 'Free French' scientists who regarded their involvement in the Anglo-Canadian nuclear program as the best way of defeating Nazi Germany, even if it meant leaving good jobs in the United States.[29] Akers found Goldschmidt's situation particularly intriguing:

He is extremely bright and also a very pleasant youth. Incidentally, the chemical team at Chicago, and also the senior physicists such as [Arthur] Compton and [Samuel] Allison, were genuinely amused at the position whereby Goldschmidt, who had been refused engagement for the American team, was able to come to Chicago and learn the whole of the chemical side [of plutonium] as a member of the British group. He obviously made himself well liked there ... It is also clear that he will be welcomed in Chicago any time and we may wish to send him for liaison work.[30]

Another promising recruit was the Italian physicist Bruno Pontecorvo, who came highly recommended by Nobel laureate Enrico Fermi. Pontecorvo was also strongly endorsed by both Pierre Auger and George Placzek 'because of the great positive contribution he would make to the slow neutron physics knowledge in the team.'[31] Although Mackenzie did not share this enthusiasm, he eventually agreed to the appointment on the understanding that Pontecorvo would join the British team that was assembled for the Montreal Laboratory. In contrast, Mackenzie was not prepared to recruit Franco Rasetti, another of Fermi's protégés, who had been appointed professor of physics at Laval University in 1940. Despite Rasetti's obvious scientific talent, and the active lobbying efforts of Halban, Canadian and British authorities decided that the Italian scientist might be a security risk.[32]

On 10 November 1942 a list of twenty-six British Tube Alloys scientists and technicians destined for Montreal was submitted to Canadian immigration officials.[33] Eighteen were British citizens,[34] the remainder were refugee scientists from a variety of backgrounds. C.J. Mackenzie, who had become quite excited over the potential of the atomic project, was enthusiastic about all members of the Tube Alloys scientific vanguard.[35] His first meeting with members of the Scientific Committee reinforced his favourable impression of the new arrivals:

> [Pierre] Auger is one of the distinguished physicists of the world and is head of the physics department at Sorbonne. He is quite able and a man of the world and should be a good influence. He says that Halban is really intuitive and while too young to be one of the really competent physicists will be one day. [George] Placzek is a most impressive fellow ... one of the best theoretical physicists in the world – speaks about a half dozen languages and has been to universities in about a half dozen universities in about a half dozen countries.[36]

However, Mackenzie was concerned about the difficulty of recruiting high-quality Canadian nuclear scientists.[37] This dearth of talent had already been determined in a May 1942 survey conducted by the British liaison scientist G.P. Thomson, who mentioned only two Canadian university physicists – Gerhard Herzberg of the University of Saskatchewan and George Volkoff of the University of British Columbia – as outstanding recruits for the Tube Alloys project.[38] Herzberg's status as a 'friendly' enemy alien made his appointment problematic, since the NRC had a policy of not employing German refugees in top-secret projects. But Volkoff had no such liabilities, and Mackenzie lost no time in acquiring his services, especially when he discovered that the UBC scientist had studied with Robert Oppenheimer. Mackenzie's letter to UBC officials was direct and compelling:[39] 'An important and internationally known group of scientists, about twenty in number, are coming from Great Britain and will be located in Canada ... I am anxious to assemble a team of our most brilliant younger workers so that when they join the group above Canada's contribution will be appreciable ... everyone agrees that Dr. Volkoff is one of the best in Canada.'[40]

Mackenzie's other prize candidate was Henry Thode, a radiochemist from McMaster University, who was highly regarded for his work with Nobel laureate Harold Urey on the separation of isotopes and the use of mass spectroscopy.[41] But Thode was not convinced that he wanted to join the Montreal Laboratory because of his unpleasant interview with NRC physicist George Lawrence, and because he had received an even more attractive offer to join Urey's Manhattan Project team at Columbia. Mackenzie was indignant when he heard about Urey's offer: 'The people in the United States offer much larger salaries than we do and if we were to permit our scientists and technical men to accept any position offered in the United States we would be in great danger of losing many men we need so badly.'[42] A compromise was reached: Thode would be a member of the Montreal Laboratory, but most of his work would be carried out at McMaster University.[43]

Younger, less established Canadian scientists such as J. Carson Mark, Phil Wallace, and Leo Yaffe were also recruited to assist division and team leaders. Mark, who had been teaching at the University of Manitoba, was assigned to George Placzek's theoretical physics team, which at a very early stage began research on 'the solutions of the neutron transport equation,' which later became critical 'for fast neutron problems ... at Los Alamos.'[44] Yaffe also welcomed the opportunity to migrate to this exciting research environment, since it also meant an escape from his 'dreary' chemical warfare experiments and the watchful eye of Otto Maass.[45]

By April 1943 the major scientific divisions were fully staffed. Not surprisingly, the British team dominated the administrative and senior scientist positions, although some Canadians, notably Volkoff, Sargent, Lawrence, and Thode, were well placed.[46] One serious problem was the NRC's failure to obtain the services of Canadian engineers with experience in the design and construction of chemical engineering plants who might assist the small contingent of overworked engineers from Industrial Chemical Industries (ICI) who had accompanied the British atomic delegation.[47] But the most serious personnel problems were not at the division level: they were at the top.

Deterioration in the relations between the leadership of the NRC and Tube Alloys was largely the result of the competing visions of Hans von Halban and C.J. Mackenzie. Almost from the day of his arrival, Halban took exception to NRC procedures and priorities. He was also outraged by what he considered unnecessary problems in getting the laboratory functional, for example, delays in obtaining the University of Montreal facilities as well as serious equipment shortages. To make matters worse, supplies of vital uranium oxide, uranium metal, heavy water, and graphite were slow to arrive from the United States.[48] The situation was not helped by Halban's tendency to 'blow up' when he felt that the NRC bureaucrats were 'being specially obstructive.' He also did not endear himself to the parsimonious Mackenzie, who soon resented Halban's constant demands for special privileges.[49] By the spring of 1943 the NRC president was referring to Halban in his diary as 'a mere child and a temperamental one at that,' an opinion that was increasingly shared by most Montreal Laboratory scientists.[50] The Canadians were particularly critical: George Lawrence remembered Halban as 'arrogant and impatient,' while Leo Yaffe regarded him as a dictator 'who ran the laboratory in an intolerable manner.'[51]

But Halban also had his defenders, none more eloquent than his Tube Alloys boss William Akers. According to Akers, most of the laboratory's shortcomings were the fault of C.J. Mackenzie, who did 'not regard the T.A. work as having any higher priority than any other of the NRC work.'[52] Akers attributed this lack of vision to several factors, the most important being Mackenzie's intense Canadian nationalism and his paranoia that British political leaders were preventing Canada from becoming 'equal ... partners in the war effort.'[53] Akers also cited Mackenzie's concern that the escalating costs of the Montreal Laboratory might threaten the NRC's postwar financial future. But the real problem, Akers claimed, was Mackenzie's sycophantic posturing toward American scientific mandarins and his fear that the Anglo-Canadian atomic project might jeopardize the NRC's 'friendly relations with OSRD.' This divided loyalty, Akers advised his superiors, should not be allowed to continue:

If cooperation is not restored with the Americans we must, as soon as possible, have a frank talk with the Canadians about how far they are prepared to go with us. We will also have to make it clear that it will be no use their saying that they want to have the research and development carried out here, unless they are prepared to fight with the Americans for essential supplies and apparatus. I would quite understand an attitude on their part that they were not prepared to jeopardize their American relations in this way; but it is essential that we should not go on with the Canadians, here, if they are always wondering if any particular action will annoy their neighbours.[54]

Towards the Quebec Agreement

The new basis of Anglo-Canadian-American atomic relations was defined in a letter from NDRC chief James Conant to C.J. Mackenzie on 2 January 1943.[55] The message was clear: the United States was now committed to a new atomic exchange policy based on the principle 'that we are to have complete interchange on design and construction of new weapons and equipment *only* if the recipient of the information is in a position to take advantage of it in this war.' Security considerations, he added, also made these changes imperative.[56]

Why was Mackenzie the first to receive this bad news? The answer was obvious: the Montreal Laboratory had a limited future, and the NRC should be given the opportunity to cut its losses. In his fiat, Conant did not mince words: 'Since it is clear that neither your [Canadian] Government nor the English can produce elements of "94" or "25" on a time schedule which will permit of their use in this conflict, we have been directed to limit the interchange accordingly.'[57] Conant then set forth the specific exchange guidelines. First, all efforts would be made 'for the development, construction and operation of the Chicago Plant, the erection of our own heavy water plants and the design of a plant making element "49" and using heavy water.' In this endeavour the OSRD envisaged a support role for the Montreal scientists, who would carry out 'the fundamental scientific work for the use of heavy water so that duPont Company could base their designs on this experience.' Montreal would not, however, be given access to 'the methods of extraction of element "49," nor the design of the plant for the use of heavy water for this purpose, nor the methods for preparing heavy water.'[58] In closing, Conant expressed confidence that his good friend C.J. Mackenzie would soon recognize that these Manhattan Project guidelines would enhance the Anglo-American-Canadian goal, 'namely, the production of a weapon to be used against our common enemy in the shortest time under the conditions of maximum secrecy.'[59]

Neither Mackenzie nor Howe derived much solace from Conant's gratuitous exhortation. They were confused and angry that the Montreal Laboratory was now redundant, especially since Bush had assured them three days earlier that 'the programme for co-operative work as between Canada, the United Kingdom, and the United States, had been finally approved and he was very pleased about it.'[60] On the other hand, Mackenzie recognized that both the Americans and the British, as they had done on other occasions, were prepared to use Canada as a pawn in their scientific chess game. He was particularly concerned that Akers and Anderson favoured the 'big stick' approach in dealing with the Americans. This was evident in Akers's 7 January 1943 memorandum that bluntly reminded Conant and Bush that by refusing 'to co-operate one hundred percent,' the Americans had violated 'the original spirit' of the agreement.[61]

Mackenzie, however, decided not to rubber stamp Akers's missive on the grounds that the arguments were redundant and pointless.[62] In his opinion the real problem was not U.S. injustice, but British naïvety: 'I can't help feeling that the United Kingdom group emphasizes the importance of their contribution as compared with the Americans and this attitude has been one of their real shortcomings.'[63] He also empathized with American apprehension over 'discussing all the details and know how with the Montreal group which is really not an Anglo-Saxon group,' but a potpourri of various nationalities whose loyalties could not 'be guaranteed for any length of time.'[64]

Mackenzie was determined not to be caught between the British and U.S. scientific establishments; at the same time, he did attempt to resolve the differences between his two allies throughout the winter and spring of 1943.[65] On 26 July, for instance, he and C.D. Howe entertained General Leslie Groves, head of the Manhattan Engineer District (MED), in Ottawa. According to Mackenzie, Groves appeared 'very favourably disposed' towards the Montreal project: 'I told him he was crazy that he did not have the Montreal group working with his people and he said he would like that very much as he realized that the workers are in a comparatively unknown field, that there are no experts really, and he is not anxious to lose any bets.'[66]

Although the French scientist Bertrand Goldschmidt had managed to obtain small samples of plutonium from his colleagues at the Chicago Metallurgical Laboratory before joining the Anglo-Canadian team,[67] the Anglo-Canadian nuclear project was dead in the water without the assistance of the Manhattan Project, especially for heavy water and uranium oxide. In August 1943 Akers and Anderson acknowledged this reality and conceded defeat.[68] The British would accept junior partner status on terms dictated by the United States.[69] Mackenzie had a preview of this capitulation when he met Anderson in Washington, just three days before the Quebec Conference. 'Sir John,' he noted, 'has a real negotiating job on his hands, but is not as rigid as he was in London, and I think he will be willing to compromise.' On this occasion, Mackenzie also met with Conant, who assured him that 'things were going to break ... although ... cooperation won't be wide open.'[70] What Mackenzie did not see was Conant's private moment of triumph[71] as he relished Anderson's acceptance of terms 'tantamount to ... our original offer.'[72] Conant was especially pleased that the British now recognized that the partnership was based exclusively on military principles: 'On that basis ... we can now proceed with an interchange which will be in the best interests of the United States and the war effort.[73]

Redefining the Role of the Montreal Laboratory, 1943–1944

In August 1943 Prime Minister Mackenzie King hosted the Quebec City summit conference. For over a week President Roosevelt and Prime Minister Churchill, with a plethora of military and civilian advisers, attempted to resolve a number of important problems that threatened to disrupt their wartime alliance, including the breakdown of cooperation between the British and American nuclear projects. On 19 August both sides agreed to a series of commitments: to develop atomic bombs at the earliest possible time; to pool British and American scientific talent and resources; and to avoid duplicating large-scale plants. Under this Quebec Agreement the two countries also promised never to use atomic weapons against each other, and never to use such weapons against third parties 'without each other's consent.' Protection of atomic secrets was to be rigorously enforced, and both sides agreed not to communicate 'any information about Tube Alloys to third parties except by mutual consent.' In the sensitive area of postwar use of atomic energy, British authorities deferred to the authority and priorities of the United States.[74]

Although the Quebec Agreement helped restore Anglo-American atomic cooperation, it did not create an equal partnership. Full exchange of information between allied scientists was confined to those 'engaged in the same section of the field' and 'between members of the Combined Policy Committee and their intermediate technical advisers.'[75] The newly created Combined Policy Committee (CPC) was responsible for coordinating the Anglo-Canadian partnership. The CPC's members were Henry Stimson, Bush, and Conant for the United States; Field Marshal Sir John Dill and Colonel J.J. Llewellin representing Britain, and C.D. Howe speaking for Canada. The CPC's scientific advisers were Groves, Chadwick, and Mackenzie. During the first meeting of the CPC on 7 September 1943, Groves made it clear that the agreement would not include operational collaboration, 'since the heads of the American sections were not allowed to know anything outside their own sections.'[76] Whether they liked it or not, British and Canadian scientists would have to accept Manhattan Project rules. In October 1943, Akers advised Halban of this grim reality:

> It is our opinion that this system of compartmentalisation has been carried, by the Americans, to a point at which efficiency is seriously hampered, but nevertheless, it is ... essential that we should not be responsible for breaking down

Atomic Research, 1942–1946   189

their system. The American division not only takes the form of preventing ... scientists and technicians, engaged in work on Slow Neutrons and the production of Element 94, from knowing that there are other methods ... but also involves a considerable degree of compartmentalisation within the Slow Neutron project itself.[77]

This was one of Akers's last Tube Alloys pronouncements. At the end of October 1943, he was fired, and his job as scientific director of the Tube Alloys project was given to Nobel laureate James Chadwick. There were many reasons for Akers's sudden dismissal: his confrontational style, his clash with Mackenzie, his unpopularity with many of the leading British scientists, and, above all, his acrimonious relations with Bush and Groves.

Almost immediately after the Quebec Conference, Vannevar Bush had begun the campaign to have Akers replaced by 'a top British scientist ... of sound judgement ... of the calibre of Sir Henry Dale or Sir Henry Tizard.'[78] But Bush was soon upstaged by General Groves, who had his own candidate – James Chadwick. The extent of Groves's lobbying efforts is evident in a letter he sent to Colonel Llewellin, one of the British members of the Combined Policy Committee: 'He [Llewellin] asked me directly whether we had any specific objections to Akers; whether we thought that he was incompetent or impossible to get along with, I told him in general that I just liked Chadwick better to deal with and that I always felt on my guard in talking with Akers just as I do in talking to our own industrialists and that I do not have that feeling about Chadwick.'[79]

Groves's experiences with Vannevar Bush, James Conant, Arthur Compton, and Robert Oppenheimer had already demonstrated that he could get along with gifted scientists. If Chadwick accepted the priorities and guidelines of the Manhattan Project, and stroked Groves's ego, there would be no problem. Fortunately for Anglo-American-Canadian atomic cooperation, Chadwick understood the rules of the game. His shrewd assessment not only of problems plaguing the Montreal Laboratory but also of the dynamics of the Manhattan Project was evident in his 31 December 1943 letter to C.J. Mackenzie:[80] 'I do not think that Groves is aware of all the troubles in his own organisation, and while I cannot tell him about them, I think I can at least help him to avoid some of the rocks in our path ... I want to assure you that I am going to work with Groves, and not against him.'[81] Mackenzie was delighted that Chadwick recognized that compromise, not confrontation, was the only viable approach in dealing with the Americans, especially since the Manhattan Project was 'one hundred times greater than any possible United Kingdom effort, that the Americans can get along if necessary without the U.K., while the U.K. can do nothing without the U.S.'[82]

Chadwick's pragmatism and quiet diplomacy soon produced results.[83] He was particularly successful in providing British atomic scientists with exposure to the Manhattan Project so that they could acquire knowledge and experience of American scientific and engineering developments. Indeed, between September 1943 and August 1945, over 150 British scientists would work in various sections of the Manhattan Project.[84] The first to come were Rudolph Peierls, Fritz Simon, Klaus Fuchs, and other members of the gaseous diffusion team, who arrived in New York City in early December 1943.[85] Another group, led by Mark Oliphant, became deeply involved with the electromagnetic project at Berkeley and Oak Ridge.[86] But the scientists who contributed most to the development of the atomic bomb were those who went to Los Alamos, the most secret part of the U.S. atomic operation.[87] Included among the British contingent were two Canadians – chemist Louis Slotin and physicist J. Carson Mark.[88]

To Montreal scientists it seemed paradoxical that other members of the Anglo-Canadian operation were given access to the innermost secrets of the Manhattan Project, while they continued to operate under severe security restrictions.[89] The explanation, however, was simple. In the eyes of Bush, Conant, and Groves, the Montreal Laboratory was a symbol of all that was problematic with Allied atomic collaboration: it did not advance the making of the bomb; it exploited American research and development for postwar military and commercial advantage; and the loyalty of some of its scientists, notably the 'Free French' contingent, was suspect. Although the situation improved considerably when John Cockcroft replaced Hans von Halban as Research Director in April 1944, Montreal's status remained uncertain as long as Manhattan Project officials regarded it as an atomic Trojan horse.[90]

Not surprisingly, this U.S. attitude intensified the serious morale problems at the Montreal Lab, and many senior British scientists talked openly about returning home.[91] This was also the advice of American nuclear administrators, especially after the Argonne heavy water reactor – Chicago Pile 3 (CP-3) – 'reached critical mass' on 14 May 1944. Was there any point, Groves and Conant now asked, in encouraging the development of another heavy water reactor at Montreal that would be both the competitor of the Chicago pile and a security threat?[92]

The only hope of the Montreal Laboratory was to convince the U.S. members of the Combined Policy Committee that it could assist the Manhattan Project. Chadwick and Mackenzie had begun this campaign on 18 September 1943, when they invited Groves and his scientific entourage to visit Montreal facilities and to discuss the merits of constructing a larger heavy water boiler that would have a heterogeneous rather than homogeneous pile. Although MED officials remained noncommittal, the meeting was a social success and reinforced the good communications that had developed between Groves, Chadwick, and Mackenzie.[93] A more profitable joint session was held at the Chicago Metallurgical Laboratory on 8 January 1944 and included ten members of the Montreal team. On this occasion, the outstanding Chicago physicist Samuel Allison gave a detailed talk on the relative merits of homogeneous and heterogeneous piles; more important, he endorsed the Montreal Laboratory's campaign, much to the relief of its scientists.[94] Chadwick was particularly pleased that Groves took the opportunity to promise that future collaboration between Chicago and Montreal would include 'engineering research and development as well as physics and chemistry,'[95] and that Montreal scientists would be given information about the technical process for the extraction of uranium 233.[96]

One reason for Groves's more positive attitude was his appreciation of the many changes that had occurred at Montreal after John Cockcroft became Research Director.[97] Shortly afterwards, heavy water experts Lew Kowarski and A.E. Kempton brought their badly needed talent to the Montreal team.[98] Relations with the NRC were also greatly improved, especially after Ned Steacie, a highly regarded NRC chemist, became Cockcroft's personal assistant. They soon became an effective administrative team.[99]

Although these were important developments, the Anglo-Canadian project still faced formidable obstacles. Powerful American scientific administrators, notably James Conant, continued to view it as a postwar project. Conant openly scoffed at Chadwick's claim that the Montreal pile could 'produce approximately two hundred grams of "49" a day,' and that the plant could be constructed within twenty months.[100] While Groves shared many of Conant's reservations, his primary concern was in developing an atomic bomb, not winning a scientific argument. He was not prepared to alienate Chadwick and thereby lose badly needed British scientific assistance at Berkeley, Oak Ridge, and, above all, at Los Alamos. Both Groves and Chadwick also had to consider the serious political difficulties that would arise if the Montreal group were disbanded.[101]

The battle shifted to Washington and the February 1944 meeting of the Combined Policy Committee. From his vantage point as C.D. Howe's scientific adviser, Mackenzie was able to assess the respective arguments of Conant and Chadwick.[102] On the one hand, he could appreciate the American position and the reasons 'they would not wish to be in the position of providing all of the material, heavy water etc., and having our group at a small fraction of the cost ... produce in quicker time ... results which they had got at such greater effort.'[103] On the other hand, he was determined that Canada's investment in the Montreal Laboratory should be rewarded. Fortunately for Canada, Mackenzie had his own opportunity to argue the merits of the Montreal Laboratory when he, along with Groves and Chadwick, was appointed as a member of a special CPC subcommittee designed to break the Anglo-American deadlock.[104] On 11 April 1944, he addressed the Combined Policy Committee:

The present proposal is to build the pilot plant ... as a joint United States, United Kingdom and Canadian effort. Our ownership of uranium ores, our early interest in the production of heavy water at Trail, and the presence of a highly expert group of workers in Canada gives us a special interest and facility for this work.

In my opinion Canada has a unique opportunity to become intimately associated in a project which is not only of the greatest immediate military importance, but which may revolutionize the future world in the same degree as did the invention of the steam engine and the discovery of electricity. It is an opportunity Canada as a nation cannot afford to turn down.[105]

Mackenzie's passionate appeal soon produced positive results.

Successful Partnership, 1944–1946

On 8 June 1944, arrangements for the Anglo-American-Canadian heavy water project were negotiated at Chicago.[106] The agreement consisted of six major objectives:

1. to maintain existing levels of production for heavy water;
2. to continue programs at both Chicago and Montreal 'for the development of fundamental information on heavy water piles';
3. to undertake the design and construction of a heterogeneous heavy water pile 'of about 50,000 K.W.' at Montreal as a joint project;

4. to review the status of the Montreal plant 'when the performance of the first large scale graphite pile at Hanford could be assessed';
5. to create the necessary administrative machinery to properly supervise the Montreal pilot pile project; and
6. to strengthen the manpower resources at Montreal 'by the inclusion of American scientists as well as British and American scientists.'[107]

The joint project was to follow clearly defined guidelines, largely determined by Groves. On the positive side was the provision for the interchange of all information 'essential to the construction and operation of the Canadian Pilot Plant,' from both the Oak Ridge graphite pile and the Argonne heavy water pile. Equally promising was the assurance that all information 'connected with the transformation of thorium to element "23" [uranium 233 isotope] and the separation and the measurement of the physical and chemical properties of thorium and 23,' would be freely exchanged. On the negative side, plutonium information and supplies were totally excluded from the agreement; this included health research into 'the toxic effects of either "49" or fission products.'[108]

To ensure that these conditions were rigorously followed, Groves assigned two American scientists – physicist H.W. Watson and chemist John Huffman – to the staff of the Montreal Laboratory with the understanding 'that they would have access to all Metallurgical Project information and they would keep themselves informed by means of reports and ... would have the responsibility of recommending visits required by the Montreal Group to Chicago.'[109] They were joined by Major Horace Benbow, a member of Groves' special security team, whose primary mission was to prevent Montreal from becoming too dependent on U.S. resources and expertise, and to protect American plutonium secrets.[110]

But in many ways the 8 June 1944 tripartite agreement rendered these tasks impossible. Not only had Montreal been given full access to uranium 233 research, it had also been promised 'a limited amount of irradiated tube alloy in the form of Clinton [Oak Ridge] slugs ... in order ... to work out independently the extraction and chemical properties of 49.'[111] This meant, according to H.W. Watson, that there was no way that Manhattan Project officials could not 'prevent the U.K.–Canadian chemists from obtaining ideas about possible 49 extraction processes from our discussion about a 23 extraction process.'[112] And he was right. In July 1944, with the arrival of the irradiated ingots, Goldschmidt's team began their 'radioactive alchemy'; by November they were ready to discuss their plutonium research results with Chicago experts.[113]

Meanwhile, at the diplomatic level, British and American officials appeared on the verge of expanding the scope of atomic cooperation. This was confirmed and reinforced during the high-level meetings between Churchill and Roosevelt at Quebec City in August 1944, and at their subsequent negotiations at Hyde Park the next month. On 19 September the two leaders signed an *aide-mémoire* stipulating that 'full collaboration between the United States and the British Government in developing Tube Alloys for military and commercial purposes should continue after the defeat of Japan unless and until terminated by joint agreement.'[114] Evidence that Churchill believed he had finally convinced Roosevelt to support Britain's postwar atomic research and development can be found in his 21 September communiqué to Lord Cherwell: 'The President and I exchanged satisfactory initialed notes about the future of T.A. [tube alloys] on the basis of indefinite collaboration in the post-war subject to termination by joint agreement.'[115] Unfortunately for the British, Roosevelt did not inform either Bush or Conant about this arrangement. Nor was the *aide-mémoire* part of the atomic briefing package Harry Truman received when he became president in April 1945.

Whether C.J. Mackenzie was aware of the Hyde Park negotiations is not clear. However, he did notice an improvement in the attitude of MED officials toward the Montreal Laboratory during the fall of 1944. He was also pleased with the encouraging progress report he received from Cockcroft and Ned Steacie in September 1944 on the future of the NRX heavy water pile: 'The Pilot Plant should operate before mid-1945. It should have a production of 300 grams per month of 49 and about 25 grams per month of 23. Provided that separation plants for 49 and 23 are built this output should enable us to build up stocks of two of the three fissionable elements, stocks which would form a useful starting point for the next stage of experimental work.' The report went on to describe how the pilot plant would provide 'an extremely intense source of slow and fast neutrons' that could be used in future technical experiments in nuclear physics, in radiation chemistry, and for tracer techniques. Such an undertaking, it was suggested, would give the Montreal team the necessary skills and experience 'to design and build a higher power plant for large scale production of 23 [uranium 233].'[116] While Mackenzie was delighted with these postwar prospects, he was also becoming increasingly concerned about whether Canada would have the necessary scientific, engineering, and financial resources to carry out such an ambitious undertaking, especially when he suspected that the British government intended to concentrate its atomic research and development operation in England rather than at the new Canadian nuclear facility being built at Chalk River, Ontario. Cockcroft summarized Mackenzie's concerns in a May 1945 letter to Chadwick:

First, he fears the U.S. may withdraw their heavy water at the end of the war and leave Canada with a derelict plant.
Second, he feels that he would be unable to justify the expenditure of the order of $2,000,000 a year ... if the expenditure were for scientific and technical purposes only.
Third, he is in some doubt as to whether Canada can afford the number of scientists required to carry on the work.[117]

In his usual diplomatic fashion, Cockcroft sought to reassure the NRC president that the existing complement of 500 scientists and technicans could be reduced to 150 once the war was over, and that Canada could easily provide '70–80 scientific and engineering staff, of whom only 20–25 need be scientists.' Mackenzie, however, was not comforted and grumbled that the British were once again ready to take 'all the cream off the last two or three years with very little expense.'[118]

Chadwick had little time to reassure his Canadian ally. Germany's defeat, and the imminent use of the atomic bomb against Japan, had created a variety of new challenges for Britain's atomic ambassador. What, for instance, was the future status of the British scientists at Los Alamos? Would they, as some were suggesting, be transferred to the Montreal Laboratory in order to accelerate the construction of the pilot plant?[119] And what would these scientists take with them when they left Los Alamos? On this final point, Chadwick's message was clear:

We cannot leave all our information behind and go with empty hands. Either British workers must take their notebooks with them or they must copy out the important abstracts. Further, they should collect copies of all technical memoranda and reports in which they have been concerned. In addition, I think all our people should be asked to give you [Dr P.B. Moon] a note about the reports which they consider important for our future work in England. You could then give me a full list and I would try and clear these reports through General Groves.[120]

Hiroshima and Nagasaki

President Truman's decision to use atomic bombs against Japan remains one of the most controversial events of the Second World War, and a subject of intense historical debate. Among the critical studies, the most exhaustive analysis is Gar Alperovitz's *The Decision to Use the Atomic Bomb, and the Architecture of an American Myth*.[121] Alperovitz provides extensive evidence that Truman's decision was based more on countering the perceived Soviet threat in Europe and the Far East than on defeating Japan. Alperovitz also shows that Churchill, as was his prerogative under the Quebec Agreement, readily agreed to the U.S. decision, despite his view, and that of most allied military officials, that Japan was on the verge of surrendering, if only the unconditional surrender terms would include protection for the Japanese emperor.[122]

Unfortunately, Alperovitz's otherwise fine study completely ignores Canada's role in nuclear research, despite its membership on the Combined Policy Committee and despite the Montreal Laboratory's quasi-partnership with the Manhattan Project. Perhaps Alperovitz assumes that the Canadian government was merely a British puppet that would automatically endorse Churchill's position, right or wrong, or that Canadian policy makers were unaware of the impending use of the atomic bomb against Japan. On neither count, of course, was this the case.

C.J. Mackenzie's diary entries for July–August 1945 are replete with references to the forthcoming use of the bomb. On 5 July, for instance, he recorded an important conversation with C.D. Howe, who had just returned from a meeting of the Combined Policy Committee where the decision to use the bomb had been announced:

> The main event [the Trinity tests] will take place in the immediate future. The Americans have all their press releases ready and it is going to be a most dramatic disclosure. They are going to tell a great deal about the project in general terms, all the money spent, where they are working, etc. Mr. Howe said we must get busy immediately and get our press releases ready as it is the biggest opportunity Canada will ever have to participate in a scientific announcement.[123]

By the end of the month, Mackenzie was supervising efforts 'in getting all three releases – U.S., U.K., and Canada harmonized.'[124] One week later he described the actual attack:

Atomic Research, 1942–1946 197

On Monday, August 6 the President of the United States announced that the first atomic bomb had been dropped on Hiroshima. The day before I telephoned to Dr. Cockcroft and Mr. Howe asking if he wanted me to come in but Howe said he had given out the first announcement ... It was an eventful week. The Russians declared war on Japan, the second bomb was dropped on Nagasaki, which was where the first bomb was to be dropped, and then the rumours regarding Japanese surrender started flying thick and fast.[125]

Prime Minister Mackenzie King's response to the advent of atomic warfare was somewhat different. While he had been kept informed of the Anglo-Canadian project at Montreal and the strategic implications of U.S. nuclear strategy, he did not take a stand on whether the atomic bomb should be used against Japan.[126] But when the possibility was transformed into a reality, King became quite agitated, as his diary entry of 4 August shows: 'It appalls me to think ... of the loss of life that it will occasion among innocent people ... It can only be justified through the knowledge that for one life destroyed, it may save hundreds of thousands and bring this terrible war to a close.'[127]

President Truman's official announcement of the Hiroshima bombing came while King was presiding over the opening session of a Dominion–Provincial Conference in Ottawa. After confiding with his aides, King told the conference that he had a world-shaking announcement to make. The conference delegates listened to his statement 'in dead silence.' In private, King brooded over the significance of Hiroshima:

It is quite remarkable that it should have been given to me to be the first in Canada to inform ... of this most amazing of all scientific discoveries and of what certainly presages the early close of the Japanese war. We now see what might have come to the British race had German scientists won the race. It is fortunate that the use of the bomb should have been upon the Japanese, rather than upon the white races of Europe. I am a little concerned about how Russia may feel, not having been told anything of this invention.[128]

The coming of the atomic age also had an enormous impact on the Department of National Defence. On 8 August, DND sent an urgent request to the Canadian Military Mission in Washington for specific information about the A-bomb attacks: 'How many were dropped? Where? What sizes of bombs? What were the expected and the observed terminal ballistic results? What other information is now available on this bomb and its design? Is desirable that these and other relevant facts be released on appropriate security grade to CGS and Minister of National Defence.'[129] Each Service also sought expert advice about the military implications of nuclear weapons. In September 1945, for example, the deputy minister of the Naval Service W.G. Mills asked if C.J. Mackenzie could arrange for a high-level briefing session, since 'the National Research Council has taken an important part in the development work of the Atomic Bomb and has become the Canadian repository of the most advanced knowledge.'[130]

During August and September 1945 the media buzzed with accounts of the devastation of Hiroshima and Nagasaki. Rumours about this terrible 'winning weapon' circulated wildly. One of the most troubling, at least from a Canadian point of view, was an inflammatory editorial in the *Washington Times-Herald* on the question of Canadian uranium and its postwar uses:

Canada has the basic raw material in quantity. We have the plants to make the atom bomb ... That means that Canada, already our ally in this war, will serve its own best interests as well as ours by making uranium in ample quantity available to us at all times, and by making it not available to other nations. In matters like this, it is a question of kill or be killed in wartime ... As to uranium, Canada should make itself our out and out and exclusive ally. If she won't do so, well ... enough patriotic Americans can probably be found to see to it that Canada does the right thing by us and by itself with its uranium.[131]

Stung by this jingoistic rhetoric, Mackenzie and Howe asked General Groves whether the U.S. War Department could issue a statement 'which would kill this sort of irresponsible statement.'[132] Their request, however, was politely ignored.

Postwar Atomic Planning, August–December 1945

Canada's postwar atomic program was strongly influenced by developments in Great Britain and the United States, most notably by changes in British policy and priorities after the war. On the political front, the Churchill government had been defeated in the June 1945 general elections by the Labour party led by Clement Attlee. Somewhat unexpectedly, given its prewar anti-militarist stance, the Labour government soon demonstrated, by the creation of the Advisory Committee on Atomic Energy (ACAE), that it regarded atomic weapons as crucial for

Britain's postwar defence. The committee's mandate was threefold: 'to investigate the implications of the use of atomic energy'; 'to advise the Government what steps should be taken for its development ... either for military or industrial purposes'; and 'to keep in close touch with the work done by the similar Committee which has been set up in the United States.'[133]

The composition and mandate of the committee were the subject of much debate in Britain, but some of the most vigorous protests came from the British atomic team at the Montreal Laboratory. On 30 August thirty-one scientists and technicians of the laboratory sent a sharply worded letter to James Chadwick, attacking the appointment of Sir John Anderson as chairman of ACAE on the grounds that he was 'one of the leading members of the cabinet which has just been rejected by the whole nation,' and 'the man who must be held ultimately responsible for all the early mistakes of policy concerning T.A.' Equally troubling to the Montreal group was the ACAE's exclusion of such outstanding nuclear scientists as John Cockcroft and Mark Oliphant, a decision they regarded as a betrayal of all those 'who had been encouraged to expect a brighter future for research in nuclear physics.'[134]

Canadian officials carefully monitored the British atomic debate, particularly C.J. Mackenzie, who remained in charge of the Chalk River reactor project, which, after a year of planning and construction, was ready for operation in September 1945. This was an expensive undertaking – costing $4 million and employing hundreds of scientists and technicians. Although the Canadian government did not contemplate developing nuclear weapons, Canada would obviously be affected by Britain's decision to proceed in that direction.[135] More immediately, Mackenzie had to adjust to the reality that the British government regarded its Canadian commitments as a temporary expedient and intended to concentrate its nuclear program at the proposed Harwell site in England. Another important issue was whether Anglo-American wartime cooperation would continue. A related factor was whether the powerful American First lobby in Washington, which was calling for stringent security regulations, would impede communication between Chalk River and American nuclear installations.

As the Manhattan Project began to wind down its operation, Ottawa was concerned that wartime exchange commitments would be terminated. Despite these concerns, General Groves continued to be generous in meeting reasonable requests from the Montreal Laboratory, especially after receiving favourable reports about the potential of the forthcoming NRX pilot plant at Chalk River and the operation of the Zero Energy Experimental Pile (ZEEP), the first atomic reactor outside of the United States, which began a 'divergent chain reaction' on 5 September 1945.[136] In November 1945, George Weil, the new MED liaison officer, recommended a free exchange of scientific information between Chalk River and U.S. nuclear research centres:[137]

It is well recognized that the effect of such free interchange will be to stimulate and advance the activities of all laboratories, rather than, as has been sometimes suggested, to discourage original work. This is especially true in the present situation, not only because of the unique facilities which will be available at Chalk River, but even more because of the high quality of the NRC scientific personnel. It would be an extremely unfortunate circumstances if the research activities at the Chalk River laboratory were to be deprived of the experience and knowledge outlined in the U.S. laboratories, and the latter, in turned, deprived of information developed at Chalk River.[138]

While Groves was interested, he first wanted the remaining 'Free French' scientists – Gueron, Kowarski, and Goldschmidt – to leave the Montreal project on the grounds that their commitment to France, and its future nuclear program, meant that they could not be relied on to protect Manhattan Project secrets. Nor was Groves moved by C.J. Mackenzie's angry protests concerning the devastating impact these expulsions would have on the Montreal Laboratory and on Canada's postwar nuclear policies.[139]

With Goldschmidt's departure, the final link with the Free French scientific team, who had contributed so much to the success of the Montreal Laboratory and Chalk River, was broken. For his part, Goldschmidt retained warm memories of his Montreal experiences, and of his involvement with the complex world of atomic diplomacy:

Reading this correspondence today, in which I am but a pawn of strong personalities wanting to assert the primacy of their own theses, I find myself filled with amazement and a certain admiration for the manner in which each of them conducted his battle. Groves fought for U.S. isolationism and an atomic monopoly, Chadwick for the survival of the Anglo-American collaboration, Cockcroft for his independence as project manager, and Mackenzie for a fair shake for Canada in this tripartite association.[140]

Conclusion

On 14 August 1945, C.J. Mackenzie wrote James Conant, congratulating him on his wartime contributions, and expressing his 'great relief and satisfaction ... when the first atomic bomb was dropped and the results of months and years of extraordinary activity were known to be successful.'[141] In reply, Conant acknowledged that his years as OSRD atomic liaison had 'been a very interesting and rather strange journey through "Alice in Wonderland,"' especially in 1942 and 1943, 'when the responsibilities were heavy and the decisions difficult.' He expressed satisfaction that the atomic relationships between Canada, the United States, and Great Britain had 'worked out so smoothly,' and praised Mackenzie for his 'understanding spirit in the difficult days and ... diplomatic help in handling international complications.'[142]

While their words were guarded, there is little doubt that Mackenzie and Conant believed that they, and their governments, had acted correctly; that it was British naïvety and inflexibility that had caused the 1943 alliance crisis. Throughout the war, American and Canadian scientific administrators often had difficulty understanding fluctuations in British defence policies in general, and atomic policies in particular. After 1943, Conant and Bush realized they held the winning cards, and they played them well. While they were not enthusiastic about the Quebec Agreement, they were prepared to live with it. They also became reconciled to the gradual emergence of the Montreal Laboratory/Chalk River project as a promising slow-neutron research centre. But what they were not prepared to accept was a permanent Anglo-American atomic partnership.

The Quebec Agreement and, even more important, the appointment of James Chadwick as Tube Alloys technical adviser, gave the Montreal Laboratory a second life. Without Chadwick's special touch with Groves – knowing how much to demand and when – the Americans might never have accepted the 8 June 1944 proposals for an Anglo-Canadian-American heavy water pilot plant at Montreal. During the next eighteen months, Chadwick and Cockcroft sought to derive the maximum benefit from the Montreal–Chicago exchange, especially in the fields of uranium 233 and plutonium extraction, and in the construction of the heavy water pilot plant at Chalk River.

The contributions of C.J. Mackenzie to the well-being of the Montreal Laboratory are more controversial. For instance, would the Canadian government have accepted British proposals in 1942 for creating the Montreal Laboratory without Mackenzie's intervention? Probably not. Moreover, Mackenzie's ability to mobilize personnel, equipment, and material in support of the laboratory was essential, especially during its first six months of existence. On the other hand, Mackenzie was quite unprepared for Conant's ultimatum of January 1943, or the subsequent bitter counterattack by Tube Alloys officials, and his political and personal response to this crisis often ignored British sensibilities. Although Akers's portrayal of Mackenzie as an American stooge is unfair, there is good reason to believe that Mackenzie's pro-American sympathies strengthened the resolve of Conant and Bush to redefine the Canada–U.S. partnership. What Akers did not appreciate, however, was that the Montreal Laboratory was only one of many vital military projects that Mackenzie had to coordinate; and for each he required American assistance.

Akers was on firmer ground when he criticized the impact of Mackenzie's open hostility toward Hans von Halban. While few found the Austrian physicist easy to like, Mackenzie, by undermining Halban's authority over his fellow scientists, contributed to the serious morale problem that afflicted the Montreal Laboratory in the 'dark' days of 1943.[143] The problem was exacerbated by the fact that, until Steacie's appointment in June 1944, no senior Canadian scientist had sufficient prestige and tact to facilitate communication between Halban and the NRC president. However, once John Cockcroft became scientific director, Mackenzie's support for the Montreal Laboratory changed dramatically, especially after June 1944, when General Groves allowed scientific exchanges between Montreal and Chicago.

What did Mackenzie expect from the Montreal Laboratory? Clearly, he did not believe that it would produce militarily significant quantities of either plutonium or uranium 233 during the war. Instead, he viewed the project as a postwar opportunity in which Canadian physicists, chemists, engineers, and physicians would take part in developing an exciting new technology – nuclear power. Most of Mackenzie's scientific colleagues shared these goals.[144]

8

Secrets, Security, and Spies, 1939–1945

It was learned that the main secrets of the British Empire War effort are contained in the vaults of the National Research Council, and many of their chemists and engineers are fully conversant with quite a number of these secrets. It seems necessary that the most extensive enquiries should be instituted with regard to the reliability of the personnel in these laboratories ... and investigate everybody who possesses vital and secret knowledge. Mr. Mackenzie is very anxious for this to be done, as he says he and other members of the Council spend many sleepless nights wondering how safe this important information is being kept.

> Solicitor General, Documents Released by the Canadian Security and Intelligence Service, Folder 141, file 11-91-99, Security Screening of NRC and the CAScW, Report of Supt. E.W. Bavin, Intelligence Officer, RCMP, 29 October 1940 (CSIS Documents)

Attempts to assess the effectiveness of the efforts of Canada's Armed Forces and the NRC to protect their scientific secrets have been impeded because much of the relevant documentation has not yet been released. Most studies of this subject have had to rely on the documents and testimony of Soviet defector Igor Gouzenko, which covered only the last three months of the war.[1] In addition, most of the relevant Canadian and British security records remain closed.[2] Still, the fragmentary evidence suggests certain patterns. One of these was the lack of systematic office, laboratory, and factory security procedures to prevent the extraction and copying of top-secret documents. Another was the inability of the Royal Canadian Mounted Police and Canadian military intelligence to monitor the activities of Soviet embassy officials or to adequately screen highly placed civil servants and scientists. The most important factor, however, was that in the Second World War Canada was committed to military alliances that included the Soviet Union. In this context, defeating fascism was deemed more important than protecting scientific secrets.

Before the outbreak of war Canadian officials had made sporadic attempts to establish security procedures to protect sensitive military information. For example, by 1935 the Canadian Armed Forces had established a procedure for circulating secret files. Three years later, as the clouds of war gathered, the Chief of the General Staff sent a memo to all district commanders instructing them to familiarize their officers with the Official Secrets Acts.[3] The memo referred especially to Section 3(2), which placed the burden of proof on the accused: 'On a prosecution under this Section, it shall not be necessary to show that the accused person was guilty of any particular act tending to show a purpose prejudicial to the safety or interests of the State, and notwithstanding that no such act is proved against him, he may be convicted if, from the circumstances of the case, or his conduct, or his known character as proved, it appears that his purpose was the purpose prejudicial to the safety of the state.'[4]

To what extent were Canadian scientists aware of the full implications of the Official Secrets Act?[5] Nuclear technician Norman Veall describes his experience of March 1946, when the full force of this draconian measure was suddenly aimed at him:

This is the first time I have seen the Canadian Act, and I must confess I am very shocked in that a person can be tried and convicted on the flimsiest of circumstantial evidence. In fact I could go even further and state that I know dozens of perfectly innocent people, including myself, who would be liable for conviction under this Act. In short, no scientist engaged in secret work is safe from arbitrary arrest and imprisonment.[6]

The Evolving National Security State, 1940–1945

Fear of enemy sabotage and subversion was widespread during the early stages of the war,[7] especially after the German Blitzkrieg of May 1940 and reports of a sinister German 'fifth column' working behind Allied lines.[8] Not surprisingly, these developments profoundly affected Canadians. Public concern, even panic, was clearly evident in the hundreds of letters and petitions sent to Prime Minister MacKenzie King during the spring and summer of 1940.[9] Part of the government's response was to take action against the different enemy alien populations, as well as potentially subversive groups, under the Defence of Canada Regulations. On 4 June various fascist organizations were declared illegal, as were Communist and anarchist organizations.[10] Although Japan and Canada were not yet at war, Canadian security officials had already made plans for the internment of Japanese Canadians who posed a security risk, if the military situation so required.[11] Another challenging task was the screening of the 2600 interned 'friendly' enemy aliens, mostly German Jews, whom Britain had sent to Canada for safekeeping during the panic of July 1940 when a German invasion appeared imminent.[12] Among the internees were a number of scientists, such as nuclear physicist Klaus Fuchs, who wished to return to Britain as soon as possible in order to participate in defence research.[13]

Canadian university teachers also came under the scrutiny of the RCMP, especially those who either were enemy aliens or had leftist sympathies. While some were dismissed from their jobs, university administrators were usually able to protect their academic colleagues. At the University of Toronto, for example, several German-born staff members received Canadian citizenship in 1940 through the personal intercession of president H.J. Cody. This select group included mathematical meteorologist B. Haurwitz, physiologist Bruno Mendel, chemist Emil Fischer, and mathematician Erich Baer.[14] Not so fortunate was geophysicist Samuel Levine, whose leftist connections made him vulnerable to internment and dismissal.[15] His troubles began in September 1940 when he was arrested for possessing subversive literature contrary to the Defence of Canada Regulations. Despite Cody's attempts to have the charges dropped, Levine was sentenced to six months in jail and placed indefinitely in the Petawawa internment camp on the grounds that he remained a security threat. When he was eventually released in the fall of 1941, he no longer had a job.[16]

Levine's 'dismissal' soon became a cause célèbre among 'progressive' scientists in Canada and the United States. His defence campaign was coordinated by his Toronto colleague mathematician Leopold Infeld, who wrote a scathing article in *The Canadian Forum* portraying Levine as an innocent victim of wartime hysteria.[17] An even more passionate appeal came from Israel Halperin, a mathematician at Queen's University:

This matter is perhaps the concern of all Canadian scientists but it is particularly disturbing to one, like myself, who studied side by side with Levine as undergraduates at the University of Toronto. I need hardly refer to the position held by Dr. Levine in the scientific world. I know that the administration of the university shares the pride which we, its graduates, feel in the high achievements of one of our number. In view of Levine's recognized position as a scientist, it is very disturbing to see that he is struggling with such difficulty to be able to continue his scientific work.[18]

The appearance of an article by Harry Grundfest and Infeld in the prestigious journal *Science* generated another wave of protest from prominent American scientists, who urged Cody to reinstate Levine 'because of his outstanding qualities as a scientist, and because of the contribution he can make to our ultimate victory over Nazi Germany and her satellites.'[19] But all to no avail.[20]

While the Levine case was an extreme example of RCMP harassment, other 'leftist' academics also faced the threat of dismissal during the Second World War. One of these was Toronto historian Frank Underhill, a prominent member of the Canadian Commonwealth Federation (CCF) and an outspoken critic of Canadian rearmament, who in 1940 almost lost his job because of his 'disloyal' public statements.[21] His CCF colleagues economist Eugene Forsey and social worker Leonard Marsh were also targeted, and were dismissed by McGill in 1940.[22] Although this type of intimidation decreased after June 1941, when the USSR became an ally of the British Commonwealth, periodic controversies over the status of political dissenters and enemy aliens still erupted. In March 1943, for example, Tommy Church, the combative member of Parliament for Toronto's Broadview riding, warned the House of Commons that it was unwise to allow 'friendly' enemy alien internees who had been released from Canada's internment camps to attend Canadian universities, where they might 'get all kinds of information from ... our science courses and laboratories.' In this instance, both Cody and McGill's principal, Cyril James, were able to refute these allegations.[23]

The issue of employing scientists with 'divided' loyalties also affected the National Research Council. In January 1943, for instance, C.J. Mackenzie was forced to respond to the demand from J.L. Synge, chairman of Toronto's Applied Mathematics Department, to follow 'the lead of Great Britain and the United States in using all available talent, provided that no risk to national security was involved.' Synge was particularly insistent that Alfred Schild, 'a gifted mathematician and physicist, who was a student at University College [London] before he was interned,' should be employed 'in ballistic or radio research with the NRC.' This proposal was not well received at NRC headquarters. Colonel F.C. Wallace, director of the Radio Branch, declared that he was 'extremely reluctant to consider anyone with enemy alien background for use in the radio field,' while Mackenzie claimed that the NRC was determined 'not to employ enemy aliens in any work having contact with secret information.' Synge countered by accusing the NRC of having a garrison mentality. To strengthen his case, he cited the more flexible approach adopted by American defence agencies. Mackenzie rebutted: 'It is true that there are a few aliens of enemy origin working in the United States on secret war projects, but they are very few and most of them have been in the country for some time ... [and] are only allowed to have knowledge of a very partial field of secret projects.'[24] What Mackenzie neglected to mention was that large numbers of European émigré scientists were working for the Manhattan Project and that his own Montreal atomic laboratory included many German, Austrian, Italian, and French scientists.

Protecting Secrets

Throughout the war C.J. Mackenzie often had difficulty convincing academic scientists to follow NRC security guidelines. At issue was whether war-related research could be published in scientific journals or whether involvement in defence projects could be used to obtain academic rewards from university administrators. In March 1942, for example, Charles Macklin, Professor of Histology and Embryology at the University of Western Ontario, cited his involvement in biological warfare research to request additional support from the university's board of governors. He also declared a willingness 'to answer any questions on the subject of War Project No. M 1000 and its relation to the University.' This breach of security outraged Mackenzie and Maass, and Macklin's involvement with the biological warfare project was sharply reduced.[25]

An even more serious problem was the tendency of many scientists to view security procedures and classification systems as arbitrary and bureaucratic, with little or no relevance to the scientific or military value of the information. Some complained about the quality of security at top-secret research installations such as the Montreal Laboratory. Chemist Leo Yaffe of McGill remembered working on nuclear experiments in 1943 while 'World War One veterans marched up and down the hall with huge Lee Enfield rifles, having much difficulty turning around.' In his opinion, the RCMP officers assigned to the laboratory were not much better: 'We had a small fire at the laboratory,' he related, 'a crowd gathered and a sight seer took a picture. The RCMP then made a big deal of confiscating the film, thereby confirming the suspicion that something suspicious was going on inside.'[26]

To be fair, the RCMP was ill prepared for the enormous responsibilities that were suddenly thrust upon it. By 1942 its resources were already stretched to the limit guarding military installations and rounding up Nazi and Communist subversives.[27] Although it had initiated rudimentary screening of newly hired civil servants in the 1930s,[28] the RCMP simply did not have the manpower or the skills to conduct a comprehensive or effective screening of the thousands of people who joined the public service, the Armed Forces, and defence agencies during the war years.[29]

That is not to say that the RCMP did not try to screen NRC scientists who had access to 'main secrets of the British Empire War effort.' Indeed, in October 1940, after consultations between C.J. Mackenzie and E.W. Bavin, Superintendent of the RCMP Intelligence Service, arrangements were made for two RCMP officers to carry out 'extensive inquiries, first of all to establish the reliability of all employees ... and investigate any other weakness that may be in the safe-guarding of secrets and work being carried out there.' As a preliminary step, the RCMP requested that all NRC employees be required to complete a questionnaire, with particular attention being focused on those involved in radar and chemical warfare research who were 'investigated individually as to their antecedents.'[30] Most defence scientists were, however, only subject to an RCMP file review, and even these were likely only perfunctory since the NRC and the Armed Services were clamouring for the services of every qualified person. Once in the system, it was possible for these scientists, particularly if they were bright and able, to secure jobs that gave them access to top-secret British and American information. This pattern was certainly evident in the case of those scientists who were named as spies by the Royal Commission on Espionage. Durnford Smith, for example, was the NRC Radio Branch liaison officer with the Rad Lab at MIT; David Shugar had access to restricted naval research laboratories in both Britain and the United States; Raymond Boyer was secretary of the Canadian–American RDX Committee; and Alan Nunn May was a key member of the Montreal Laboratory and a frequent visitor to the Argonne nuclear research laboratory near Chicago.[31]

During the war, however, the Canadian military made various

attempts to improve security procedures. This was evident in October 1940, when General H.D.G. Crerar, Chief of the General Staff, issued a warning to all personnel 'not to discuss information, the knowledge of which comes to them in the course of their duty, and that this applies to members of their families as well as other civilians.'[32] Of particular concern was the effectiveness of security arrangements in Ottawa-based research centres. For example, it was reported that during an October 1944 fire drill in Ottawa, Army Ordnance personnel, despite repeated warnings, fled from the building, leaving 'classified papers and files ... exposed, the doors of offices ... left open and even hot plates ... left on.'[33]

Protecting defence secrets became an important priority for C.J. Mackenzie after British and U.S. officials complained about the effectiveness of Canadian security procedures. In September 1941, for example, P.B. Linstead of the British Supply Council charged that top-secret radar equipment was not being sufficiently safeguarded 'to prevent ... any discoveries made by men working with the N.R.C. being used against us.'[34] However, it was not until November 1942 that the NRC and the Department of National Defence made attempts to systematically protect radar equipment, components, and data.[35] An even more demanding undertaking – and one that was of special concern to Mackenzie – was adequate military and RCMP protection of the isolated weapons research centres. By 1943 this included the chemical warfare experimental station at Suffield, Alberta; the biological warfare research centre at Grosse Île, Quebec; the RDX and ballistic production and testing facilities at Shawinigan Falls and Valcartier, Quebec; and the Anglo-Canadian atomic research project in Montreal.

The establishment of the Suffield Experimental Station in the fall of 1941 presented the Canadian Army with its first major challenge in protecting a top-secret research site. Such an undertaking involved more than just providing guards, checking visitors, and ensuring that tests were conducted in secrecy.[36] It also necessitated the protection of top-secret reports, equipment, and poison gas supplies moving between research stations in Canada and the United States. While the system of chemical warfare exchanges generally worked well, there were a few notable exceptions. One of these occurred in September 1942, when 570 tons of toxic mustard gas en route by train from Huntsville, Alabama, to Suffield was left unattended at the border for almost twenty-four hours because U.S. authorities had not informed the Directorate of Chemical Warfare and Smoke of the train's itinerary.[37] One of the most embarrassing lapses in security occurred in February 1944, when a Suffield scientist lost a suitcase full of top-secret documents in a crowded Chicago railway station while on his way to attend a joint Canada–U.S. chemical warfare committee meeting.[38]

Protecting biological warfare secrets was deemed even more important. The first meeting of the M-1000 Committee in November 1941, for example, was preoccupied with speculation that enemy saboteurs 'could take advantage of the opportunity afforded by a position of trust in a laboratory to prepare material which could be conveyed to enemy agents for distribution.' It therefore recommended that directors of important laboratories be given authority 'to investigate the antecedents of all employees.'[39] Even more stringent security measures were adopted for the top-secret rinderpest and anthrax projects at Grosse Île. All personnel were given a full RCMP screening, as revealed by the following report on J.W. Stevenson and R.W. Reed: 'Our investigation is now completed, and we have to advise that reliable persons interviewed, in this connection, do not hesitate to vouch for the reliability and trustworthiness of either of the candidates. Both are considered to be loyal subjects, and there is no record of their having been connected with radical or subversive organizations.'[40] In addition, Maass insisted that the existence of the research station be a secret and that the full powers of wartime censorship be applied.[41] Another important consideration was how to safeguard top-secret biological warfare information that had been obtained from American defence agencies. This was often a frustrating business since most of these documents were channelled through the Canadian Legation in Washington.[42]

Problems of maintaining an effective and consistent Canadian security system were greatly complicated by the exigencies of alliance warfare. D.W.R. McKinley, head of the NRC's microwave research, described some of these problems in a 1946 report: 'The security and classification of [secret] information is a complex and interlocking problem requiring simultaneous action by the appropriate bodies in U.K., U.S.A. and Canada. The picture is very cloudy due to disagreements, slow action and enforced alterations in status caused by the seemingly endless leakages.'[43] Nor was the system of classifying documents any more consistent or careful within the Canadian armed forces, as was revealed in this April 1946 exchange between George Fawcett, the RCAF director of electronics research and development, and the lawyer for Squadron Leader Matt Nightingale:

Q. I thought it was the originator who was supposed to [decide] ... whether or not he would put it within a security classification?

A. That is correct; but the definitions of the security classifications are clearly laid down ...

Q. Was it done by a joint board of the army, navy and air force?

A. They were originally laid down by the RCAF but subsequently the security classifications were reviewed with the express purpose of protecting United States procedure, bringing it into line with the British Commonwealth ...

Q. And the different armed services of Canada seem to have followed the British, and some followed the Americans; is that right?

A. To the best of my knowledge, the RCAF ... followed the British system ...[44]

By 1944 it was the United States that determined what military scientific information would be exchanged within the alliance. A November 1945 memorandum of the U.S. Joint Chiefs of Staff provides a useful example of how specific weapons systems were categorized:

(a) *Chemical warfare*: top secret ... no information regarding it should be released, except as defensive measure;
(b) *Explosives-RDX*: secret: Limited Access.
(c) *Guided missiles*: top secret: No information should be released.
(d) *VT, Proximity Fuse*: top secret: Information regarding it should not be released under any circumstances.
(e) *Radar*: top secret and secret: Clearance by the Armed Forces and OSRD is required.[45]

Canada was also affected by American plans to begin declassification of defence science information as the war in Europe drew to a close. On 17 November 1944, President Roosevelt instructed Vannevar Bush to investigate how 'the information, the techniques, and the research experiences' associated with the war effort could now be used for 'the creation of new enterprises, bringing new jobs, and the betterment of the national standard of living ... consistent with military security, and with the prior approval of the military.'[46] This movement toward openness was accelerated in June 1945, when Harry S. Truman established a publications board to review what information could be released to the American public. It was also authorized 'to deal with duly accredited representatives of foreign governments with regard to declassification and publication of scientific information.'[47]

Protecting Atomic Secrets

The debate over security was most pronounced in the case of the Anglo-American-Canadian atomic energy alliance. In some ways the security problems surrounding the development of the atomic bomb were similar to those associated with other Allied weapons systems. But important differences set the Manhattan Project apart.[48] The most crucial distinction was the Anglo-American refusal to share atomic secrets with the USSR. This policy decision had been codified in article five of the 1943 Quebec Agreement, which stipulated that neither country would 'communicate any information about Tube Alloys except by mutual consent.'[49]

The task of protecting the secrets of the Manhattan Project rested with General Leslie Groves and his staff. Although the elaborate security procedures were set up ostensibly as a defence against German espionage, in Groves's mind the Soviet Union posed the greater threat. As a result, every scientist with a left-leaning background was put under surveillance by the Manhattan Project's security section.[50] The major mechanism for ensuring secrecy was a rigid compartmentalization system in which scientific information could be exchanged only on a 'need to know' basis. In his own folksy way, Groves explained what he meant by compartmentalization: 'Just as outfielders should not think about the manager's job of changing pitchers, and the blocker should not worry about the ball carrier fumbling, each scientist had to be made to do his own work.'[51] This narrowly defined 'need to know' approach infuriated most American nuclear scientists and was a continual source of tension at the three major Manhattan Project sites of Chicago, Oak Ridge, and Los Alamos.[52]

No such rigorous security system prevailed at the Montreal atomic laboratory, which by the end of 1943 employed over 175 scientists. But Montreal paid a high price for its relaxed research environment. Groves considered the Montreal Laboratory a serious security risk, ensuring its virtual exclusion from the secrets and technology of the Manhattan Project, at least until June 1944, when provision was made to include Montreal scientists in some of the important plutonium and reactor research projects. However, Montreal would now have to adopt the same strict security standards of the Manhattan Project. To ensure that this happened, two MED atomic scientists, H.W. Watson and John Huffman, were appointed as special 'consultants' to the laboratory's director, John Cockcroft, to monitor the liaison arrangements and to recommend visits 'required by the Montreal Group to Chicago.' Major Horace Benbow had the added responsibility of ensuring that security procedures at Montreal met Groves's standards. It was a daunting task.[53] By August 1944 Benbow was complaining that many of the British scientists working on the project had 'never been cleared by British Intelligence.' He was even more agitated that Cockcroft insisted on keeping 'all records and clearances for security of the U.K. personnel under the control of his immediate office instead of turning this information over to RCMP to whom this office looks for information.'[54]

From the perspective of Canadian scientists, many of these U.S.-dictated security procedures were vexatious and insulting. One of the most outspoken critics was physicist B.W. Sargent, who was outraged when his former mentor, J.A. Gray of Queen's University, was denied access to the Montreal Laboratory. 'I am sorry,' he wrote Gray, 'that you were invited to the lab. This has happened to many others. A week ago a very distinguished physicist travelling under the name of Nicholas Baker [Niels Bohr] came to Montreal and was not permitted to visit the lab ... Our collaboration with the Americans must hang on a slender thread.'[55]

Security problems associated with collaborative research continued to disturb Manhattan Project officials. They were particularly concerned about unauthorized visits by Montreal scientists to the Chicago and Oak Ridge facilities. In January 1945, for example, Walter Zinn, Director of the Argonne National Laboratory, forwarded a report to Groves complaining about the problems of having visiting scientists around his laboratory for extended periods of time:

As an example of his point, Zinn described an incident which occurred during the visit of an A.N. May. The results of experiments performed at Hanford were being given to May who was led to believe the work had been done at Argonne. All went well until a visitor from the Metallurgical Laboratory, who did not realize May's status, remarked to a group that the work had been done at Hanford.[56]

Once alerted, Groves made a point of discussing this breach in MED security with James Chadwick, who assured him that May 'was exceptionally reliable and close-mouthed.'[57] John Cockcroft also endorsed his colleague's loyalty to the point of proposing that May be allowed to visit Oak Ridge and examine its 'control system and experimental facilities' before his return to England. While Chadwick was anxious to acquire additional Manhattan Project secrets, he correctly assumed that General Groves would regard May's visit with great suspicion, especially since the information being sought was really not 'connected with the Montreal project, but with our future work in England.'[58]

An even more disturbing factor, as far as MED officials were concerned, were the activities of the Montreal Laboratory's Free French scientists. Although Hans von Halban had long been suspected of sending Manhattan Project secrets to General Charles de Gaulle, it was his unauthorized meeting in Paris with left-wing physicist Frédéric Joliot-Curie in December 1944 that brought matters to a head.[59] Groves was outraged by this breach of security and demanded that all four French scientists be immediately sacked.[60] Fortunately for the Montreal team, C.J. Mackenzie managed to convince Groves that at least Bertrand Goldschmidt and Lew Kowarski be allowed to complete their important research projects. They remained associated with the National Research Council until the fall of 1945.

Anglo-Soviet-Canadian Relations, 1943–1945

The alliance warfare relationship that proved to be the most difficult to control was Canada's involvement with the Soviet Union. In part, this stemmed from the fact that in July 1941 Canada had little say in Britain's decision to enter into a wartime alliance with the USSR and was later drawn by default into bilateral arrangements such as the Anglo-Soviet Technical Accord of September 1942. Under this agreement, Britain and the USSR agreed to share information on all 'weapons, devices, or processes which ... are ... or in future may be deployed ... for the prosecution of the war against the common enemy.' A rider clause made it possible for either country to withhold specific requested material, a decision that had to be justified.[61] Until the fall of 1944 the British government felt obliged to fill most Soviet requests for military technology.[62]

Britain's appreciation of the Red Army's military prowess was enhanced after the November 1942 Battle of Stalingrad. This positive stance was reflected in the February 1943 decision of the War Cabinet to send a scientific mission under Sir Henry Tizard to Moscow with a mandate 'to discuss anything which ... would be useful to the Russian [war effort] and to the general development and liaison of confidence between the two Governments.'[63] In particular, Tizard and his two key advisers, John Cockcroft and P.M.S. Blackett, felt that it was essential that the Russian Armed Forces receive advanced forms of British radar (both ground and airborne) and communications equipment. In return, they expected the Kremlin to provide information about Soviet incendiary bombs, rockets, and chemical weapons.[64]

None of this would materialize. By the summer of 1943 Tizard's Moscow mission had been scuttled, largely because of fierce opposition from the U.S. Joint Chiefs of Staff.[65] The American military had three major complaints: they had not been properly consulted about the mission; they had not been given full details of the Anglo-Soviet Technical Accord; and they believed 'that the Russians should not get any information on equipment they did not already possess.'[66] Despite its attempts to convince the Joint Chiefs to consider a possible extension of the accord, the British government eventually had to back down. Indeed, it was not until after the second Quebec Conference of August 1944 that Britain and the United States adopted a common exchange policy with the Soviets.[67]

Ironically, by this stage in the war, with Germany's defeat imminent, British officials increasingly regarded the Technical Accord as more of a liability than an asset.[68] Although political considerations militated against outright cancellation, a variety of administrative changes were implemented in order to reduce the number of Soviet requests for top-secret British weapons.[69] In addition, all Russian diplomatic and trade personnel were required to obtain from the Ministry of Supply a special permit that was designed to drastically limit 'any secret equipment which can be shown.'[70]

Canada was directly affected by these complex wartime negotiations. Although British policy makers generally regarded Canada as the most assertive of the Dominions, they correctly assumed that the King government would accept the 1941–2 arrangements with the Soviet Union as a necessary part of Britain's wartime strategy.[71] Indeed, by 1943, public support in English Canada for the Russian war effort was high. The King government arranged a number of bilateral agreements with the USSR, the most important of which was the Mutual Aid Act of May 1943, which provided for the delivery of 'planes, tanks, wheat, bacon and lumber to all the Allies,' including the Soviet Union. An earlier initiative was the February 1942 agreement to establish diplomatic relations between the two countries.[72] Within eight months, a Soviet mission, headed by ambassador Fedor Gusev, arrived in Ottawa.[73] In early 1943 Colonel

Nikolai Zabotin of the Red Army intelligence (GRU) assumed his official duties as military attaché, a job that included streamlining the delivery of appropriate military technology to the USSR. By the end of the war Canada was supplying the Soviets with approximately $167.3 million worth of food, medical supplies, and war materiel.[74]

The Soviet Espionage System

Soviet intelligence saw obvious advantages to operating from Ottawa between 1943 and 1945. Most important was that Canada, because of its extensive involvement with major Anglo-American weapons systems, possessed valuable military secrets. In addition, unlike its more powerful western allies, Canada did not have the same capacity for mounting an effective counterintelligence operation. Another factor was the appearance of a number of organizations that championed the advantages of the Canadian–Soviet connection. This optimistic outlook was reflected in a June 1943 Canadian poll asking Canadians whether they thought Russia could be trusted 'to cooperate with us when the war is over'; 62 per cent of Ontario respondents answered in the affirmative.[75]

Soviet requests for war materiel and information about important weapon systems were normally forwarded through Canadian diplomatic and military channels under guidelines determined by the Canadian War Cabinet and by the liaison division of the Department of National Defence.[76] In addition, the Mutual Aid program provided Soviet officials such as Colonel Zabotin and Trade Commissioner Ivan Krotov with other means of obtaining military technology and supplies.[77] And when all else failed, Zabotin could argue that Canadian officials must respect both the spirit and the terms of the 1942 Anglo-Soviet Technical Accord.[78]

Radar secrets were high on Zabotin's list of demands. This was evident in June 1943, when a special meeting of NRC radar experts was called to discuss whether it was possible to release 'all available documents concerning the GL Mark III C and ... GL Mark III C equipment to the USSR.' After considerable debate, the committee concluded 'that a recommendation be sent to the Chiefs of Staff Committee that such a release should be agreed to.'[79] But the Soviets wanted more. On 21 August the Mutual Aid Board received requests for 'three or four full sets or Radio Locators GL 3' as well as a demand that two or three Soviet engineers be placed 'in the plant of Research Enterprises Limited for the obtaining of practical experience.'[80] On this occasion, the RDF Committee decided that the issue was so politically sensitive that it should seek the approval of C.D. Howe, Minister of Munitions and Supply, before proceeding. Howe's response, however, was equivocal:

(1) We can make available to USSR three full sets of Radio Locators GL 3 provided we are authorized to do so officially by the British government. These are of British design, and can only be released by consent of the designers.
(2) In the event the British government officially authorizes shipment of GL sets to Russia we will arrange to place two or three USSR engineers in the plant of Research Enterprises to learn the operation of these sets.[81]

Even more controversial were negotiations over Soviet access to RDX secrets.[82] This saga began in February 1943, when the British government received a request from Moscow for information about the production methods and technical characteristics of RDX. Significantly, the note did not specify which process the Soviets wanted – the older Woolwich method or the more effective U.S.–Canadian Bachmann continuous process. British military and scientific officials assumed the former, which fell under the 1942 Technical Accord. But to be safe, they decided to consult Washington. Much to their surprise and chagrin, the U.S. response was negative. According to the Joint Chiefs of Staff, no RDX secrets could be released without their approval since it was a joint research project of great military importance.[83] While British authorities reluctantly accepted this veto, they had difficulty justifying this exclusion both to the Kremlin and to pro-Russian groups at home.

The Americans' position on RDX had ramifications for the Canadian government, particularly since many groups in Canada viewed any attempt to keep advanced weapons from the Red Army as counterproductive.[84] Some of the most intense criticism came from the National Council for Canadian–Soviet Friendship and from organizers of the pro-Communist Labour Progressive party such as Fred Rose.[85] Indeed, it was Rose who managed to convince McGill chemist Raymond Boyer, a major participant in the development of the Bachmann system, to provide the USSR with classified RDX information during this February 1943 controversy.

Yet it should also be acknowledged that the Kremlin obtained information about the Bachmann system officially through the Department of Munitions and Supply and the Mutual Aid Board, in part because officials of these agencies appeared confused about what RDX information could be divulged. For example, in September 1944, permission was granted for a Soviet engineering delegation to tour the Shawinigan Chemicals plant without any guidelines or supervision. Munitions and Supply authorities left the matter 'entirely ... for Shawinigan Chemicals to decide as to how far they should go in allowing foreign visitors through their plant.'[86] From all accounts, the Soviet engineers found their tour profitable.[87]

As military attaché, Colonel Zabotin was a regular visitor to the offices of Munitions and Supply, National Defence, and the NRC. He was also a familiar presence in Ottawa's social world, especially at the many social events hosted by the Soviet embassy.[88] Located in a mansion formerly owned by an Ottawa Valley lumber baron, the embassy was 'larger and more luxurious than Laurier House where the Prime Minister of Canada lived.'[89] Although ambassador George Zaroubin nominally presided here, the real star was Colonel Zabotin. An impressive looking man, handsome and tall, with an engaging sense of humour, the GRU colonel was a natural *bon vivant* who charmed many members of Ottawa's political establishment. Zabotin was also a frequent visitor to the hunting and fishing camps outside Ottawa.[90] His most rewarding fishing trip occurred during the spring of 1945, when he managed to convince a Canadian Army officer to take him to a popular site on the Ottawa River, which just happened to be located exactly where the Chalk River atomic plant was being built. The officer, apparently, 'obligingly sat steady in the canoe while Zabotin took pictures of the lovely scenery.'[91]

As the coordinator of the GRU's intelligence system, Zabotin had onerous responsibilities, and he knew the price of failure.[92] The actual task of managing Red Army agents was, however, delegated to his subordinates: Colonel Vassili Rogov, Colonel Motinov, and Major Vsevolod Sokolov.[93] While each had demanding assignments, Rogov's responsibilities were the greatest since he orchestrated the Fred Rose/Gordon Lunan cell, whose job it was to recruit and manage engineers Ned Mazerall and Durnford Smith, mathematician Israel Halperin, and, for a time, physicist David Shugar. Rogov's rather heavy-handed approach did not win him many friends among the Canadian recruits. Gordon Lunan has described how the Soviet colonel often acted 'like a goon' and offered money to his informants.[94]

According to Igor Gouzenko, other Soviet espionage during these years was carried out by the People's Commissariat of State Security (NKVD), Naval Intelligence, the Commercial Attaché, and by representatives of the Soviet news agency TASS. But the NKVD was the oldest, best staffed, and most ruthless of these groups and had the additional authority to check 'on internal Embassy conditions – [including] spying on other Soviet organizations from within.'[95] The NKVD operations in Ottawa were directed by Vasail Pavlov, the official Second Secretary of the Embassy, who carried out his own recruitment activities.

By the time of Japan's defeat, Soviet spy masters had good reason to feel confident about their achievements.[96] In the case of the GRU, Colonel Zabotin's organization had by September 1945 secured the services of at least seventeen agents, many of whom were 'highly regarded' civil servants and scientists.[97] Although it is difficult to re-create the full scope of this operation, since relevant Russian records remain closed, an analysis of the kinds of documents Gouzenko removed from the Soviet embassy show certain patterns. For instance, during January 1945 Zabotin forwarded 109 separate reports to Moscow. Seventy were from an agent 'Foster,' who provided an exhaustive account of the operations of the Department of Munitions and Supply, citing production levels of munitions, aeroplanes, and ships.[98] During the next seven months, even more valuable information was obtained, especially from leftist scientists involved with radar, sonar, explosives, and atomic energy.

In recruiting Canadian scientists who might be favourably disposed to the USSR, Colonel Zabotin had the assistance of two leading members of the Canadian Communist party: Sam Carr, its national secretary, and Fred Rose, Labour Progressive MP for Montréal-Cartier. By 1943 both men had achieved national status as champions of the radical left and were deeply involved with organizations such as the National Council for Canadian–Soviet Friendship and Marxist study groups in Montreal, Toronto, and Ottawa.[99] In Ottawa such groups appealed primarily to middle-class civil servants, academics, and scientists.

The issue of whether these study groups were conspiratorial Soviet cells remains problematic. For those who participated in them, such allegations are considered ridiculous.[100] Most of their recollections stress that the vast majority of members were 'CCF people' who engaged in 'good, yeasty talk about creating a better Canada.'[101] Other accounts are less charitable, with the most damning indictment coming from the Royal Commission on Espionage,[102] which called these groups 'recruiting centres for agents.'[103] According to the commission, Soviet spies had created a carefully orchestrated system, in which Fred Rose or Sam Carr would recommend certain study group members as possible espionage agents: 'Col. Zabotin would get details about the "candidate" including his "possibilities" – that is, place of work and the kind of information to which he had access – and would send this to Moscow. Moscow would then telegraph Zabotin permission or refusal to use this particular "candidate."' If the message was positive, Zabotin would instruct his subordinates to contact the new recruit, arrange code names, establish meeting places, define assignments, and offer small sums of money 'for blackmail purposes should the agent's enthusiasm for the cause ... wane.'[104] According to the commission, some of those recruited would be given the responsibility of managing the activities of other cell members. One of those named in this category was Eric Adams of the Research Division of the Bank of Canada; another was Gordon Lunan, editor of the Department of National Defence publication *Military Affairs*, who had a wide range of military and scientific contacts.

During the spring of 1945, Lunan (or Back, as he was known to the GRU) was given the challenging task of first recruiting and then coordinating the espionage activities of engineers Ned Mazerall and Durnford Smith of the NRC's Radio Branch.[105] Mazerall's work involved research on 10 cm and microwave radar, especially the airborne distance indicator.[106] Smith was also involved in the microwave field and had been given responsibility for the design and construction of specialized radio equipment 'for a new project being undertaken for the British Admiralty.'[107] Other scientists who attracted GRU attention were Israel Halperin (ballistics), David Shugar (sonar),[108] and Matt Nightingale (loran navigation systems). They were deemed the most valuable sources of information not only for what they could obtain in Canada, but also because they had access to top-secret laboratories in Britain and the United States.[109]

Zabotin's real prize, however, was Alan Nunn May, a highly placed nuclear physicist at the Montreal Laboratory. His task, as directed by Colonel Motinov, was to obtain top-secret nuclear data and samples from the Montreal and Chicago laboratories, which May visited on a number of occasions in 1944 and 1945. Before he returned to Britain, May provided the GRU with small quantities of enriched uranium 233 and 235 and plutonium, as well as a general analysis of the explosive power of the Hiroshima and Nagasaki bombs.[110]

Pro-Soviet Organizations in Wartime Canada

It is difficult to understand the actions of May, Boyer, Mazerall, and Smith without appreciating the Zeitgeist of the war years, when the Soviet Union was a valued ally.[111] The most revealing indicator of pro-Russian sentiment was the operation of the National Council for Canadian–Soviet Friendship (NCCSF).[112] It was created on 22 June 1943 when an overflow crowd jammed Toronto's Maple Leaf Gardens and 'cheered enthusiastically almost every mention of Soviet deeds.' Prime Minister King was on hand to introduce the two keynote speakers: Joseph Davies, former U.S. ambassador to Moscow, and a leading exponent of American–Soviet friendship; and Feodor Gousev of the Soviet Legation. King took the opportunity to declare that his government 'warmly approved' of the formation of the new organization, and sought even closer relations with the government of the USSR. Similar endorsements came from John Bracken, leader of the Progressive Conservative party, and from M.J. Coldwell of the CCF. Gousev's reply was greeted 'with an uproarious burst of applause.'[113]

This founding convention also elected a national executive that included the influential financier Sir Ellsworth Flavelle as president and John David Eaton as vice-president.[114] Within a short time, the NCCSF established branches in eighteen centres across the country, actively disseminating 'authentic information ... about Russia.' Efforts were also made to create specialized committees to establish 'working relationships with similar groups in Russia.' The NCCSF's recruitment and organizational work was greatly facilitated by the establishment in January 1944 of the *Council Bulletin*, which during the next two years provided information about the NCCSF executive, study groups, special committees, and major events.[115] The May 1944 edition, for example, described how Victoria, British Columbia, was planning to celebrate the recapture of its 'sister city' Sevastopol with 'a special concert of Russian music ... and a formal presentation of the Soviet flag to the city ... by officers of the Red army.'[116]

One of the major goals of the NCCSF was to establish close links with specialized agencies in the USSR. This goal had been discussed at the December 1943 NCCSF committee meetings, where representatives of committees on science, medicine, agriculture, education, trade, labour, and the arts reported on their ability to develop links with their Soviet counterparts.[117] On this occasion, the work of the science committee received special praise from the NCCSF executive because of its efforts 'to apply information discovered by Russian scientists to Canadian problems.' This splendid beginning, it was suggested, should be followed by renewed efforts 'to secure the free exchange of scientific data between Canada and the U.S.S.R.'[118]

The NCCSF leadership were particularly pleased that a number of high-profile scientists had joined the organization. For example, the executive of the science committee included Leopold Infeld (mathematics, University of Toronto), B.H. Speakman (Director of the Ontario Research Foundation), Christian Sivertz (chemistry, University of Western Ontario), and David Shugar of the Royal Canadian Navy.[119] An even bigger catch was Wilder Penfield, director of the prestigious Montreal Neurological Institute and head of the NRC medical committees on wounds, plastic surgery, and motion sickness. Penfield had also participated in the well-published 1943 tour of Soviet medical centres as a member of the British-American-Canadian surgical mission to the USSR.[120]

In the summer of 1944 the Canadian Association of Scientific Workers (CAScW) joined the NCCSF in calling for closer scientific ties with the USSR. The CAScW shared the goals and aspirations of its parent organization, the British Association of Scientific Workers, which had developed an extensive campaign to improve working conditions for British scientific workers. This included collective bargaining rights, additional government support for research, and international cooperation between scientists.[121] The CAScW was the product of Canadian wartime conditions, which facilitated a collectivist response. For example, defence scientists were required to work on weapons systems that required large scientific teams, an effective division of labour, and a cooperative spirit.[122] It is not surprising then that the first branch of the CAScW was founded in July 1944 by scientists working at the Montreal Atomic Laboratory and McGill University. The formation of this branch was described by nuclear technician Norman Veall, one of the CAScW's founders:

The majority of people in our Lab, including Dr. Cockcroft, are members, and the suggestion arose ... that we should form a branch of the British Association here so that we would not lose touch with developments in England. At the same time we had been talking to several Canadians and found they were all interested ... so we thought it would be more helpful if we helped the Canadians to form a Canadian Association ... Dr. [Carl] Wallace, Dr. [George] Volkoff, Dr. [Leo] Yaffe, Dr. J. Carson Mark ... these people and myself ... took the initiative. The idea caught on immediately and within a week or so practically everybody we spoke to thought it was a fine idea.[123]

The CAScW shared many of the goals of its British counterpart, namely:[124] 'to promote the interests and economic welfare of scientific workers'; 'to secure the widest application of science and scientific methods for the welfare of society'; 'to expose pseudo-scientific theories'; and 'to combat all tendencies to limit scientific investigation to suppress scientific discoveries.'[125]

By the spring of 1945, the CAScW had branches in Montreal, Ottawa, and Toronto; a national executive that included prominent scientists such as Raymond Boyer,[126] and Alan Nunn May; and its own journal, *The Canadian Scientist*, which published important articles on the social responsibility of science.[127]

The Cold War and Science

The defeat of Nazi Germany brought about major changes in allied cooperation in defence science. For example, the British government lost no time in phasing out its commitments under the Anglo-Soviet Technical Accord.[128] In contrast, the United States continued to provide the USSR with weapons and technical information with the expectation that the Red Army would enter the war against Japan and thereby save thousands of American lives.[129]

In June 1945 an event occurred that demonstrated the gulf that had developed between the British government and many of its scientists. The occasion was the 220th anniversary celebrations of the Soviet Academy of Science in Moscow. Because the Kremlin was determined to make this a showcase event, it invited prominent scientists from twenty-two countries to attend this unique gathering.[130] The British delegation, initially consisting of twenty-two scientists, was the largest national group to be invited. However, at the last moment, the Churchill government, in consultation with American authorities, decided that eight of those selected, including Patrick Blackett, J.D. Bernal, P. Dirac, N.F. Mott, and Sir Charles Darwin, could not go to Moscow for reasons of national security.[131] This directive brought an immediate and vehement protest from British scientists across the country, who condemned their government's arbitrary action[132] and 'the discourtesy' shown to the Soviet Academy of Science.[133] For some, including Henry Tizard, it was an illustration of how wartime conditions could breed 'totalitarian' attitudes, while A.V. Hill wondered 'how long Winston's [Churchill's] Gestapo will want to prevent scientific men, with a knowledge of war secrets, from coming into contact with neutrals.'[134]

It was a rather subdued group of British scientists who arrived in Moscow on 14 June. Nor did spirits rise when members of the Soviet

Academy of Science began to ask questions about the 'missing eight.' B.M.H. Tripp of the Cavendish Laboratory described his embarrassment at the opening reception: 'Acad. Kapitza asked me personally why Prof. Bernal had not come. When I gave the official reply that Britain was still at war and needed his services, he said, "Why did they only discover they needed his services a few hours before he should have left?" ... Later in Leningrad a member of the Soviet Information Bureau said to me: "So Churchill wouldn't let the physicists come here?"'[135]

No such controversy surrounded the Canadian scholars who attended the Moscow celebrations: physiologist Hans Selye of McGill, political economist Harold Innis of the University of Toronto, and A.E. Porsild of the Canadian Geographical Society.[136] All three seemed to have enjoyed the proceedings, particularly the keynote address by physicist Peter Kapitza, who spoke of how the Academy of Science intended to encourage international scientific cooperation.[137]

The CAScW and Postwar Challenges

The defeat of Nazi Germany and Imperial Japan brought about many changes in the status of Canada's defence scientists. Most were concerned with returning to civilian life and resuming their involvement in pure research. The end of the war also placed the Canadian Association of Scientific Workers in a new political and economic context. The organization now focused its energies on ensuring stable postwar funding of scientific research and achieving representation on all government agencies that 'involved the application of science.'[138] In particular, there was support for the creation of a parliamentary science committee along the lines of the British organization that had been founded in 1929. More controversial was the CAScW's attempt to provide its members with collective bargaining rights in accordance with recent changes in Canadian labour law.[139]

In these initiatives the CAScW was affected by scientific developments in the United States. By August 1945 the American Association of Scientific Workers (AAScW) was calling on its members to endorse an ambitious postwar program that included the need 'to safeguard the intellectual freedom and professional interests of scientists.'[140] But while the AAScW still had a loyal following, its antiwar stance of 1940 and allegations that it was infiltrated by Communists greatly hampered its attempts to become a viable national organization.[141] By 1946 it was displaced by a more powerful and prestigious organization – the newly formed Federation of American Scientists (FAS). The FAS had two major advantages: it included the elite of American science, and it led the campaign for civilian control of atomic energy.[142]

However, the most important factor that shaped the CAScW's scientific activism was the British Association of Scientific Workers (BAScW), on which the CAScW modelled its constitution, organization, and goals. The two organizations also shared many members – indeed, most of the eighty-four British atomic scientists at the Montreal atomic research laboratory belonged to both the CAScW and the BAScW.[143] By the end of the war, the British organization had a membership of 16,000, including such high-profile defence scientists as Patrick Blackett, J.D. Bernal, and Robert Watson-Watt.

The BAScW's finest moment was its February 1946 conference entitled 'Science and the Welfare of Mankind,' attended by delegates from the United States, Canada, Australia, South Africa, France, Holland, and China. According to Grant Lathe, the Canadian representative, the meetings were a great success, despite the absence of Soviet scientists. He was particularly impressed by the speech of BAScW president Patrick Blackett, who called for 'the complete abolition of weapons for mass destruction as the only solution for maintaining world peace.'[144] Equally well received was the BAScW's endorsement of a proposal to create the World Federation of Scientific Workers, with J.D. Bernal and Frédéric Joliot-Curie as co-organizers for the July 1946 founding conference.[145]

Peace, disarmament, and nuclear arms control were among the key priorities of the Canadian Association of Scientific Workers. After the bombings of Hiroshima and Nagasaki in August 1945, the CAScW demanded that the Canadian government work toward 'control of the atomic bomb by the United Nations Organization.'[146] On 19 November 1945, the CAScW announced its own policy on atomic energy. Its ten-point statement began by lamenting the fact that 'international relations had deteriorated since the use of the atomic bomb through suspicion and distrust.' The brief called on the Canadian government to provide a confidence-building model by returning 'to the normal practice in the publication and dissemination of scientific and technical information [and] ... proving to the world that we do not intend to use our possession of atomic weapons as a bargaining instrument in international politics.'[147]

By the end of 1945 the CAScW had become a dynamic organization with a promising future.[148] It now had over 280 members, with branches in Montreal, Ottawa, Kingston, Toronto, Winnipeg, Edmonton, Fredericton, Guelph, and London. Moreover, its membership had expanded from an original complement of nuclear scientists to include chemists, physicists, and engineers.[149] By February 1946 the CAScW had also established links with two prestigious scientific organizations: the Federation of American Scientists and the British Atomic Scientists Association (BASA), through the personal efforts of Alan Nunn May, now resident in London.[150] In order to promote the ideals of scientific internationalism,[151] the CAScW's secretary, Norman Veall, began an active correspondence with Soviet scientists in February 1945 through the USSR Society for Cultural Relations with Foreign Countries (VOKS), in addition to circulating scientific papers, books, and films provided by the Soviet embassy.[152]

Conclusion

How effective were the GRU and NKVD espionage operations in Canada? Did they obtain vital defence secrets? This question remains difficult to answer because most Canadian security records for this period remain closed. But from what evidence is available, it seems clear that the Soviets were able to obtain important information about advanced radar and sonar systems, RDX, and certain nuclear data and samples in Canada, Great Britain, and the United States.

Why were Soviet agents able to gain access to some allied weapons, but not to others? For instance, neither the GRU nor the NKVD were able to obtain secret information about the proximity fuse, although they were certainly aware of its development. One factor, of course, was that Canada and Britain were not involved with either the final stages of research or the mass production of the weapon, which was carried out exclusively in the United States. In addition, the Americans gave the project the highest security possible.[153]

However, Canadian scientists were extensively involved in all aspects of research, development, and testing of chemical and biological weapons. Why did the Soviets not obtain any of those secrets? One explanation was provided by Otto Maass, who in July 1945 claimed that security measures adopted by the Directorate of Chemical Warfare and Smoke were so effective 'that it excited the admiration of the United States Chemical Warfare Service.'[154] A more credible explanation is that Soviet scientists were well advanced in their own chemical and biological research and, even more important, would soon be in a position to profit from the activities of their vanquished enemies. From Germany they would inherit the nerve gas facilities in east Prussia, which were later moved to the USSR. The Red Army also acquired some of Japan's biological warfare expertise when it invaded Manchuria in August 1945.

Historical studies dealing with Soviet espionage in Canada during the Second World War generally give the RCMP low marks for efficiency, especially for their failure to anticipate the Soviet threat. This viewpoint is reinforced by the assessment of former RCMP commissioner William Kelly: 'Although the RCMP had suspected espionage ... they never, until Gouzenko, had come to grips with this kind of thing.'[155] The force had no postwar plans to improve its screening of the Canadian public service, which meant that NRC scientists, unless they were working on 'extra secret work' would only be subject to a general file check.[156]

Another issue is whether wartime security measures created more problems than they solved – that excessive security measures caused severe delays in the development of sophisticated weapons. This was certainly the viewpoint of many American scientists. One of the most outspoken critics was the nuclear physicist Leo Szilard, who charged that General Groves's rigid system of compartmentalization delayed the development of the atomic bomb by at least one year. Less well known are the complaints of American biological warfare scientists that the U.S. program 'had been seriously limited ... by the drastic security regulations ... and by attempts to subject the research and development program to a military system.'[157] Whether these accusations are valid remains uncertain. But one thing is clear: most Canadian scientists, like their counterparts in Britain and the United States, found security systems imposed by the military to be vexatious and unwelcome.

9

Scientists, National Security, and the Cold War

Dr. Maass phoned me. Apparently the long awaited developments have taken place and a series of arrests were made last night. He appreciated my having informed him of what might happen as otherwise he would have been in a very embarrassing position ... by defending Professor Boyer.

C.J. Mackenzie Diary, 15 February 1946

In the early morning of 15 February 1946, twelve Canadian public servants were arrested in their homes under the authority of the War Measures Act.[1] The charge: violation of Canada's Official Secrets Act. That same day the Royal Commission on Espionage, headed by Supreme Court judges R.L. Kellock and Robert Taschereau, began its public hearings. By the end of June the inquiry was complete.[2] Twenty-two Canadians, including seven scientists, were accused of spying for the Russians; eleven would later be sent to jail.

Between September 1945, when the federal government began its own inquiry into Soviet espionage, and June 1946, when the Royal Commission on Espionage finished its work, Canadians' perceptions of the USSR changed dramatically. But the most important changes occurred within the government itself. By the summer of 1946 Canadian military planners no longer considered war with Russia a remote possibility, and plans were made to reactivate defensive arrangements with Britain and the United States. This included plans to carry out systematic weapons research, which in Canada was coordinated by the newly established Defence Research Board. Yet another initiative was the creation of the Security Panel, which was given authority to monitor the activities of potentially subversive elements within the country.[3]

The Royal Commission on Espionage and the Cold War

On 5 September 1945 Igor Gouzenko, an obscure cipher clerk at the Soviet legation in Ottawa, walked out of the Soviet embassy with a number of confidential GRU documents. These included a series of telegrams between Colonel Nikolai Zabotin and GRU headquarters in Moscow; pages from Zabotin's personal notebook; and a series of documents dealing with top-secret allied weapons systems, including the atomic bomb. Gouzenko was well placed to obtain this information because of his pivotal role in communicating with the GRU 'centre' through special codes and ciphers,[4] and because of his special access to the safe where Colonel Zabotin's papers were kept.[5] Gouzenko had been planning his defection for some time, using the documents to expose Soviet espionage activities and thereby obtain asylum in Canada.[6]

Gouzenko's motives for seeking sanctuary in Canada remain controversial. In his testimony, both before the Royal Commission on Espionage and in subsequent 'treason' trials, Gouzenko espoused noble intentions: his acquired love of Canada, his growing disdain for communism, and his determination to expose the nefarious Soviet espionage network that was 'preparing to deliver a stab in the back of Canada.'[7] Other accounts, both contemporary and historical, point to Gouzenko's survival instincts, particularly since he was about to be recalled to Russia with a negative performance record.[8]

Reg Whitaker and Gary Marcuse, in their sweeping study *Cold War Canada: The Making of a National Insecurity State, 1945–1957*, have provided the most comprehensive analysis to date of how Mackenzie King's government responded to Gouzenko's defection and of the subsequent crisis in Canadian–Soviet relations.[9] Of paramount importance was the Canadian government's verification of Gouzenko's story, and its decision to grant him asylum in return for his documents and testimony.[10] Top-ranking British political and security officials participated in this debriefing,[11] with agents of the U.S. Federal Bureau of Investigation not far behind.[12] This chapter focuses on the substance of Gouzenko's testimony and the relevance of his documents from the perspective of the Royal Commission on Espionage.

The commission had its beginnings in a special Cabinet working group, formed in September 1945, and code named 'Corby,' that consisted of Prime Minister Mackenzie King; Norman Robertson and Hume Wrong of External Affairs; Minister of Justice Louis St Laurent;

and RCMP Superintendent Charles Rivett-Carnac.[13] On 6 October 1945, Parliament passed a secret order-in-council (PC 6444) giving the RCMP authority, under the War Measures Act, to arrest and interrogate thirteen Canadian suspected of being part of a Soviet spy ring. Department of Justice argued that convictions against the accused could only be obtained by using the powers of a royal commission to force them to testify under the Canada Inquiries Act. In addition, the War Measures Act would be used to deny those arrested the rights of habeas corpus. What was still not determined, however, was when the suspects would be detained. That decision rested largely in Washington and London.

One of the crucial questions of late 1945 was whether the United States and the USSR could negotiate a viable bilateral agreement to prevent a nuclear arms race.[14] As a result, American State Department officials were determined that revelations about Soviet espionage would not complicate their nuclear diplomacy, even if it meant a delay in apprehending known agents. This flexible response would change dramatically after the December 1945 Moscow meeting of foreign ministers, which ended in failure.[15] Evidence of a new U.S. hard-line policy on wartime espionage was forthcoming on 1 February 1946, when Admiral William Leahy, Truman's security adviser, came to Ottawa on a personal consultative mission. Three days later came the dramatic announcement by American radio commentator Drew Pearson that 'the biggest story of espionage and intrigue since the war is about to break.'[16] The Canadian government was now forced to act. On 6 February the Royal Commission began its investigation of Soviet espionage behind closed doors, reviewing the documentary evidence and listening to Gouzenko's testimony. After a week, the Commission was prepared to authorize a series of arrests.

On 15 February at 6:00 a.m. squads of RCMP officers descended on the homes of eleven suspects; another two, Gordon Lunan and Eric Adams, were lured to Ottawa and likewise detained.[17] All were held incommunicado for weeks at the RCMP headquarters in Rockcliffe and subjected to intense interrogation. They were not allowed to contact their families or obtain legal advice. These draconian measures were based, in part, on the advice that lawyer E.K. Williams, an outside consultant, had submitted to the special Cabinet working group on 9 December 1945 emphasizing the advantages of using the broad powers of the Inquiries Act to extract testimony. It is, however, not entirely clear whether the Cabinet was informed about these provisions when it approved PC411 on 5 February, thereby officially creating the royal commission, under Supreme Court justices Roy Lindsay Kellock and Robert Taschereau.[18]

Official reaction to the creation and activity of the Royal Commission on Espionage varied. The most flamboyant response came from American newspapers, which vied with each other in embellishing Drew Pearson's allegations 'that a Russian spy ring involving as many as 1700 people in Canada and the U.S., was known to exist ... and that the Russians probably already had the atomic bomb.'[19] British press reaction was described by the office of the High Commissioner as generally 'brief and cautious.'[20] They did indicate, however, that the British government was concerned about the strategic ramifications of a complete diplomatic break with the USSR, particularly since influential groups in the country were calling for closer Anglo–Soviet relations:

You may be interested to know that one cause of difficulty for the Foreign Office is that they are under great pressure at the moment from the Royal Society to invite a group of Soviet scientists to England for some Newton Memorial celebrations. You will remember the painful incident last June when a group of British scientists were about to leave for Moscow and the atomic specialists were taken off the plane at the last moment. The Foreign Office considers this a most inopportune time for a return visit, but they do not know whether they can beat off the demands of Professor A.V. Hill, who incidentally is the leading spokesman of the view that Soviet espionage in Canada can be blamed entirely on the refusal to give up atomic secrets.[21]

The Kremlin, however, remained strangely quiet. Indeed, it was not until until 20 February, six days after the story broke, that the Canadian chargé d'affaires in Moscow, Robert Ford, was invited to meet with Deputy Commissar of Foreign Affairs Lozovski and was given a brief official statement.[22] It was a peculiar document. While it acknowledged 'that the Soviet Government knew that some information relative to radio location and other matters had been irregularly obtained by a member of the Soviet Embassy in Ottawa,' this was not an admission of wrongdoing. Indeed, the communiqué emphasized that the 'information was of little interest to Soviet authorities, who already possessed superior data at home.' The Russian response focused mainly on Canada's lack of diplomatic propriety. Why had Moscow not been contacted 'before the public statement of February 15 was made'?[23] Had not the USSR voluntarily recalled Colonel Zabotin? What more did Canada want? These same issues were raised in the 21 February edition of *Pravda*, which launched a bitter personal attack on Mackenzie King, calling him a puppet of British Foreign Secretary Ernest Bevin and a willing accomplice of British militarists.

Although King was troubled by these accusations and by the deterioration in Canadian–Soviet relations,[24] he was reassured that the Kremlin 'probably would not wish to break relations provided results of inquiry are not too scandalous for them.' King was not sure that Canada could avoid a serious diplomatic rupture with Moscow, particulary as tensions between the western democracies and the USSR increased both in Europe and in the United Nations.[25] In addition, there were further revelations about Soviet espionage activities.[26] On 3 March 1946, the British atomic scientist Alan Nunn May was arrested;[27] the following day the interim report of the Royal Commission on Espionage was released;[28] and on 5 March former prime minister Winston Churchill gave his famous Iron Curtain speech at Fulton, Missouri, in which he called for a common front of English-speaking peoples against 'the spread of Soviet-dominated police government in Eastern Europe.'[29]

In Ottawa, the commission was continuing its secret hearings in the offices of the Department of Justice. Emma Woikin, a clerk in the department of External Affairs, was the first 'witness' to appear (22 February), followed by Kathleen Willsher (25 February) of the office of the British High Commissioner. Two members of the so-called NRC spy ring followed: Edward Mazerall (27 February) and David Gordon Lunan (28 February). Four other scientists were scheduled for March: Raymond Boyer (7 March), David Shugar (8 March), Durnford Smith (18 March), and Israel Halperin (22 March).[30]

On the surface, the commission appeared to have a strong case to indict those charged with compromising Canadian defence secrets. Igor Gouzenko's testimony and documents, especially the reports of Colonel Zabotin and Colonel Rogov, were particularly damning, as were the notebooks, telephone diaries, and other personal paraphernalia seized from the homes of the accused during the 15 February RCMP raids. There was also considerable circumstantial evidence about alleged pro-Communist tendencies of the accused, including membership in Marxist study clubs, involvement in organizations such as the Canadian Association of Scientific Workers, and connections with Communist party organizers Fred Rose and Sam Carr. However, neither the Department of Justice nor the commission's lawyers considered this evidence sufficient to obtain criminal convictions. They needed self-incriminating testimony, which, they argued, could be obtained if the accused were told they were 'witnesses' who were required to testify under the provisions of the Canada Evidence Act and the Ontario Public Inquiries Act. This strategy, however, would only work if Justices Kellock and Taschereau ordered all those who appeared before them to answer questions posed by the Crown or to face contempt charges.[31]

In defending themselves before the commission, the accused adopted several tactics. Some, like Boyer, Lunan, and Smith, acknowledged that they had committed a 'technical' violation of the Official Secrets Act[32] but emphasized that their actions were consistent with Canada's wartime commitments since the information they provided could have been officially transmitted through either the Anglo-Soviet Technical Accord or the Mutual Aid Board.[33] Others were more defiant and refused to be intimidated, even when threatened with contempt charges.[34] David Shugar, for example, dismissed allegations that he had shown an inappropriate interest in the American sonar research being carried out in Washington, DC, and Orlando, Florida, by pointing out that this liaison work was part of his job with the Royal Canadian Navy.[35] Halperin was equally contemptuous of how the commission interpreted illegal disclosures: 'My own principal [R.C. Wallace, Queen's University],' he stated, 'had lunch with me once and asked me what I was doing. He did not ask for secret information, and I did not understand the question that way. I told him I was helping to organize an army research and development establishment which I told him would play an important role.'[36] Although Halperin's retort did little to convince the commission of his innocence, he was eventually acquitted during his 'treason' trial, as were Shugar and Nightingale. By contrast, Boyer, Mazerall, Smith, and Lunan were found guilty at their trials for violating the Official Secrets Act, and were sent to jail.[37]

By June 1946, when the Royal Commission on Espionage submitted its final report, Canada perceived the Soviet Union as an adversary. In many ways, therefore, the commission's report was more of a Cold War document than a comprehensive analysis of Canada's wartime relationship with the Soviet Union. For instance, in its discussion of the wartime context, the report made virtually no mention of either the Anglo-Soviet Technical Accord or the Canadian Mutual Aid Board. Nor did it address the fact that Soviet diplomats, trade officials, and military advisers were able to obtain radar sets, explosives, airplanes, and other war materiel through official Canadian channels. Instead, the commission saw only conspiracy: allied defence science secrets had been obtained by GRU spies, and by Soviet agents 'working along the same lines in the United Kingdom, the United States and elsewhere.'[38] One of the best examples of the commission's narrow focus was its negative portrayal of the Canadian Association of Scientific Workers.[39]

Scientific Activism and the Cold War, 1946–1947

In its final report, the Royal Commission on Espionage branded the CAScW as a pro-Communist organization that 'could be used in a variety of ways not only for propaganda purposes but eventually as a base for recruiting adherents to that Party from among scientists'[40] This assessment was based, in part, on evidence provided by the RCMP and by the fact that three prominent members of the CAScW's national executive – Raymond Boyer, David Shugar, and Alan Nunn May – were in custody.

The Mounties' dossier on the CAScW was extensive and of long standing. As early as January 1945 the Intelligence Section of the RCMP Criminal Investigation Branch had denounced CAScW attempts to organize scientific workers as being under 'Communist ... direction and therefore control.' Their suspicions were intensified when the CAScW expanded its membership and its range of activities. By early 1946 RCMP headquarters were convinced that the CAScW was formed 'not for the purposes of maintaining the standards and ethics of men of science of the highest possible level, but as a means of furnishing the revolutionary movement a source of information heretofore denied it.' The RCMP was particularly concerned that 'some dozen of [CAScW] members in Montreal worked on some phases of the atomic bomb.'[41]

The most critical period for the Canadian Association of Scientific Workers was between February and July 1946, during the hearings of the Royal Commission on Espionage. Also during this period the American controversy over who would control atomic energy – the military or civilian authorities – reached its most intense stage.[42] The degree to which these two events were related is shown by the extensive correspondence between the international secretary of the CAScW, Norman Veall, and the executive of the Federation of American Scientists (FAS).[43] This exchange reveals a number of important trends about the relationship between these two organizations, both of which were caught up in the evolving 'red scare.'

The dialogue began in earnest on 16 February 1946, when Veall advised W.A. Higinbotham, president of the FAS, about rumours that the USSR had obtained vital atomic secrets from its Canadian-based spies. Would it be possible, he asked, for the FAS to advise the CAScW whether Soviet agents would have been able to penetrate the Manhattan Project's security system? 'Specifically, I find it difficult to visualize General Groves letting a Russian spy walk out of the U.S. with "blueprints of the atomic bomb and a sample of the metal," and I would like your views on this article as soon as possible as I intend to bring this matter up when our National Executive meets next Wednesday [18 February].'[44] Since the Canadian case 'was still *sub judice*,' Veall indicated that he was unable to challenge evidence the Royal Commission on Espionage had obtained. Although he did not mention the impending appearance of Raymond Boyer before the commission, Veall was obviously concerned about the arrest of the CAScW president.[45]

Higinbotham's reply of 21 February was guarded, in part because his information on the Canadian situation was incomplete:

The situation is very confusing. The press and the public in this country are hysterical and, as you say, no real information exists as to what is going on. There seems to be little doubt that information was actually stolen from the Canadian government and by this time Russia has admitted that her agents had obtained it. It also seems true that this was not atomic information. I am sure that General Groves or anyone else would not have allowed spies to escape with blueprints of a bomb and a sample of metal. Furthermore, Byrnes [U.S. Secretary of State] has stated that the secret of the bomb is not known in Canada.[46]

At this stage, Higinbotham's greatest concern was that anti-Soviet 'hawks' and militarists in Washington would manipulate the Canadian spy scare 'to stir up nationalistic feeling' in Washington, a strategy he felt would have disastrous effects on FAS attempts to ensure effective civilian control over atomic energy:

We have not been able to think of a good line of defence. The Oak Ridge Association has released a statement asserting that Army sources are trying to discredit scientists and trying to block the McMahon Bill for national control of atomic energy which specifies civilian control. I suspect this may be the case but do not think it is possible to attack this issue until it becomes more clear of what actually has happened and who is behind the campaign.[47]

On 20 February 1946, the national executive of the CAScW issued a public declaration entitled 'The Spy Scare and Atomic Energy' which pointed out that since Prime Minister King's official statement five days earlier, a rash of inflammatory stories in Canadian newspapers had tried to link Soviet espionage with the Canadian atomic project. These allegations, the CAScW claimed, were the work of powerful right-wing groups in Washington who wished 'to stampede both the public and the legislature into supporting legislation such as the May-Johnson Bill which restricts academic freedom of scientists, the possibility of fruitful international cooperation, and the chances of developing atomic energy for peaceful purposes.'[48]

The arrest and confession of Alan Nunn May in Britain stunned both the Federation of American Scientists and the Canadian Association of Scientific Workers. In order to deal with the crisis, Veall proposed a joint emergency meeting of the two organizations in Montreal 'to pool our information, and perhaps develop some ideas for action to meet what appears to be a serious situation.'[49] The prospect of such a meeting, however, did not appeal to FAS leaders, who were beginning to feel the heat from the growing U.S. red scare.[50] On 23 March, as part of its own damage control, the FAS national executive revised its position on the possible involvement of some Canadian scientists with Soviet agents. Instead, in an official statement it emphasized that most of the 'leaks' had occurred during the war 'when Russia and Canada were allied in a death struggle against Germany.'[51] This admission was accompanied by a warning to its own members to remain vigilant in their support of FAS principles, especially given 'the unfortunate effect of the Canadian spy scare on atomic energy legislation in this country' and the possibility that security agencies in the United States would demand powers similar to those utilized by the Royal Commission on Espionage.

Concern over 'justice Canadian style' was reinforced on 28 March, when Veall told FAS leaders about the sweeping scope of the Official Secrets Act: 'I know dozens of perfectly innocent people, including myself, who would be liable for conviction under this act; in short, no scientist engaged in secret work is safe from arbitrary arrest and imprisonment.'[52] Veall's apprehension about his own safety became even more evident when he described the testimony of Igor Gouzenko at the preliminary stages of Fred Rose's Montreal trial:

He [Gouzenko] presented the court with a list of 35 names and 'cover names' of people involved in the spy ring. Among these names was Norman A. Veall – cover name not known. The first I knew of this was when I found my name listed as a spy on every front page in Canada. This is apparently due to the fact that as International Correspondent of the C.A.S.W. I have had some perfectly innocuous correspondence with the Embassy [Soviet]. Cockcroft tells me not to worry, but it means I have to appear before this Royal Commission in a week or so to explain why my name appears on Embassy documents and get the whole business straightened out officially.[53]

Veall's bad news staggered the FAS executive. Was their Canadian correspondent also a spy? Had they exposed their organization and movement to irrevocable damage through association with Veall and the CAScW? Would J. Edgar Hoover and his FBI bloodhounds try to implicate the FAS executive in this 'international' A-bomb conspiracy?

While the FAS anguished over its growing vulnerability, the Canadian organization was fighting for its life. On 3 April the CAScW launched a desperate broadside against the Royal Commission on Espionage, denouncing its violation of fundamental civil liberties and its complicity in the 'orgy of red-baiting and anti-Russian propaganda.' It also repeated its warnings that the Canadian red scare was being manipulated by 'sinister' elements in the United States 'to discredit scientists in general, and atomic scientists in particular, who support the McMahon Bill.'[54] But this counterattack was too little, too late. By the end of April 1946 a number of the CAScW's national leaders were either in custody or forced to justify their actions before the commission. Fearing their careers would be jeopardized, many Canadian scientists left the organization; those who remained were dispirited.[55] Their siege mentality was evident when the FAS asked Canadian atomic scientists for their views on the recent Acheson-Lilienthal Report, which set forth an imaginative agenda for the international control of atomic energy. This request elicited virtually no response from Canadian scientists, a situation Veall attributed to the pervasive political conservatism that now enveloped Chalk River scientists.[56]

Despite this rebuff, FAS scientists did not abandon their beleaguered Canadian colleagues, since they knew that this communist 'witch-hunt' knew no borders and was poised to destroy their organization. One of the more outspoken critics of the actions of the Canadian government came from Harvard physicist Wendell Furry, who denounced the commission's hidden agenda:

It seems plausible that the decision to act on the case instead of dropping it ... was politically motivated: (a) against Russia (b) against Canadian radicals (c) against scientists in Canada and the U.S. Also it is not unreasonable to suppose that holding the accused incommunicado so long served two purposes: to make possible the lurid and bogus prolonged publicity, and to force out confessions.[57]

One of the last efforts by the FAS executive to assist the CAScW came in late May 1946, when it tried to convince four high-profile American nuclear scientists, along with U.S. Senator Glen Taylor, their ally in the campaign against the May-Johnson Bill, to attend the CAScW's second annual conference.[58] But there were no takers.

One reason for their reticence might have been Alan Nunn May's recent conviction by a British court, a case that threatened double jeopardy for the scientists' movement: not only was May a self-confessed Soviet spy, he was also a former member of the CAScW national executive.[59] The severity of May's sentence (ten years) was also interpreted by representatives of the British Association of Atomic scientists as an attempt 'to try and frighten us into silence,' particularly since the trial had coincided with the introduction of the British Atomic Energy Bill, which the government was 'trying to force ... through.'[60]

By the time the CAScW held its second convention (in Toronto, 25–27 May 1946), the scientists movement in Britain and the United States was in decline. It was a strange affair. Despite the intense anti-Soviet sentiments in the country, the CAScW leadership made a point of reading to the assembly the fraternal greetings from the Soviet Academy of Science, which stressed 'the necessity of international cooperation in science.'[61] Delegates passed a variety of resolutions calling for substantial changes in Canada's scientific policies, notably a commitment to use science for peaceful purposes. The high point of the meeting was the speech by the CAScW's new president, Christian Sivertz of the University of Western Ontario, who predicted a positive future for the organization:

We feel proud of ourselves ... for having come through a year which has tried us all. But it seems to have brought out the best in those who were good; it seems to have sharpened their conceptions with regard to the whole object and meaning of our Association. It seems to have welded them more closely into a group determined to build their Association as it should be built – the pioneer of something which is becoming a new age.[62]

Fortunately for the CAScW, before the royal commission's final report of June 1946, few reporters linked the organization with espionage. As a result, the *Globe and Mail*'s coverage of the association's convention was relatively benign, and did not differentiate between that convention and the Toronto gathering of the apolitical Canadian Association of Professional Physicists (CAPP). Indeed, the newspaper was more interested in the CAPP meeting, where two fascinating proposals about possible future civilian use of atomic bombs were presented. One was by meteorologist P.D. McTaggart-Cowan, who speculated on the possibility that potentially dangerous hurricanes 'might be scattered by dropping an atomic bomb into the middle of it.' Not to be outdone, NRC physicist R.W. Boyle envisaged dropping a bomb on some of the Rocky Mountain glaciers: 'Think of the increased water supply for western Canada if one of those glaciers were melted by a bomb.' It was not recorded whether Boyle volunteered to drink the first glass of this radioactive glacier water.[63]

Meanwhile, in Washington, the FAS was making plans to distance itself from the Canadian Association of Scientific Workers. By the fall of 1946 the national executive claimed that their previous brief of 23 March in defence of the CAScW was nothing more than 'a compilation of excerpts from press reports,' and was 'not an expression of FAS opinion except as our interests may have coloured our selection of material.'[64] In addition, members of the FAS executive were being coached on how to respond to hostile press inquiries, including such leading questions as 'Aren't scientists likely to let secrets leak by mistake? Who knows Russian scientists? What are the Russians doing on the bomb?'[65]

Within Canada, the impact of the espionage inquiry was felt by federal government departments and agencies. High on the list was the National Research Council. Its president, C.J. Mackenzie, had become involved in the inquiry on 8 September 1945 when he had been informed of Gouzenko's defection and asked to verify what NRC secrets had been compromised. Moreover, four of those charged with violating the Official Secrets Act – Boyer, Mazerall, Smith, and May – had all worked on major NRC projects.[66] Although he had maintained his composure during the Royal Commission on Espionage hearings, Mackenzie deeply resented that his 'boys' had betrayed his trust and, even worse, jeopardized the future of the Council. He was also concerned that General Groves, Vannevar Bush, and other American science mandarins would view lax Canadian security as justification for not sharing defence secrets in the future.[67]

The pressure on Mackenzie was certainly evident when in March 1946 Bruno Pontecorvo, a Chalk River physicist, asked permission to visit his family in Italy. Under normal circumstances, the request would have been immediately granted. That March, however, was the most intense month of the Canadian spy stories, and American security agencies were watching. So Mackenzie took advantage of the fact that Pontecorvo had been a member of the British atomic team, and referred the decision to James Chadwick. Pontecorvo was eventually allowed to leave the country, but only after Chadwick consulted with General Groves, and on the condition that the Italian physicist not visit Frédéric Joliot-Curie in France.[68]

McGill University was another institution whose image was adversely affected by the spy trials, since so many of the accused had some connection with the university. Raymond Boyer was the most controversial. On 15 February 1946, immediately after the RCMP raids, Mackenzie had briefed Otto Maass about his colleague's involvement in the spy ring so that he would not place himself 'in an embarrassing position ... by defending Professor Boyer.'[69] Once Boyer had been arrested, it was McGill Principal Cyril James's turn to be in the hot seat. First, he had to convince powerful members of the University's board of governors, such as Walter Molson, that the university was not at fault in hiring Boyer as a lecturer in 1940 despite his reputation as a political radical.[70] In his rebuttal, James claimed that it was 'unjust and unfair to blame McGill University either for his [Boyer's] Communistic views or for having him in their employ ... [since] he was ... put here by the director of a government agency to do government war work and we accepted him unquestionably on this account.' Nor was there any concern over Boyer's loyalty, James asserted, until he became national president of the Canadian Association of Scientific Workers.[71]

But for McGill the worst was yet to come. On 20 March 1946, Solon Low, national leader of the Social Credit party, condemned the university in a widely quoted speech in the House of Commons: 'It must have struck thousands of Canadians as peculiar that so many of the suspects already charged with serious offenses were either on the staff or were graduates from McGill University.' Low confessed that he personally was not surprised by this subversive behaviour since he had long regarded the institution as a 'red hotbed.' What most concerned Cyril James about the Socred leader's speech was not its content, but the fact that prominent McGill graduates, such as Minister of Defence Brooke Claxton and Finance Minister Douglas Abbott, 'did not contradict him in the House.'[72] Left to his own devices, James prepared a measured, but scornful rebuttal:[73]

So far as the 'heads' of McGill University are concerned, he [Low] has not produced any evidence that Sir Arthur Currie, who was Principal for thirteen years after the last war, was a Communist, nor that either of my two predecessors belonged to that party. If I am a Communist, Mr. Low has discovered it before it has been revealed to me ... At the moment there are 750 members of the teaching staff and more than 6,000 students at McGill University. Of these students, approximately 3000 have just returned from active service, many of them with gallant records, and wearing decorations conferred upon them by His Majesty. Are these the Communists to whom Mr. Low refers?[74]

James also attempted, with only mixed success, to convince newspapers such as the Quebec City *Chronicle-Telegram* that McGill did not deserve their insinuations.[75] Although the controversy gradually abated, McGill administrators remained vigilant throughout the late 1940s in suppressing communist-type views on the part of faculty and students.[76]

Queen's University was also affected by the spy scare. This was most pronounced after mathematician Israel Halperin resumed his teaching duties there in the fall of 1946. Although Halperin had been named in the royal commission's report, and had been brought to trial in 1946, three factors helped him weather the storm: the Crown's case was almost entirely circumstantial; he had not incriminated himself; and he was well connected.[77] Indeed, almost immediately after Halperin's arrest, twenty-six of his Queen's colleagues had circulated a petition denouncing the commission's actions as 'alien to the basic principles ... of freedom under the law.'[78] Scientists at other universities rallied behind Halperin, one of the most outspoken being University of Toronto physicist Leo Infeld.

Frustrated in their attempts to convict Halperin, the RCMP resorted to other tactics. In 1947 a member of Queen's board of trustees was told that Halperin's acquittal did not 'invalidate the conclusion reached by the Royal Commission,' and that for the good of the university he should be fired. It was only after the timely intervention of chancellor Charles Dunning, who dismissed the campaign as a witch-hunt, unworthy of Queen's democratic traditions, that Halperin was allowed to continue his distinguished career.[79]

In contrast to McGill and Queen's, the University of Toronto emerged virtually untarnished from the espionage inquiries. But this was more good luck than good planning. Certainly, the RCMP had bulging files on a number of Toronto faculty members,[80] notably Leo Infeld, who had first come to their attention during the 1941 Levine case. Infeld's involvement with the National Council for Canadian–Soviet Friendship, the Canadian Association of Scientific Workers, and the Halperin defence campaign increased his vulnerability.[81] However, the university made no attempt to discipline Infeld until the spring of 1950, when he was dismissed because of his alleged pro-Communist activities.[82]

The Military Value of the Scientific Secrets

Did the 'secrets' obtained by the GRU's and NKVD's espionage operations enhance postwar Soviet military capabilities? This question has both contemporary and historical dimensions. This section of the chapter focuses on the debate about the military significance of the information Soviet agents obtained from Canadian sources in the year following Gouzenko's September 1945 defection.

The first public analysis of the military value of the scientific secrets came from the Kremlin. On 20 February 1946, Soviet authorities issued a communiqué claiming that any information obtained through Colonel Zabotin's activities was of little interest, given the 'more advanced technical attainment in the USSR.' The communiqué also denied that any of the British or U.S. radar and atomic data would be of use to the Soviet armed forces, since 'the information in question could be found in published works on radio location, etc., and also in the well-known brochure of the American, J.D. Smyth, *Atomic Energy*.'[83]

The Royal Commission on Espionage came to a different conclusion. At the very minimum, it asserted, the secrets the Soviets had acquired in Canada complemented the 'body of data' they had already obtained 'in England and ... in the United States.'[84] Moreover, the extensive activities of the GRU in themselves confirmed that 'the information sought was considered of the greatest importance by the Russian espionage leaders, and that alone might be a fair test of the question of value.' But the most important factor, according to the commission, was 'that the bulk of the technical information sought by the espionage leaders related to research developments, which would play an important part in the post-war defense of Canada, Great Britain and the United States.'[85]

In building its case against Colonel Zabotin's spy network, the commission utilized a number of documents that Igor Gouzenko had extracted from the Soviet embassy. Particular emphasis was placed on the activities of those NRC scientists recruited by Colonel Rogov. On 5 July 1945, for example, Rogov informed Zabotin that he had given

Durnford Smith (code named Badeau) a long list of requests, including data about 'the newest types of aerial photo apparatus of the RCAF' and 'particulars about the Electron Shells [proximity fuse].'[86] On 27 August an elated Zabotin informed GRU headquarters that Badeau had fulfilled some of his mission, delivering

17 top secret and secret documents [English, American, and Canadian] on the question of magnicoustics, radio-locators for field artillery; three secret scientific-research journals of the year 1945. Altogether about 700 pages. In the course of the day we were able to photograph all of the documents with the help of the Leica and the photo-filter. In the next few days we will receive almost the same amount of documents for 3 to 5 hours ... I consider it essential to examine the whole library of the scientific Research Council.[87]

The commission's documents showed that the GRU was also seeking important scientific information from the Royal Canadian Air Force, more specifically, data about the 'land-lines communications, network and allocations of aerodromes, maps of the R.C.A.F. and possibly the Gander project in Newfoundland.'[88] Another set of documents described the tasks assigned to agent 'Prometheus,' whom the commission identified as David Shugar. These documents included:

Tactical and technical facts of the naval and coastal hydrophonic acoustic stations working in ultra-sound diapason ... type of 'Asdic' which is used in a new submarines [sic] and other ships ... [and] Plants, workshops, Scientific Research Institutes and laboratories in England and in the U.S.A. which are making and planning the hydrophonic apparatus.[89]

But did Colonel Zabotin ever receive any of this information about the RCAF communications systems or allied sonar? The commission could not be sure. Perhaps it was for this reason that the commission placed so much emphasis on the RDX secrets Raymond Boyer admitted giving Fred Rose in early 1943:

I told him that RDX was used as a high explosive in the form of what is known as Composition A, which is a composition of RDX and beeswax. I told him that RDX was used in the form of Composition B, which was RDX, TNT and beeswax. I told him that RDX was used in the form of torpex, which is the same as Composition B with aluminum dust added. I told him that RDX was used in the form of plastic explosives.[90]

244   The Science of War

It is uncertain, however, whether Boyer provided data about the older Woolwich process or the joint U.S.–Canadian Bachmann continuous process. Moreover, in 1944 the USSR had been given much more relevant engineering information about RDX under the Mutual Aid program, and the Red Army had captured German RDX production facilities, which were similar to those in the United States.[91]

The commission was most concerned about any atomic bomb secrets Alan Nunn May might have conveyed to Moscow. Their apprehension was justified, since several of the GRU documents stolen by Gouzenko related to May's espionage activities. On 9 August 1945, for example, Colonel Zabotin sent the following report to his Director:

Facts given by Alek [May]: (1) The test of the atomic bomb was conducted in New Mexico, (with '49,' 94–239'). The bomb dropped on Japan was made of uranium 235. It is known that the output of uranium 235 amounts to 400 grams daily at the magnetic separation plant at Clinton [Tennessee]. The output of '49' [plutonium] is likely two times greater ... The scientific research work in this field is scheduled to be published, but without the technical details, The Americans already have a published book [the Smyth Report] on this subject. (2) Alec handed over to us a platinum with 162 micrograms of uranium 233 in the form of oxide in a thin lamina.[92]

The GRU documents also showed that before his return to England, May had been instructed to check out the Zeep reactor experiments at Chalk River and to acquire whatever other information he deemed relevant.[93]

At the same time that the royal commission was building its case against members of the Soviet spy ring, military experts in Canada, Great Britain, and the United States were carrying out extensive damage control assessments. Which weapon systems had been most compromised? How had this been done? And what long-term strategic advantages could the Soviet Union derive from these Allied secrets?

In September 1945 Canada's wartime allies were first informed of Gouzenko's startling revelations. However, only a small group of political and security officials in the United States and Britain were given any particulars. One of these was J. Edgar Hoover, director of the U.S. Federal Bureau of Investigation, who told President Truman that Canadian evidence confirmed the existence of a Soviet spy network among American scientists. Hoover, however, was more disturbed by reports that

May had provided the Russians with 'a small quantity of uranium 233 ... [and] specimens of uranium 235.'[94] Another interested party was General Groves, who, as head of the Manhattan Project, had the wartime responsibility of protecting U.S. atomic secrets.

After the Royal Commission on Espionage had completed its report, a second round of inquiries began. On 14 June 1946, Canadian military intelligence received a communication from Colonel R.E.S. Williamson, the American military attaché in Ottawa, asking what top-secret military information Soviet agents in Canada had turned over to the USSR. Specifically, he wanted to know whether the Soviet Union had been able to obtain information about the Manhattan Project, 'any samples of uranium 235 or plutonium ... [or] any plans or drawings showing bomb construction.' In addition, Williamson wanted to know what information the Soviets had gained about other advanced weapons systems: for example, explosives such as RDX; U.S. or British developments in the field of guided missiles; 'the U.S. Navy electronic shell,' or proximity fuse; and German nerve gases.[95]

The Americans were not alone in feeling that their national security had been compromised by Soviet espionage. Throughout May and June 1946, the British War Office and British intelligence agencies sent urgent messages to Canadian officials asking about 'any British military material or documents that may have been compromised to Russia.'[96]

These official inquiries forced Canadian scientific and military administrators to provide a detailed assessment of what secrets had been compromised. Given his wartime role as Canada's leading defence scientist, it was appropriate that C.J. Mackenzie should submit the most comprehensive analysis. In a 19 September 1946 letter to O.M. Solandt, Director General of Defence Research (Army), Mackenzie responded point by point to Colonel Williamson's letter. On the release of atomic secrets, Mackenzie was reassuring: Alan Nunn May could only have transmitted 'general information about the construction of the small pile at Chalk River but nothing of the operation of ... any full scale piles – in fact little more than was published in the Smyth report.' With reference to the methods of separating uranium 235 from uranium 238, Mackenzie claimed that May could not have disclosed any useful information, 'since Canada did not work on this project.' Nor could the British scientist have been able to disclose anything about the 'construction, method of assembly, and operating features of an atomic bomb.' Mackenzie did, however, admit that it was 'possible' that May had given the GRU 'a minute quantity of plutonium.'[97]

An earlier analysis of what atomic secrets had been compromised was provided by John Cockcroft, scientific director of Chalk River, during the fall of 1945. He placed more emphasis on May's pirate samples of uranium 235 and 233, especially the latter, since, in his opinion, this would alert the Russians 'that we were in the future going to use thorium.' Cockcroft also suggested that although May did not have access to either the weapons assembly plant in Los Alamos or the Hanford plutonium plants, he still could have provided Soviet scientists with 'valuable information ... if he had written out a report of everything he knew.'[98]

General Leslie Groves provided the most definitive American assessment of May's espionage activities. In his testimony before the U.S. Senate, Groves acknowledged that May 'must have become familiar with the work then going on at the Argonne Laboratory ... [and] probably acquired knowledge of some technical problems which we encountered in the operation of the first Hanford pile.' On the other hand, he was quick to reassure the senators that the Soviet spy had virtually no contact with Los Alamos and that none of his Manhattan Project scientists would have given May any information of value:

He [May] had few social contacts with the other scientists although he was generally well liked by them. They have described him as a charming, shy, little man with a dry sense of humour. The American scientists with whom he was in most intimate contact are in my opinion men of unquestioned loyalty and integrity. The revelation of his activities came as a complete shock to them.[99]

A more recent scholarly critique about the value of Soviet atomic intelligence has been provided by historian David Holloway.[100] Drawing on recently released Kremlin documents and extensive interviews with former Soviet atomic scientists, Holloway claims that while 'espionage played a key role in the Soviet atomic project,' once Stalin had decided to mobilize all Soviet resources behind the atomic project, there is little doubt that a bomb would have been built by 1952,[101] even without the information supplied by its spies. Nor does he consider Alan Nunn May's contribution[102] in any way comparable to that provided by Klaus Fuchs.[103]

Although atomic espionage most concerned American and British defence officials, they were also anxious to learn what other weapons systems might have been compromised. Once again, C.J. Mackenzie's assessment was comforting, if somewhat evasive:

Scientists, National Security, and the Cold War 247

Explosives: We have no further knowledge than that disclosed in the Commission report.
Guided Missiles: As far as we know, nothing.
Electronics and Proximity Fuses: As far as we know, nothing.
Chemical Warfare: Nothing as far as we know.

Mackenzie did, however, acknowledge that there was a divergence of opinion among the various technical experts about the level of damage the GRU espionage ring had caused.[104] The most disturbing report came from Robert N. Battles of the Royal Canadian Navy Directorate of Electrical Engineering, who listed a number of important antisubmarine secrets that might have been jeopardized:

(1) The function of individual A/S Research and Development Establishments, testing plants and trial bases, in the United States and United Kingdom, together with their relation to the overall Anti-Submarine organization.
(2) General description of the function of audio equipment installed in Naval craft and use of type 1479 as a depth determining equipment.
(3) General description of the application of audio equipment in the defence of harbours and Naval bases.
(4) Information as to general purpose of production test laboratories and location in Canada at Renfrew, Ont.
(5) General information on functions of audio recorder paper and experiments under investigation for its improvement.[105]

Loss of radar secrets was the focus of the appraisal prepared by D.W.R. McKinley of the NRC Radio Branch. But his message was mixed. McKinley conceded that Mazerall and Smith had access to top-secret radar information, including 'some details essential to the production of successful and practical [microwave] radar equipment for military purposes.'[106] This might mean, he admitted, that certain features of the British Admiralty's top-secret 931 naval radar set, which operated on 'a narrow wavelength band about 1.25 cms.,' could have been divulged.[107] McKinley stressed that the vast majority of NRC research activity was involved with 'non-secret and well established' information and techniques. He also indicated that in most instances, restrictive procedures were in place only because the Canadian Armed Forces automatically kept 'details of these techniques on wave lengths shorter than 3 centimetres ... secret.'[108] In summary, McKinley claimed that NRC radar secrets could only be considered vital 'in case of war or in anticipation of war.'[109]

By the summer of 1946, the prospect of another world war no longer seemed such a remote possibility.

Continuity and Change in Canadian Defence Science, 1945–1947

The end of the Second World War marked the end of the defence science relationship between Canada's National Research Council and the U.S. Office of Scientific Research and Development. The OSRD was dismantled and much of its work was taken over by organizations such as the Joint Committee on New Weapons, and by the ever-expanding research laboratories of the Army, Navy, and Air Force. In Canada, the NRC happily returned to its prewar civilian status, leaving the Defence Research Board (DRB) with the difficult task of coordinating the work of the country's scientific and military communities.[110] C.J. Mackenzie and Charles Foulkes, Chief of General Staff, were key participants in creating the DRB, along with Colonel Omond Solandt, its first director-general. All benefited from the 'lessons' of the war years, as well as from postwar British and American debates over what direction defence research should take.[111] The views of Vannevar Bush and Henry Tizard carried special weight for Canadian defence scientists after 1945, as they had during the Second World War.

The terms of reference for the new office of Director-General of the Defence Research Board were set forth in the December 1945 memorandum of the Cabinet Committee on Research for Defence, which stressed the importance of having one agency 'to coordinate research activities between the Armed Services with a view to economy of effort in fields of common interest.'[112] The selection of Solandt to occupy this powerful position, with a rank equivalent to the other Service chiefs, was based on his reputation as an outstanding defence scientist who, despite his youth and Canadian background, had moved up the slippery ladder of the British military hierarchy during the Second World War.[113] Another asset was his involvement with the British scientific team that analysed the physical damage and casualty levels in Hiroshima and Nagasaki in September 1945.[114]

On 15 February 1946, Solandt assumed his formal duties as director-general and immediately ordered a complete survey of Canadian defence research facilities and problems. On 3 May this report, entitled 'Policy and Plans for Defence Research in Canada,' was sent to the Cabinet for approval. On 28 November, after a delay of six months, the Defence Research Board was officially created under the terms of order-in-council PC 4316 and reinforced by the 1 April 1947 amendment of the National Defence Act.[115] Under this legislation, the board was given the mandate to act 'as a medium through which all the scientific resources of Canada and of other friendly nations can be made available to the Canadian Armed Forces and to make plans and preparations for the rapid mobilization of these resources in the event of war.' In addition, the board took over the operation of the research establishments involved with 'applied research on weapons and explosives and on chemical and biological warfare.'[116] As a liaison organization, it was assumed that the Defence Research Board would have 'the closest collaboration with the Services on the one hand and the National Research Council and the Universities on the other.'[117]

In keeping with the legacy of the Second World War, planning for Canadian defence occurred within the context of the evolving Anglo-American alliance.[118] This was outlined in the May 1946 report of the Cabinet Defence Committee (CDC). In the case of Britain, the CDC report stressed that while there were 'no specific arrangements of an inter-governmental nature involving post-war defence commitments,' there was mutual understanding 'that each country will regard the other as a potential ally in the event of a general war.'[119] On 22 May 1946, the Cabinet summarized the scope of postwar Canadian defence science priorities as follows:

In some fields of defence research, Canada has developed original ideas, while in others it has been largely dependent upon the United Kingdom and the United States. Canada, in the interests of all three nations, should gradually assume a greater degree of independence in defence research while, at the same time, seeking ever closer collaboration with the other two countries. Canadian activities should be limited to fields in which Canada has special interests, facilities or resources.[120]

It was assumed, therefore, that Canada would concentrate on those weapons systems where it had a relative advantage. This meant, in turn, that the Department of National Defence would continue to operate most of the major research stations: the Suffield Experimental Station, the Chemical Warfare Laboratory (Ottawa), the Canadian Signal Research and Development Establishment (Ottawa), and the Canadian Armament Research and Development Establishment (Valcartier). There were also encouraging signs that British military planners were interested in helping to fund some of these operations.

250 The Science of War

By the summer of 1946 British plans to continue utilizing Canadian defence science resources were well developed.[121] Discussions of this topic had been included in the July 1945 report of Sir Henry Tizard's Joint Technical Warfare Committee,[122] and the special Advisory Committee on Atomic Energy, which had been created in August 1945.[123] Even more important were the special June 1946 meetings of Commonwealth defence scientists held in London. Henry Tizard, who had organized the conference, urged the assembled scientists to focus their discussion on five weapons systems: guided missiles, supersonic aircraft, sonic research, and biological and chemical warfare.[124] Tizard was also instrumental in convincing the delegates of the advantages of creating the Commonwealth Advisory Committee on Defence Science. The actual work of the committee began in November 1947, when twenty-six Dominion scientists and military officials gathered in London.[125] Although a number of defence issues were discussed, future developments in the field of biological warfare were of special interest to the delegates.[126]

While these Commonwealth discussions progressed, the Canadian government was considering the nature of its postwar arrangements with the United States. In December 1945 the Cabinet discussed a report that recommended an extension of the ABC-22 continental defence plan, which had originally been negotiated in October 1941.[127] These tentative discussions were expanded by the Cabinet Committee report of 7 May 1946, which advised that Canadian interests would be advanced by coordinating specific defence programs with the United States through the auspices of the Permanent Joint Board of Defence 'in all matters relating to the security of the northern part of the western hemisphere.' The report also anticipated that cooperative defence arrangements with the United States would include

(a) interchange of personnel;
(b) encouragement of the adoption, as far as practicable, of common designs and standards in arms, equipment, organization, etc.;
(c) cooperation and representation at joint manoeuvres and tests;
(d) reciprocal provision of military, naval and air facilities, and transit rights to the other country;
(e) free and comprehensive exchange of military information affecting the security of the two countries.[128]

The Cabinet report also noted that a number of joint defence science projects had already been initiated by the Canadian Armed Forces since the end of hostilities. These ranged from the Navy's involvement in oceanographic research at Wood's Hole, to the RCAF–U.S. Army Air Force loran trials, which had been carried out in connection with Exercise Musk-Ox.[129]

American defence officials had their own agenda, especially in regard to how the Canadian North fit in with their plans for continental defence. In October 1946 the Joint Chiefs of Staff accepted the view 'that foreseeable new weapons enlarge the necessity of keeping any prospective enemy at the maximum possible distance ... [and] projecting our advance bases into areas well removed from our vital areas.'[130] But Ottawa was reluctant to accept any proposal that called for a comprehensive defence alliance with the United States.[131] Residual commitments to Commonwealth defence, suspicion about U.S. intentions, depressed military budgets, interservice rivalries, and the absence of an immediate Soviet atomic threat were sufficient reasons to justify a cautious approach. In addition, the new Minister of Defence, Brooke Claxton, did not consider North American defence critical, claiming 'that any attack on this continent would be diversionary,' with the main Soviet thrust coming in Europe.[132]

Alliance Planning and Weapons of Mass Destruction, 1945–1947

Postwar attempts to establish a system of defence arrangements between Britain and the United States strongly affected Canada, especially in the case of weapons of mass destruction, where the United States was in a position to dictate policies to its two wartime allies.

The decline in Britain's wartime bargaining position was most evident in the discussions over Anglo-American atomic energy exchanges. From the British perspective, there were three sacred covenants: the 1943 Quebec Agreement, the 1944 Hyde Park Aide-Mémoire of September 1944, and the arrangements worked out at the November 1945 Washington summit. In the latter case, both sides agreed to exchange scientific information; pledged not to divulge atomic secrets until 'effective enforceable safeguards against its use for destructive purposes can be devised'; and concurred on the need for an international commission under the United Nations to control the future of atomic energy.[133] This understanding was also endorsed by Prime Minister Mackenzie King, whose inclusion in these high-level discussions reflected Canada's unique wartime contributions in this field.[134]

Five months later, the Americans changed the rules of the game. In April 1946 President Truman announced a new interpretation of the Washington Agreement, especially in relation to the clause 'full and effective cooperation.' According to Truman, not only had the British authorities misunderstood the U.S. position as set forth at the November 1945 meetings, but they had also misinterpreted the essentials of the wartime arrangement. In a letter to the British prime minister, Clement Attlee, Truman wrote: 'Under the Quebec Agreement the United States was not obligated to furnish to the United Kingdom in the post war period the design and assistance in construction and operation of plants necessary to the building of a [UK] plant.'[135] Nor, Truman claimed, did the United States ever consider encouraging the proliferation of atomic weapons plants, even in the case of its wartime allies. Why else would he, as American president, have joined with the UK and Canada 'to request the United Nations to establish a Commission to control the production of Atomic Energy so as to prevent its use for military purposes'?[136] Although Attlee sent a spirited rebuttal, Truman showed no interest in continuing the debate.[137]

Although its exclusion from American technology severely hampered Britain's attempt to develop the atomic bomb, the Attlee government pushed resolutely toward that goal. On 6 November 1946 the British Atomic Energy Act, which set forth the country's nuclear priorities and strategies, received royal assent. Canadian High Commissioner Norman Robertson, who had carefully observed the parliamentary and public debates on the issue, noted the high priority Britain placed on the military uses of atomic energy. This situation, he mused, was summarized by a phrase used in the House of Commons to justify nuclear military preparedness: 'If there is one prospect more alarming than the prospects of an armaments race, it is the prospect of an armaments race in which we come in last.'[138] Throughout 1946 Robertson was also busy trying to avoid having Canada caught in the middle of this confrontation between its two major allies. The situation was particularly tense in April, and again in November, when Prime Minister Attlee tried to enlist Canadian support in his campaign against U.S. restrictive policies.[139] On both occasions, Mackenzie King's response was characteristically cautious: instead of taking sides, he emphasized Canada's record of atomic cooperation with both its allies.[140]

Canada's role as a nuclear ally of Britain and the United States also depended on the extent of its participation in United Nations Atomic Energy Commission. Although Ottawa wanted to be involved with the

UN agency, there was one complication: only members of the Security Council could participate in the discussions. Since Canada had not been elected as a nonpermanent member of the Security Council, some way had to be found to ensure that its views could be expressed during the complex debate about effective international control of atomic energy. After much negotiation, a compromise solution was reached,[141] and Canada's representative, General Andrew McNaughton, participated in the prolonged and acrimonious negotiations between the U.S. and Soviet delegations. But it was a frustrating experience for the General and his scientific advisers.[142] Part of the problem was the rigid and provocative proposal submitted by Bernard Baruch, the chief American negotiator. The so-called Baruch Plan had three major components: the Soviet Union would have to renounce any plans to develop nuclear weapons before the United States would even begin dismantling its atomic bomb arsenal; the Soviets would have to accept international inspection; and the USSR would have to abandon its Security Council veto during these UN Atomic Energy Commission (UNAEC) deliberations before any discussion could take place. While the Canadian delegation accepted the theory of U.S. exceptionalism 'as the only possessor of atomic weapons,' they attempted to modify Baruch's unworkable proposal through UNAEC's Scientific and Technical Committee, which had been formed to discuss the logistical aspects of inspection.[143] Indeed, one of the more imaginative attempts to break the American–Soviet log jam came from NRC physicist George Lawerence, who tried to convince Richard Tolman, the U.S. scientific representative on the Scientific and Technical Committee, that since the United States had already disclosed so much information about the Manhattan Project, 'further release of basic and technical information sufficient to permit Russia to build experimental fission plants for peaceful purposes would not shorten the time that country would require to complete bomb material plants and make atomic bombs.'[144] Tolman vigorously disagreed.

In the end, McNaughton's hand was forced by Baruch's insistence that the UNAEC vote on the acceptance of his virtually unaltered proposal. And despite its reservations, Canada joined with Britain and other U.S. allies to endorse the Baruch Plan. It was a high price to pay for Alliance solidarity.[145]

Atomic bombs were not the only weapon of mass destruction whose future control and development concerned Canadian military planners. The newly created biological munitions were deemed the most destructive weapon next to the atomic bomb, because 'mass casualties could be caused by relatively light weights of attack.'[146] In Canada, the Suffield Experimental Station remained the focal point for Canada's postwar chemical and biological warfare research and development, although the formidable costs of running it raised doubts about its future after Germany's defeat.[147] Otto Maass attempted to deal with these problems at a meeting of the CW Inter-Service Board on 11 September 1945. Most of the discussion centred on the past performance and future role of the Experimental Station, and on the necessity of maintaining a staff of almost 400 people. Suffield's scientific director H.M. Barrett emphasized the station's unique qualities as 'the only well-equipped experimental establishment in the British Empire with an area large enough to permit the carrying out of full scale trials with long range artillery, high altitude bombing, or large scale experiments with CW or BW agents.' He also predicted that the station would continue to attract British and American chemical and biological weapons experiments, which would keep Canada in the forefront 'of new methods of offense and defense.'[148]

The potential of biological weapons was the focus of discussion during a December 1945 meeting of British defence officials. On this occasion, scientists predicted that future research could create biological weapons '100 times more efficient, in terms of bomb load, than it is now. In other words, a bomb or warhead of 1,000 lb. suitably designed, might cause a 50% risk of death over an area of about 3 square miles.'[149] Similar discussions were taking place at Camp Detrick, which remained under the control of the U.S. Army's Chemical Warfare Service at an annual budget of almost $1 million.[150] The impact of these British and American initiatives on Canadian defence planning was evident in May 1946, when Omond Solandt endorsed Otto Maass's proposal that biological and chemical warfare 'be considered a priority in the Canadian Defence Research programme,' since Canada had much wartime expertise in these fields, and was 'well equipped, with existing facilities and skilled personnel.' Canadian policies would, he assured, adhere to international protocols prohibiting the use of war gases and biological agents, although there was 'the possibility that the Government might decide to employ these agents as a retaliatory measure.' This position, he added, was consistent with British and U.S. policies; indeed, certain projects had already been 'allocated to Canada by various combined committees and conferences.'[151]

The deterioration of Soviet–American relations between 1947 and 1950 provided an additional impetus for closer cooperation between Detrick and Suffield. In October 1950, the United States Joint Chiefs of

Staff, in preparing their nation for possible war with the USSR, acknowledged that in their 'efforts to make the United States capable of effectively employing toxic and chemical agents and biological weapons at the outset of a war, we have received valuable assistance from our Canadian ally.'[152]

This was not an isolated example of Canada's participation in the development of weapons of mass destruction. Although Chalk River was committed to the peaceful use of atomic energy, by 1948 C.J. Mackenzie had made plans to sell 'sixty kilograms of plutonium annually,' derived from the spent fuel rods of the NRX reactor, to the U.S. Atomic Energy Commission.[153] This profitable exchange system would soon become institutionalized and provide Canada with an adjunct role in the nuclear arms race. In addition, it should be remembered that Britain, France, and the Soviet Union gained some of their expertise required to develop atomic bombs through their involvement, legal or otherwise, with the Montreal Laboratory and Chalk River.[154]

The Korean War (1950–3) provided additional justification for Canada's collaboration with the United States in the design and production of new weapons systems. In turn, this meant an enormous increase in defence spending, which in Canada reached $2,100 million by 1952–3. Most of the major weapons projects were coordinated by the Defence Research Board and by the Department of Defence Production, a newly created agency under the control of C.D. Howe.[155] Even the National Research Council was involved, despite its reduced involvement in defence research after August 1945. C.J. Mackenzie, in one of his final reports as president of the NRC, described how the intensification of the Cold War had affected his organization:

With the growing uneasiness in the world situation that has been so marked during the past year, pressure on research for defence purposes has been rising, and once again the facilities of the National Research Laboratories are being turned more and more to projects arising from the requirements of the Armed Forces. At the present time, in addition to the atomic energy project, which always has had a dual character, the division of electrical engineering, mechanical engineering, and to some extent building research, applied physics and applied chemistry *have turned almost exclusively to war work.*[156]

Mackenzie must have wondered if it was September 1939 all over again![157]

Conclusion

During the Second World War Canada was involved in an epic struggle against Nazi Germany, fascist Italy, and imperial Japan, and for the second time in the twentieth century, the country was forced to commit itself to total war. Between 1939 and 1945 millions of Canadians served either in the Armed Forces or in the war economy. Others, as this study has outlined, made unique contributions through their involvement in Canada's scientific war, which was carried out in conjunction with the country's primary wartime allies: Great Britain and the United States. Within the context of alliance warfare, the role Canadian scientists played was often pivotal, but at other times less than crucial.

While many actors participated in this wartime drama of mobilizing Canadian science for war, it was the scientific mandarins who shaped and directed Canada's defence science effort. The most important, of course, was Dean C.J. Mackenzie, president of the National Research Council, assisted by Otto Maass of McGill University, who coordinated the country's chemical warfare and explosives program. The Canadian legend and quintessential hero Frederick Banting was the catalyst behind Canada's biological warfare planning and aviation medicine until his death in February 1941. This is just to mention the prominent few. But as this study makes clear other scientists made important contributions to the military victory over the Axis powers.

What distinguished the top scientific managers from their scientific colleagues was administrative authority, access to substantial funding, support of the military, and communication with the leading British and American defence mandarins. Although final authority for weapons research and development rested with the Canadian War Committee, these scientific administrators were given great latitude in making arrangements with American and British scientists for developing and testing various weapons. Canada was a junior partner in the Anglo-American alliance, and in most cases had neither the capability nor the responsibility of bringing most of these weapons into battlefield use. It is a tribute to Canadian scientific administrators that their country's contribution to the Allied war effort was much more than merely being a source of scientific resources, both material and human, to be mined at will by British and American defence officials.

The story of this book began in September 1939, when Canada, because of its military, scientific, and industrial resources, became Britain's most important Commonwealth defence partner. This Anglo-Canadian connection was greatly expanded in June 1940, when Britain stood virtually alone against the Nazi behemoth. While the wartime North Atlantic defence alliance was gradually taking shape, so was Anglo-American-Canadian cooperation in military technology. The secret mission of Henry Tizard set up a three-way dialogue among the three leading defence science administrators – Henry Tizard, C.J. Mackenzie, and Vannevar Bush, head of the U.S. Office of Scientific Research and Development.

The successes and occasional failures of this exchange system have been extensively documented throughout the book. For example, extensive cooperation between the NRC Radar Branch and the massive research and development facilities at the MIT Radiation Laboratory resulted in great achievements in the field of radar. No less important were the many links that existed between proximity fuse scientists at the University of Toronto and Johns Hopkins, and between RDX scientists at McGill and their counterparts at the University of Michigan. Chemical and biological warfare exchanges linked Canadian scientists at Suffield with British scientists at Porton and U.S. scientists at Edgewood and Camp Detrick. By 1944, the Anglo-Canadian Montreal Laboratory had become an adjunct of the powerful U.S. atomic bomb program, largely due to the efforts of C.J. Mackenzie and James Chadwick, who were able to negotiate a deal with General Leslie Groves, the czar of the Manhattan Project.

Since most of Canada's defence science contributions came about within the framework of alliance warfare, this study has analysed some of the major debates and controversies that affected defence science in Britain and the United States during the war years. In the case of Britain, the rivalry between Henry Tizard and Frederick Lindemann, which began during the mid-1930s rearmament period, had severe ramifications for Commonwealth defence in general and British defence science in particular. Lacking a centralized organization such as existed in Canada and in the United States, British defence science often failed to utilize effectively its own scientific talent.

In contrast, American scientists such as Vannevar Bush and James Conant enjoyed a special relationship with President Franklin Roosevelt and Defence Secretary Henry Stimson, enabling them to create the NDRC and the OSRD, and giving them powerful cards to play in dealing with the U.S. military. Their influence was further enhanced in 1942 with the creation of the Joint Committee on New Weapons, with Bush as its chairman.

The sometimes troubled relationship between Canada's National Research Council and the Canadian Armed Forces has received attention throughout this study. Although the book has made extensive use of NRC sources and the correspondence and diaries of C.J. Mackenzie, it has also taken other interpretations into account. For example, the criticisms of Royal Canadian Navy officials about NRC inefficiency in radar research and development has some legitimacy, as do the fulminations of British Tube Alloys officials against C.J. Mackenzie. Still, deficiencies in NRC performance need to be assessed not just from the perspective of what should have happened, but rather from the vantage point of what options were available to a country suddenly thrust into total war. After all, in 1939 Canada was seriously underdeveloped militarily, lacking an effective industrial infrastructure and technological expertise. And not to be forgotten is the fact that Canada was only a junior partner in the North Atlantic military alliance, and was taken seriously only when it advanced either British or U.S. strategic interests.

As this book demonstrates, C.J. Mackenzie developed many successful strategies to cope with Canada's subordinate military status, as did General Maurice Pope, head of the Canadian military mission in Washington. Both were ever watchful to assure a fair deal for the Canadian war effort. Otto Maass, head of the Army's Directorate of Chemical Warfare and Smoke, which coordinated Canada's biological and chemical warfare programs, was perhaps the most successful in negotiating partnerships that were favourable to Canada. He was able to accomplish this because he held two major assets: the Suffield and Grosse Île experimental testing stations. Nevertheless, these installations were only as good as the teams of scientists who contributed their services, both during and after the war.

Although most academic scientists resumed their university careers after V-J Day, many continued to be involved with Canadian military research through extramural contracts. Others opted for careers with the Defence Research Board or with the Chalk River atomic complex. This continuity in research staff allowed the DRB to concentrate its efforts on certain weapons systems in which Canada had relative advantages – particularly in the fields of explosives and ballistics and chemical and biological warfare.

Allied Weapons Systems: Did Canada Make a Difference?

In a provocative recent collection of essays, writers from a variety of scholarly backgrounds speculate on whether Canadian foreign policy has 'made a difference in the world.'[1] This question could also be asked about Canada's role during the Second World War. Did Canada make a difference? In terms of the country's overall military contribution, the consensus among historians is that, despite shortcomings of leadership, Canadian soldiers, sailors, and air crews performed admirably. A similar assessment emerges from the numerous studies that have examined the performance of Canada's wartime economy. But what about the country's contribution to the development of the major weapons systems of the Second World War? Did Canada make a difference?

These are not easy questions to answer, since, as Colonel C.P. Stacey so accurately observed, it is 'impossible to produce any definite quantitative assessment of the value of Canadian research and development as a contribution to Allied victory ... simply [by] the fact that it was a *contribution* – a share, and necessarily in most cases not a major share.'[2] Given the degree of collaboration that went into developing complex weapons systems, it is impossible to construct a precise credit ledger of Canadian defence science accomplishments. Yet, as this study has shown, it is possible to discern definite patterns that reveal the extent of Canadian expertise and contribution.

One approach to this complex subject is to assess how the war changed Canada's involvement with weapons research relative to the efforts of its allies. The greatest progress occurred in the United States, which grew, under the leadership of Vannevar Bush, from a state of relative military underdevelopment in 1940 to become a military superpower by 1945. By contrast, Canadian advancements in the field of defence science seem rather meagre. But there are good reasons why Canada lagged behind the United States. At the beginning of the war, Canada had a dislocated economy, with limited scientific and technological resources, and its political leaders wished to avoid large-scale military commitments. Moreover, during the war Canada's scientific organizations most often functioned within the larger framework of Anglo-American scientific and military planning. On the other hand, if one compares Canada's performance in the field of military technology with that of the other 'middle powers,' or even with that of Great Britain, a much different conclusion emerges. Canada's achievements were most notable in the fields of explosives and propellants, atomic energy, and chemical and biological warfare, while Canadian research and development in aviation medicine, radar, and the proximity fuse were also important to the total Allied war effort.

While many Canadian scientists proved exceptionally talented and energetic in their contribution to military science, the person most responsible for the coordination of these varied and complex research endeavours was C.J. Mackenzie. In many ways Mackenzie shared some of the qualities of his U.S. counterpart Vannevar Bush, although their situations were quite different. Bush directed the research of the most powerful nation in the world, which had large and wealthy universities, staffed by gifted scientists and engineers, as well as a highly developed industrial sector with its own impressive research and development capacity. While Canadian universities and industries were willing wartime partners of the National Research Council, they were simply not in the same league as their U.S. colleagues. The two men also operated within different political environments. Bush had ready access to President Roosevelt, the main source of power in the United States. This was particularly important when he had to deal with the American military on contentious issues. Mackenzie was not as fortunate. Although he had a powerful political patron in C.D. Howe, the Minister of Munitions and Supply, there was no assurance that Howe could carry either the War Committee or the Prime Minister with him. Moreover, while U.S. President Roosevelt took a personal interest in military technology, Canada's Prime Minister Mackenzie King was singularly uninterested in and uninformed about scientific matters.[3]

Did Canadian defence scientists feel accepted by their British and American counterparts? While it is difficult to provide an exact assessment, certain trends are evident. For example, it is clear that Canadian scientific administrators such as Mackenzie, Banting, Maass, and Murray had difficulty dealing with the arrogance of certain British defence officials and elite scientists. This attitude was particularly problematic in situations where the NRC was expected to provide unswerving loyalty, as was the case during the 1942–3 Anglo-American nuclear dispute, or submit to British illusions of superiority, such as occurred during the 1944 controversy over Canada's membership on the U.S. Biological Warfare Committee. But the source of greatest frustration for Mackenzie and his associates was not being consulted about new directions in British defence science policy, which were changed either unilaterally or through the Combined Chiefs of Staff, from which Canada was excluded.

In contrast, Canada's relations with U.S. defence scientists were generally harmonious. This was certainly the case in the Mackenzie–Bush relationship, as it was in the complex negotiations Otto Maass and E.G.D. Murray had with American chemical and biological warfare administrators, and in the interactions of George Wright and Arnold Pitt with U.S. explosives and proximity fuse experts. Likewise, NRC liaison teams had no problems negotiating with American military officials such as General William Porter, head of the Army's Chemical Warfare Service. In his dealings with General Leslie Groves, Mackenzie had much better luck in resolving bilateral issues than did his British counterparts. It helped, of course, that by 1943 both Mackenzie and Maass recognized that the United States was the major superpower and that American priorities would determine the allied scientific exchange system and security guidelines for the duration of the war.

Perhaps the best measure of Canada's military science contributions was the level of its involvement in postwar defence science planning through the auspices of the Defence Research Board. Between 1945 and 1950 the DRB was involved in a number of high-level meetings and consultations with its British and American counterparts. Of particular importance was the June 1946 meeting of the Commonwealth Defence Science Committee, where Omond Solandt, the DRB's director-general, assumed an important role in determining which weapons system should receive priority. Canada's expertise in chemical and biological warfare – especially given the Defence department's decision to continue the Suffield operation, which kept close contact with Porton and Camp Detrick – ensured the country a role in the 1947 BW-CW tripartite agreement, which remains in effect today.[4]

Canada's contributions to allied weapons systems also made a difference in the field of nuclear research and development. Evidence for this can be seen in the establishment of the Chalk River nuclear facilities, Canada's technological achievements in heavy water reactor design, the training of atomic scientists, and the sale of plutonium to the United States. In addition, Canadian nuclear scientists were active participants in the United Nations' efforts to control atomic energy and prevent the proliferation of nuclear weapons.

The Impact of the War on Canadian Scientists

While this book has examined the institutional framework of Canada's defence science achievements, it has also brought to light the goals and insights of individual scientists. The way in which scientific mandarins like Mackenzie, Maass, Banting, and Murray reacted to the multitude of scientific, technical, administrative, and political problems demonstrates their allegiance to Canada's war effort and their commitment to the allied cause. These men served their country often at high personal cost. Otto Maass's health, for instance, was severely undermined by his intense wartime schedule. Nor was he alone in this regard. Frederick Banting's diary entry for 17 April 1940 captures the physical and mental strains these men experienced:

I am so tired that I cannot sleep nor do I want to go to bed. I feel that I want time alone to think what is best to do ... On top of all these things I have to draft a memorandum on the Medical equipment required for our troops when they go to Greenland. Diet, high fat, vitamins, infectious diseases, clothing, frost bites, lice, drugs, etc. etc. are circulating through my head ... It is hardly possible for one small brain to accomplish the various tasks I have set myself. But it is war. My greatest worry is that I make mistakes & because of my negligence that our troops may suffer, or even that men may lose their lives.[5]

Fortunately, Banting and Mackenzie kept copious wartime diaries, which provide a wealth of information about many important issues. Mackenzie's entries were particularly valuable for this study because of his central role in defence planning, and because of the scope of his contacts. In any given week throughout the war years, Mackenzie consulted with cabinet ministers, the hierarchy of the Armed Forces, high-ranking civil servants, visiting British and American officials, and an array of Canadian scientific experts.

While Canada's scientific administrators assigned the projects and issued directives, most of the actual research and development of weapons systems was carried out in the laboratories of the NRC and the country's universities. Indeed, another way of assessing the impact of the war on Canada's scientific community can be obtained by examining the experiences of important project directors: George Wright and

Conclusion 263

Raymond Boyer for RDX; Arnold Pitt for the proximity fuse; J.S. Foster and Donald McKinley for microwave radar; Andy Gordon and H.M. Barrett for chemical warfare; Guilford Reed and James Craigie for biological weapons; and B.W. Sargent and Harry Thode for nuclear energy. In carrying out their duties, these scientists usually worked in close cooperation with American and British experts, with whom they often developed close personal friendships.

Most of Canada's leading defence scientists in this period were men, a trend that reflected the gender profile of physics, chemistry, and the medical sciences. There were, however, some notable exceptions such as microbiologist Carol Rice of the University of Rochester, who helped Guilford Reed prepare the Esoid protective toxoid during the spring of 1944, and Madge Macklin, who assisted her husband's biological research at the University of Western Ontario.

For most Canadian scientists the end of the war meant a return to their campus laboratories and to the world of pure science. But the experiences of the Second World War were not easily forgotten. Many remembered the war years as a time of excitement and opportunity when, through their work in advanced weapons systems, they had a chance to play in the scientific 'big leagues.' Harry Thode, a prominent member of the Montreal-based Anglo-Canadian atomic team, recalled the exhilaration of collaborative research: 'You couldn't go into the laboratory without some new scientific discovery being made each day, each week, each month ... and there would be scientists from all over the world.'[6]

The postwar years offered unprecedented opportunities for Canadian scientists to accept lucrative positions in the rapidly expanding research centres in the United States. As scientists increasingly moved between the worlds of state-directed research and the university laboratory, a new type of university scientist began to emerge. Needless to say, the consequences of this development for the universities were profound. As the famous American physicist Isidore Rabi has written: 'The whole picture changed as a result of government money ... It became no longer a matter of the department or the university. Each professor who was any good wore a knapsack, he could travel anywhere.' In sum, the war had, for better or for worse, ushered in a new age in state–university relations, and in the role of science in Canadian society.[7]

Scientific Activism and the Security State

During the First World War scientists of all the major powers participated in the development of a deadly array of new weapons.[8] With peace, however, came questions about the militarization of science, or more specifically, about whether scientists should allow patriotism to supersede their sense of scientific ethics. Criticism of the military gained further momentum with the signing of international peace covenants such as the Kellogg-Briand Pact of 1928, and with the advent of the Great Depression, which demonstrated the vulnerability of capitalist economic systems. In contrast, many peace activists and social reformers were impressed by the Soviet experiment, in which scientists appeared to be highly valued, and their talents used for social uplift, not for making war.

These were noble thoughts, but they were difficult to sustain in a world threatened by fascism.[9] By 1939 scientists in Great Britain, Canada, and the United States were prepared to offer their services in defence of their homelands. This meant that until the Axis powers were defeated the social activism of the scientific communities in the three countries had to be placed on hold. During the war scientists did express periodic concerns over security procedures imposed by the military and over the exclusion of the Soviet Union from important military technology. These protests became more pronounced in 1945 with the advent of the atomic age. During the next year, organizations such as the British Association of Scientific Workers, the Federation of American Scientists, and the Canadian Association of Scientific Workers began to speak up against the militarization of science and the barriers to international scientific cooperation.

The fact that some scientists were beginning to challenge state policies they viewed as unacceptable became a subject of much debate during the hearings of the Canadian Royal Commission on Espionage. The commission's obsession with conspiracy and betrayal was evident in its final report, which failed to appreciate the social, political, and military ramifications of the various Anglo-Soviet-Canadian arrangements. This narrow outlook has also influenced many writers and historians who have written on the subject.

The Science of War has attempted to understand the actions of certain 'disloyal' scientists within the context of the war years, not from a Cold War perspective.[10] If, as the royal commission implied, left-wing scientists were continuously suspect, why did C.J. Mackenzie and C.D. Howe allow Raymond Boyer, with his well known pro-Communist views and affiliations, to work on one of Canada's most secret scientific projects, and to represent Canada on the joint RDX Committee?[11] Nor is this an isolated example. In Britain, Professor J.D. Bernal, one of that country's most outspoken Marxist scientists, first became a principal scientific consultant for the Minister of Home Security and later personal scientific adviser to Lord Louis Mountbatten, Chief of Combined Operations. When Sir John Anderson, the Lord Privy Seal, was warned in 1940 that Bernal was a security risk, 'Sir John commented, "even if he is as red as the flames of hell," he wanted him as an additional adviser on civil defence.'[12] Similarly, General Leslie Groves explained his appointment of the left-leaning physicist J. Robert Oppenheimer as director of the Los Alamos atomic weapons laboratory in 1942: 'My careful study made me feel that, in spite of the record, he was fundamentally a loyal American citizen and that, in view of his potential overall value to the project, he should be employed.'[13]

Thus, military pragmatism, not ideological correctness, was the guiding principle of defence science research in Canada, Great Britain, and the United States during the Second World War. This outlook also coloured negotiations between the western allies and the Soviet Union. By 1944 all three countries had adopted a weapons exchange system with the USSR on the assumption that each new weapon would help the Red Army kill more Germans. Thus, sporadic evidence of Soviet espionage did not unduly concern the Roosevelt, Churchill, or King administrations. What mattered most was a speedy end to the war; and that meant retaining Soviet cooperation. It is in this climate of opinion that the willingness of some Canadian scientists, 'persons of marked ability and intelligence,' to supply Colonel Zabotin with certain 'secrets' must be considered.[14]

The espionage inquiry and trials of 1946-7 brought the Canadian Association of Scientific Workers crashing to the ground, effectively destroying the cause of scientific activism in Canada during the early Cold War.[15] This did not mean, however, that Canadian scientists did not protest against military-funded 'big science' or against the escalating nuclear arms race. Indeed, in the late 1950s many former CAScW members participated in the international science for peace Pugwash movement,[16] or mobilized themselves when the United States and the Soviet Union appeared on a collision course toward outright thermonuclear war. Canadian scientists also assumed an important role in the complex international negotiations that resulted in the 1972 Biological Weapons Convention and the 1992 Chemical Warfare Convention on the Prohibition of the Development, Production, Stockpiling and Use of Chemical Weapons, and on their Destruction.[17]

One reason why Canada has assumed a significant role in attempts to ban weapons of mass destruction is directly linked to the degree of expertise it possesses in these fields. At the same time, Canadian defence policies have been criticized as both hypocritical and dangerous to world peace. In particular, serious questions have been raised about the Suffield Experimental Station, which continues to conduct experiments with deadly chemical and biological weapons. On the other hand, as our political and military leaders often remind us, the world remains a dangerous place, even with the end of the Cold War. And, as the Gulf War demonstrated, 'rogue' nations such as Iraq can acquire operational nerve gas and biological weapons despite international conventions and an inferior industrial infrastructure.[18]

During the Second World War Canadian scientists did not hesitate to become involved with military technology, even with weapons of mass destruction. Their justification was simple: Canada and its allies could only defend themselves by having a retaliatory capacity; it was, therefore, their duty as scientists to ensure their country's survival. Frederick Banting captured the essence of this challenge in his diary entry of 20 October 1939: 'This war is one of brains and scientific ingenuity rather than one of man power, mass infantry and massacre. Our job is to lick the Germans under Hitler.'[19]

Appendix 1

Major Military, Political, and Scientific Events

1915 Poison gas used against Canadian troops at Ypres.

1917 National Research Council Act passed by Parliament.

1917 Colonel A.G.L. McNaughton directs artillery assault on Vimy Ridge.

1918 Over 124,200 tons of poison gas used during the war.

1923 Frederick Banting shares Nobel prize for medicine.

1924 H.M. Tory made honorary chairman of NRC; becomes president in 1928.

1925 Geneva Protocol bans military use of chemical and biological weapons.

1929 General McNaughton becomes Chief of the General Staff.

1932 Opening of the NRC Sussex Street (Ottawa) laboratories.

1935 General McNaughton appointed president of NRC. C.J. Mackenzie appointed to NRC Council. First stages in the development of British radar equipment. Italy invades Ethiopia, uses poison gas.

1936 Outbreak of the Spanish Civil War. Germans occupy the Rhineland.

1937   Imperial defence conference takes place in London.

1938   Canada accepts role in Commonwealth Air Training Plan. NRC Associate Committee on Medical Research created. Frederick Banting warns of German biological warfare intentions. British Biological Warfare Sub-Committee. Munich crisis. Otto Hahn achieves successful atomic fission experiment.

1939   Germany annexes Czechoslovakia. Germany invades Poland. Britain and Canada declare war on Germany. General McNaughton leads First Division to England. C.J. Mackenzie named acting president of NRC. NRC's Radio Branch becomes involved in radar research.

1940   Germany invades France and western Europe. Winston Churchill leads wartime coalition government. Italy enters the war on the side of Germany. Luftwaffe strategic bombing of England begins. Liberal government of Mackenzie King is re-elected. Canadian government enacts National Resources Mobilization Act. United States creates National Defense Research Committee. Canada–U.S. Permanent Joint Board on Defence created. Henry Tizard leads scientific mission to United States. Canadian War Technical & Scientific Development Fund. Research Enterprises Ltd. created as Crown corporation. Bletchley Park experts break German Ultra codes. Radiation Laboratory (MIT) becomes centre for U.S. radar work. British consider using gas against German invasion. Otto Maass and J.R. Donald begin RDX research. Formation of British MAUD Committee to study atomic bomb. Einstein-Szilard letter to President Roosevelt about atomic bomb.

1941   Germany invades the Soviet Union. British military defeats in North Africa. Formation of the Combined Chiefs of Staff. Hyde Park Agreement coordinates U.S.–Canada defence production. Japan attacks Pearl Harbor; the United States declares war. Germany declares war on the United States. Canadian troops captured at Hong Kong. Sir Frederick Banting killed in air crash. James Conant leads U.S. scientific mission to England. Office of Scientific Research and Development (OSRD) founded. OSRD directs U.S. atomic research. Canada–U.S. Technical Sub-Committee on Explosives. British Central Research Office in Washington. Joint U.K. Canadian Suffield Experimental Station. Otto Maass appointed head of Directorate of Chemical Warfare. E.G.D. Murray head of NRC M-1000 Committee on Biological Warfare. Merle Tuve directs NDRC proximity fuse work. Arnold Pitt heads Toronto proximity fuse team. J.S. Foster becomes NRC liaison at Radiation Laboratory.

1942 British defeat German forces in Libya. Joint U.S.–British invasion of North Africa. Russian victory at Stalingrad. American naval forces defeat Japanese at Midway. Canadian troops suffer heavy losses during Dieppe raid. Relocation and internment of Japanese Canadians. Conscription crisis divides Canadians. Anglo-Soviet Technical Accord. General Pope head of Joint Staff Mission in Washington. Formation of U.S. Joint Committee on New Weapons. Canadian and American military consider gas as defensive weapon. Roosevelt and Churchill warn Germany and Japan not to use gas. Rinderpest and anthrax research at Grosse Île. George Merck heads U.S. biological research. Establishment of the Anglo-Canadian Montreal Laboratory. Army's Manhattan Project takes over U.S. atomic bomb development. Successful fission experiments at University of Chicago.

1943 Allied invasion of Sicily and Italy. Allied continuous bombing of German cities. Russians drive German forces westward. Recapture of strategic islands from Japanese. Diplomatic exchange between Canada and the USSR. Quebec Conference agreements between Churchill and Roosevelt. Henry Tizard's unsuccessful mission to Russia. American SCR 584 gun-laying radar replaces Canadian GL IIIC. Mass production of RDX by Tennessee Eastman. Creation of U.S.–Canada Chemical Warfare Advisory Committee. Formation of the Combined Policy Committee for joint atomic research. Robert Oppenheimer heads U.S. nuclear team at Los Alamos. James Chadwick directs British atomic program.

1944 U.S. forces regain Marshall Islands, Philippines. D-Day landings in Normandy. German V-1 and V-2 rockets attack London. President Roosevelt wins fourth term; Harry Truman vice-president. Conscription crisis divides government of Mackenzie King.

GIN anthrax research transferred to Suffield. Major chemical warfare, biological warfare, flame-thrower tests at Suffield. Debate over German use of biological warfare weapons in Normandy. British–U.S. operational studies on use of chemical weapons. British War Office assumes control of biological warfare research. Camp Detrick able to weaponize a number of biological warfare agents. Japanese balloon attacks on Pacific North West. Proximity fuse weapons used in Britain and France.

1945 Yalta conference decides U.S., British, Russian spheres of interest. Death of President Roosevelt; Harry Truman succeeds as president. Germany's unconditional surrender; death of Hitler. Labour party triumphs in British general election. Mackenzie King wins a majority government. San Francisco founding conference of the United Nations. Truman informs Stalin about atomic bomb at Potsdam Conference. Kremlin orders massive atomic bomb research and development. United States attacks Hiroshima and Nagasaki. Japan surrenders; Second World War ends. Igor Gouzenko defects, revealing Soviet spy ring in Canada. Washington Agreement: U.S., U.K., Canadian atomic partnership. Moscow conference fails to prevent nuclear arms race. American military propose using gas against Japan. Tizard Committee studies Britain's postwar weapons. Discovery of Japanese biological warfare research and German nerve gas. Alan Nunn May provides Soviet spies with nuclear data. Opening of the Chalk River nuclear facility; test of the ZEEP. Formation of Canadian armament and explosives research centre. Canada included in British–U.S. postwar code-breaking plans. Defence Minister McNaughton supports continuation of Suffield.

1946 Relations between the West and the Soviet bloc deteriorate. United Nations fails to prevent nuclear arms race. McMahon Bill prevents sharing of U.S. atomic secrets. U.S. tests atomic bombs at Bikini. Britain decides to develop its own nuclear weapons. Britain, Canada, and U.S. continue chemical warfare and biological warfare research and development. Canadian Royal Commission on Espionage conducts its hearings. Defence Research Board created, with Omond Solandt as director. National Research Council withdraws from defence research. McNaughton made head of Atomic Energy Control Board. McNaughton leads Canadian team at UN Atomic Energy Commission talks. Collapse of the Canadian Association of Scientific Workers. Conviction of Alan Nunn May and other Canadian scientists. Defence Research Board involved with Commonwealth defence science. C.J. Mackenzie supervises Canada's nuclear program.

Appendix 2
Brief Biographical Sketches

Frederick Banting (1891–1941) was Canada's most famous scientist during the interwar years because of his Nobel Prize–winning work in the discovery of insulin. He and his colleagues at Toronto also conducted pioneering research in aviation medicine and biological warfare. During the war he was one of Canada's leading defence scientists, until his death in February 1941.

H.M. Barrett (1908–78) joined the University of Toronto's chemistry department in 1932. Between 1941 and 1946 he was research director at the Suffield Experimental Station.

J.D. Bernal (1901–71) of Cambridge University was an authority in the field of crystallography. He was better known as a leftist writer, social activist, and defence scientist. He was a member of BAScW.

Norman Bethune (1890–1939), like Banting, was involved in military medicine on the western front during the First World War. He gained notoriety because of his blood transfusion work during the Spanish Civil War, and for his social activism in Montreal.

Patrick Blackett (1897–1974) was one of Ernest Rutherford's many former students who made outstanding contributions to Britain's war effort. He was most famous for his operational research work, and for his criticism of British postwar defence policy. He, like Bernal, was an active member of the BAScW.

Raymond Boyer (1906–93) was a McGill-trained chemist who was active in various progressive organizations during the 1930s. During the war he was a leading member of Canada's RDX team; in 1944 he became the first president of the CAScW. In 1946 he was charged with violation of the Official Secrets Act.

R.W. Boyle (1883–1955) was McGill's first physics Ph.D. (1909). He was director of the NRC Physics Division during the Depression and the war years.

Eli Burton (1879–1948), a graduate of the University of Toronto, became the head of its physics department after the 1932 departure of John McLennan. He carried out various war-related projects.

Vannevar Bush (1890–1974), an engineer and administrator, directed American defence science between 1940 and 1946. In 1940, he was made chair of the National Defense Research Committee. In coordinating the activities of the Office of Scientific Research and Development (OSRD), Bush was able to draw on the support of President Roosevelt and the high prestige he enjoyed among American scientists. He also established a good working relationship with British and Canadian defence scientists.

James Chadwick (1891–1974) was assistant director of radioactive research in the Cavendish Laboratory from 1923 to 1935. He was awarded the 1935 Nobel Prize in physics for his discovery of the neutron. Between 1943 and 1945 he directed Britain's Tube Alloy's project, working closely with C.J. Mackenzie of the NRC and General Groves of the Manhattan Project.

John Cockcroft (1897–1967) trained at the Cavendish Laboratory. He received the 1951 Nobel Prize in physics. During the war he was an expert in operational research, scientific director of the Anglo-Canadian atomic projects at Montreal and Chalk River between 1944 and 1946. He later directed British atomic energy research at Harwell, England.

James Conant (1893–1978) was an internationally acclaimed chemist before becoming president of Harvard. During the war, he was second in command of the Office of Scientific Research and Development, and played an important role in nuclear diplomacy. Conant was an important defence science consultant during the early stages of the Cold War.

Major General Harry Crerar (1888–1965) became a career officer with the Canadian Army after the First World War, which he continued to serve until his retirement in 1946. His approach to war was greatly influenced by General McNaughton, and he shared his mentor's interest in defence science.

E.L. Davies (b. 1897) occupied a senior position at Porton Down before becoming superintendent of the Suffield Experimental Station in 1941. After the war, he became a prominent member of the Defence Research Board.

Raymond Dearle (1891–1970) studied with Toronto's J.C. McLennan; he taught physics at the University of Western Ontario throughout the interwar and war years. His most important research contribution was in the field of microwave radar.

Paul G. Fildes (1882–1971), a highly respected microbiologist, was director of British biological warfare research between 1940 and 1945.

J.S. Foster (1890–1964) was one of Canada's leading physicists when he left McGill University to work at the MIT Radiation Laboratory between 1942 and 1945. During the postwar years he resumed his work on nuclear physics, particularly after McGill opened its new cyclotron in 1946.

Ralph H. Fowler (1889–1944), an accomplished physicist, served as the British scientific liaison officer in Ottawa and Washington in 1940–1. Through his efforts, an Allied weapons exchange system evolved.

Bertrand Goldschmidt, a French nuclear chemist, was a prominent member of the Free French scientific team at the Montreal Laboratory between 1942 and 1946. He later assumed an important role in France's atomic energy program.

Charles Frederick Goodeve (1904–80), a graduate of the University of Manitoba, was one of many Canadian-born scientists who gained prominence in Britain during the interwar years. His most outstanding wartime achievement was in the field of naval operational research.

A.R. Gordon (1896–1967) was a Toronto chemist who played an active role in chemical warfare and explosives research.

Joseph A. Gray (1884–1966), a student of Ernest Rutherford, came to

Queen's University in 1922. He quickly gained a reputation for inspiring young physicists to pursue academic careers.

Hans von Halban (1908–64), an Austrian-born physicist, arrived in Canada in 1942 as the scientific director of the Montreal Laboratory. Before the fall of France in 1940 he had been a key member of Frédéric Joliot-Curie's atomic research team.

Israel Halperin (b. 1911) obtained his Ph.D. in mathematics at Yale in 1937; he then taught at Queen's University before joining the Canadian Army in 1942. He was implicated in the 1946 espionage inquiry.

George H. Henderson (1892–1949) studied at Cavendish before joining Dalhousie's physics department in 1922. He became actively involved in naval war research both through the NRC and in the Naval Research Establishment laboratory in Halifax.

John T. Henderson (1905–83), a physicist, was trained in Britain before joining the NRC Radio Branch in 1934. He became one of Canada's leading radar experts during the war, first with the NRC and then with the Operational Research Centre of the RCAF.

Gerhard Herzberg (b. 1904) was an outstanding German physicist who fled to Canada from Nazi Germany in 1935. He taught at the University of Saskatchewan between 1935 and 1945, and worked on explosives during the war. In 1948 he was appointed director of the NRC's physics division; in 1971 he was awarded the Nobel Prize in chemistry for his molecular spectroscopy research.

A.V. Hill (1886–1977), a Nobel laureate in physiology, was a prominent member of the Royal Society and one of Britian's leading defence scientists. His 1940 personal lobbying in North America helped prepare the stage for the Tizard scientific mission.

Clarence Decatur Howe (1886–1960) came to Canada in 1913 after graduating from the Massachusetts Institute of Technology. During the federal election of 1935 he was elected as the Liberal MP for Thunder Bay, and became minister of transport in Mackenzie King's Cabinet. In 1940 he became minister of the new department of Munitions and Supply with the responsibility of directing Canada's war production and manufacturing new weapons for the Canadian Armed Forces. In 1944 he was asked to assume responsibility for the Department of Reconstruction.

Leopold Infeld (1898–1968), a Polish-born physicist, came to the University of Toronto in 1939 on the recommendation of Albert Einstein. He was dismissed in 1950 for alleged pro-Communist activities.

George Lawrence (1905–87), a nuclear physicist, was educated at Dalhousie and Cambridge universities. A member of the NRC, he was one of the senior Canadian scientists at the Montreal Laboratory, and McNaughton's scientific adviser at the 1946 meetings of the United Nations Atomic Energy Commission.

Brigadier Charles Lindemann (elder), brother of Lord Cherwell, was a career military officer who served as a scientific adviser of the British Army in Paris until the spring of 1940, when he was sent to the British Military Mission in Washington to help arrange Anglo-American military cooperation.

Frederick Lindemann (Lord Cherwell; 1886–1957) was an Oxford University physicist who worked closely with Winston Churchill both during the 1930s and when Churchill became prime minister in May 1940. His bitter rivalry with Henry Tizard hampered British wartime defence science planning.

Otto Maass (1890–1961) was born in the United States, and did his undergraduate work at McGill. After studying in Germany and at Harvard, he became a member of McGill's chemistry department. During the war he directed Canada's chemical warfare and explosives programs. In 1947 he was appointed director of the biological and chemical warfare research division of the Defence Research Board.

C.J. Mackenzie (1888–1984) was trained as an engineer at Dalhousie and Harvard before serving overseas in the First World War. During the interwar years he was dean of engineering at the University of Saskatchewan. He became acting president of the NRC in 1939 and full-time head in 1944. After the war, he supervised Canada's atomic energy program.

A.G.L. McNaughton (1887–1966) was a McGill engineering graduate who achieved great distinction during the First World War. During the interwar years he served as Chief of the Defence Staff and president of the NRC, which he helped shape into a defence science organization. Between September 1939 and 1943 he was Canada's senior military officer in Britain. In 1944 McNaughton became Minister of Defence, and in 1946 he represented Canada on the United Nations Atomic Energy Commission.

Alan Nunn May came to the Montreal Laboratory in 1943 as part of the British nuclear team. In September 1945, shortly after his return to Britain he was revealed to be a Soviet spy. He was sentenced to a ten-year prison term by a British court in March 1946.

E.G.D. Murray (1890–1964) came to McGill via South Africa and Cambridge. An internationally recognized bacteriologist, he presided over Canada's biological warfare program between 1941 and 1945. After the war he returned to his McGill laboratory.

Wilder Penfield (1891–1976), a neurosurgeon and scientist, established the Montreal Neurological Institute in the early 1930s. His work on war medicine was highly regarded by Canada's allies.

Arnold Pitt was a Toronto physicist who directed Canada's proximity fuze program between 1941 and 1944. His research team were responsible for a number of important innovations that helped American scientists make this weapon operational by 1944. By this stage, Pitt had accepted an academic position in the United States.

Bruno Pontecorvo (b. 1913), a nuclear physicist, had worked with Enrico Fermi before coming to the Montreal laboratory in 1942. His defection to the Soviet Union in 1950 raised questions, which have yet to be answered, about his possible involvement in espionage.

Major General Maurice Pope (1889–1978) was another protégé of General McNaughton. His career in the Canadian Army assumed many different forms. His most important wartime assignment was to direct the activities of the Canadian military mission in Washington between 1942 and 1944.

Guilford Reed (1897–1955), of Queen's University, was one of Canada's foremost bacteriologists during the interwar years. During the war Reed carried out important experiments on the military potential of botulinum toxin both in his laboratory and at the weapon-testing facilities at Suffield, Alberta. He continued this research with the Defence Research Board after 1945.

Ernest Rutherford (1871–1937) is recognized as one of the outstanding scientists of the twentieth century for his work on radioactivity and his discovery of the atomic nucleus. His influence on Canadian science continued after he left McGill to become director of the Cavendish Laboratory at Cambridge.

Christian Sivertz (1897–1983), a McGill-trained chemist, taught at the University of Western Ontario between 1929 and 1962. During the war he worked on chemical warfare, and in 1946 he was elected president of the CAScW.

Louis Slotin (1911–46), a graduate of the University of Manitoba and the University of London, joined the Los Alamos atomic weapons centre in 1944. He died in May 1946 after being exposed to a fatal dose of radiation, a tragedy that caused much debate among scientists over laboratory safety standards.

Omond Solandt (1909–93) pursued his studies in physiology at Toronto and Cambridge. During the war he became one of Britain's leading experts in operational research. In 1946 he became director general of the newly created Defence Research Board.

E.W.R. Steacie (1900–62), a McGill-trained chemist, was director of the NRC's Chemistry Division at the outbreak of the Second World War. After 1942 he was closely associated with the Montreal Laboratory, and worked with John Cockcroft in developing the first experimental reactors.

John L. Synge (1897–1995) was a member of Toronto's applied mathematics department between 1922 and 1943. He was often critical of the NRC's reluctance to use émigré scientists. He too left Canada for an academic position before the end of the war.

Harry Thode (1910–97), a graduate of the universities of Saskatchewan and Chicago, joined McMaster University as a nuclear chemist in 1939. He was a key member of the Montreal Laboratory atomic team, although most of his wartime work was carried out at McMaster.

Henry Tizard (1885–1959) gained acclaim as a physicist and scientific adviser during the 1920s. In the interwar years he supported the development of radar. His famous mission to Washington in 1940 resulted in greater cooperation between British, American, and Canadian scientists. His involvement with Commonwealth defence science planning continued after the war.

Merle Tuve (1901–82), a prominent American physicist, directed the NDRC's proximity fuse program. By the end of the war this weapon had greatly improved anti-aircraft and artillery warfare.

Norman Veall was a British technician who worked at the Montreal Laboratory between 1943 and 1946. He was better known, however, as international secretary of the CAScW, and for his appearance before the Royal Commission on Espionage.

George Volkoff (b. 1914), a Canadian who received his nuclear physics training in the United States, returned to teach at the University of British Columbia, and then to work in the Montreal Laboratory until 1945. He remained associated with Canada's nuclear program in the postwar years.

George Wright (1904–76) was a chemist who came to Toronto in 1937. During the war he was a major force in Canada's explosives and propellants programs, continuing his involvement after 1945.

Leo Yaffe (1916–97) trained as a chemist with Otto Maass during the early 1940s. During the war, he worked on chemical weapons and at the Montreal Laboratory. After 1945 he was actively involved with nuclear research both at Chalk River and in his McGill laboratory.

Notes

Abbreviations

AVHP	A.V. Hill Papers
CAB	Cabinet Records (British)
DCW	Directorate of Chemical Warfare Records
DEA	Department of External Affairs
DHist	Directorate of History, National Defence
DND	Department of National Defence Records
FAS	Federation of American Scientists Papers
HQS	Headquarters DND Records
IWM	Imperial War Museum
JCS	Joint Chiefs of Staff Papers (U.S.)
JNW	Joint Committee on New Weapons Papers (U.S.)
MUA	McGill University Archives
MED	Manhattan Engineer District Records
NAC	National Archives of Canada
NAW	National Archives, Washington
NRC	National Research Council Papers
OSRD	Office of Research and Development Papers (U.S.)
PAM	Public Archives of Manitoba
PAO	Public Archives of Ontario
PCO	Privy Council Papers
PREM	Prime Minister Records (U.K.)
PRO	Public Records Office (U.K.)
RC	Royal Commission
UMA	University of Manitoba Archives
UTA	University of Toronto Archives

282 Notes to pages 3–5

UTRB   University of Toronto Rare Book Room
UWO   University of Western Ontario Archives
WO   War Office Records (U.K.)

Introduction

1 Library of Congress, Vannevar Bush Papers, Box 68, C.J. Mackenzie File, Bush to Mackenzie, 19 March 1947; ibid., Mackenzie to Bush, 24 March 1947.
2 National Archives of Canada (NAC), Department of National Defence Records (DND), file TS 711-270-16-1, Colonel R.E.S. Williamson to Colonel W.A.B. Anderson, director of intelligence, Canadian Army, 14 June 1946.
3 David Cassidy, 'Controlling German Science: U.S. and Allied Forces in Germany, 1945–1947,' *Historical Studies in the Physical and Biological Sciences*, vol. 24, pt 2 (1994), 197.
4 Solly Zuckerman, *Scientists and War: The Impact of Science on Military and Civil Affairs* (New York, 1967), 113.
5 Ibid., 115.
6 Larry Owens, 'The Counterproductive Management of Science in the Second World War: Vannevar Bush and the Office of Scientific Research and Development,' *Business History Review*, 68 (Winter 1994), 516.
7 Ibid., 519; See also Carol Gruber, *Mars and Minerva: World War I and the Uses of Higher Learning in America* (Baton Rouge, 1975).
8 Owens, 'The Counterproductive Management of Science,' 526.
9 Carroll Pursell, 'Science Agencies in World War Two: The OSRD and Its Challengers,' in Nathan Reingold, ed., *The Sciences in the American Context: New Perspectives* (Washington, 1979), 359–78; Peter Galison, 'Physics between War and Peace,' and Paul Hoch, 'The Crystallization of a Strategic Alliance: The American Physics Elite and the Military in the 1940s,' in Everett Mendelsohn et al., eds, *Science, Technology and the Military*, 2 vols (Boston, 1988), 87–117; Gregory Hooks, *Forging the Military-Industrial Complex: World War II's Battle of the Potomac* (Urbana, IL, 1991); Roger Geiger, *Research and Relevant Knowledge: American Research Universities since World War II* (New York, 1993); Stuart Leslie, *The Cold War and American Science* (New York, 1993).
10 Carol Gruber, 'The Overhead System in Government-Sponsored Academic Science: Origins and Early Development,' *Historical Studies in the Physical and Biological Sciences*, vol. 25, pt 2 (1995), 242.
11 Donald Phillipson, 'International Research Co-ordination in Wartime: The Anglo-American Alliance, 1939–1945' (unpublished paper, 1985); Guy Hart-

Notes to pages 5–8 283

cup, *The Challenge of War: Britain's Scientific and Engineering Contributions to World War Two* (New York, 1970), 17–53; R.V. Jones, *Most Secret War: British Scientific Intelligence, 1939–1945* (London, 1978), 1–107.
12 Ronald Clark, *Tizard* (London, 1965), 230–85; Hartcup, *The Challenge of War*; R.V. Jones, *Most Secret War*, 1–107.
13 James Phinney Baxter III, *Scientists against Time* (Boston, 1948), 17–19, 127.
14 Baxter, *Scientists*, 31–2; Irwin Stewart, *Organizing Scientific Research for War: The Administrative History of the Office of Scientific Research and Development* (New York, 1948), 158–9.
15 J.L. Granatstein, *The Generals: The Canadian Army's Senior Commanders in the Second World War* (Toronto, 1993), 57. See also Stephen Harris, *Canadian Brass: The Making of a Professional Army, 1860–1939* (Toronto, 1988), 152–62.
16 M. Christine King, *E.W.R. Steacie and Science in Canada* (Toronto, 1989), 57.
17 Mackenzie was born in St Stephen, New Brunswick, on 10 July 1888; he died in Ottawa 26 February 1984.
18 Maass was born in 1890 and died in 1961.
19 Banting was born 14 November 1891 in Allison, Ontario, the youngest of six children. He died 21 February 1941.
20 Michael Bliss, *Banting: A Biography* (Toronto, 1984), 38–40.
21 The importance of science and military technology during the First World War has been effectively analysed in many recent books. See Williamson Murray and Allan Millett, eds, *Military Innovation in the Interwar Period* (Cambridge, MA, 1996); Timothy Travers, *How the War Was Won: Command and Technology in the British Army on the Western Front, 1917–1918* (London, 1992).
22 Bliss, *Banting*, 42.
23 E.A. Flood, 'Otto Maass, 1890–1961,' *Biographical Memoirs of Fellows of the Royal Society*, vol. 8 (1962), 183–204.
24 C.J. Mackenzie taped interview, 1967; transcript NRC Library.
25 Freda Fraser, Mary Ross, Edith Taylor, Helen Plummer, Jessie Ridout, and Ruth Partridge were original faculty members of the University of Toronto's School of Hygiene. Paul Bator with Andrew Rhodes, *Within Reach of Everyone: A History of the University of Toronto School of Hygiene and the Connaught Laboratories*, vol. 1: *1927 to 1955* (Ottawa, 1990), 79–81.
26 During the interwar years the NRC Chemistry Division employed two women chemists: Helen Cathaway (1930–7), a McGill graduate, and Audrey Tweedie (1930–8), who had an M.Sc. in physics from the University of Manitoba. In terms of NRC grants, women scientists received 18 per cent of the total in 1920–1, but only about 10 per cent in 1930–1. Marianne Gosztonyi Ainley and Catherine Millar, 'A Select Few: Women and the National Research Council of Canada, 1916–1991,' in Richard Jarrell and Yves Gin-

gras, eds, *Building Canadian Sciencee: The Role of the National Research Council* (Ottawa, 1992), 71–87.
27 Surveys of Canada's war effort include David Bercuson, *Maple Leaf against the Axis* (Toronto, 1995); W.A.B. Douglas, *Out of the Shadows: Canada in the Second World War* (Toronto, 1977); and C.P. Stacey, *Official History of the Canadian Army in the Second World War,* vol. 1: *Six Years of War: The Army in Canada, Britain, and the Pacific* (Ottawa, 1966).
28 Stacey also presents a brief but useful commentary on the evolution of Operational Research within the three Services and on the activity of the Army Technical Development Board, which recorded upwards of 450 different projects. In addition, Stacey refers to the Anglo-Canadian Montreal atomic laboratory. C.P. Stacey, *Arms, Men, and Governments: The War Policies of Canada, 1939–1945* (Ottawa, 1970), 528.
29 Ibid., 512.
30 John Swettenham, *McNaughton,* 3 vols (Toronto, 1969).
31 General Maurice Pope, *Soldiers and Politicians: The Memoirs of Lt. Gen. Maurice A. Pope* (Toronto, 1962).
32 Granatstein, *The Generals.*
33 J.L. Granatstein, *How Britain's Weakness Forced Canada into the Arms of the United States* (Toronto, 1989), 4. See also John English, *Shadow of Heaven: The Life of Lester Pearson,* vol. 1: *1897–1948* (Toronto, 1989), and Hector Mackenzie, 'Finance and Functionalism: Canada's War Effort as Justification for Its Participation in Wartime and Post-War Organizations,' a paper presented at The International Second Quebec Conference: A 50th Anniversary Commemoration (Quebec City, October 1994).
34 See the review article by Alex Roland, 'Technology and War: The Historiographical Revolution in the 1980s,' *Technology and Culture* 2 (1993), 117–35.
35 Cited in J.W. Grove, *In Defence of Science: Science, Technology and Politics in Modern Society* (Toronto, 1989), 70–1.
36 Spencer Klaw, *The New Brahmins* (New York, 1968), 15.
37 Grove, *In Defence of Science,* 25–47.
38 Cited in Wilfrid Eggleston, *National Research in Canada: The NRC, 1916–1966* (Toronto, 1978), 186
39 Canadian backwardness in aeronautical design would dramatically change after 1945, when Canadian aircraft companies and the RCAF 'embraced high-speed aeronautics in the hopes of gaining a place in the forefront of technological progress.' Julius Lukasiewicz, 'Canada's Encounter with High-Speed Aeronautics,' *Technology and Culture* (1986), 223.
40 D.J. Goodspeed, *DRB: A History of the Defence Research Board of Canada* (Ottawa, 1958), 165.

41 The work of the Examination Unit was greatly enhanced by the May 1943 agreement between Britain and the United States that arranged for the full exchange of intelligence data and scientists. This Anglo-American-Canadian communications and intelligence cooperation was reaffirmed in September 1945. Bradley Smith, *The Ultra-Magic Deals: The Most Secret Special Relationship, 1940–1946* (Novato, CA, 1993), 205–30; Wesley Wark, 'Cryptographic Innocence: The Origins of Signals Intelligence in Canada in the Second World War,' *Journals of Contemporary History* 22 (1987), 642–58; John Bryden, *Best-Kept Secret: Canadian Secret Intelligence from the Second World War to the Cold War* (Toronto, 1993), 120–210.

42 For an overview of Canadian military medicine, see W.R. Feasby, *Official History of the Canadian Medical Services, 1939–1945*, 2 vols. (Ottawa, 1953), and Terry Copp and Bill McAndrew, *Battle Exhaustion: Soldiers and Psychiatrists in the Canadian Army, 1939–1945* (Montreal/Kingston, 1990).

1 Canada's Defence Scientists: Organizing for War, 1938–1940

1 Annual Report of the President, Saskatchewan University, 1916–17, cited in Yves Gingras, 'The Institutionalization of Scientific Research in Canadian Universities: The Case of Physics,' *Canadian Historical Review* 67(2) (June 1986), 191.

2 This was part of Rutherford's letter to *Nature* (9 September 1915) lamenting the death of his brilliant colleague H.G.J. Mosely. Mark Oliphant, *Rutherford, Recollections of the Cambridge Days* (London, 1972), 60.

3 Kristie Macrakis, *Surviving the Swastika: Scientific Research in Nazi Germany* (New York, 1993), 24–7.

4 L.F. Haber, *The Poisonous Cloud: Chemical Warfare in the First World War* (Oxford, 1986), 107.

5 Estimates of gas casualties during the First World War range from 1,009,038 to 1,297,000. According to one American study, gas caused 26.8 per cent of the casualties in the U.S. Army, but only 11.3 per cent of the deaths. The British Army had a mortality rate of 4.3 percent, the French Army 4.2, and the German Army 4.5. The poorly equipped Russian Army suffered more casualties (475,000) and fatalities (56,000). Edward Spiers, *Chemical Warfare* (London, 1986), 13–33; Haber, *Poisonous Cloud*, 39–137.

6 Haber, *Poisonous Cloud*, 305–6.

7 The United States government did not, however, ratify the Protocol; nor did Japan. Robert Harris and Jeremy Paxman, *A Higher Form of Killing: The Secret Story of Gas and Germ Warfare* (London, 1982), 46–7.

8 McGill physicist J.A. Gray worked with British artillery sound-finding teams, while his colleague Robert W. Boyle was involved with submarine detection work. University of Toronto physicist J.C. McLennan was seconded by the Royal Navy in 1916 as a scientific adviser. Yves Gingras, *Physics and the Rise of Scientific Research in Canada* (Montreal, Kingston 1991), 71.
9 The Dominion Bureau of Statistics listed over 10,000 industrial firms in 1917.
10 Eggleston, *National Research*, 8.
11 Michael Pattison, 'Scientists, Inventors and the Military in Britain, 1915–19: The Munitions Inventions Department,' *Social Studies of Science* 13 (1983), 521–68.
12 Eggleston, *National Research*, 12, 34, 53–8.
13 The NRC budget also increased from a low of $50,000 in 1917 to $140,000 after 1923. Ibid., 34.
14 Stacey, *Arms, Men and Governments*, 3.
15 Ibid., 3–4; James Eayrs, *In Defence of Canada*, vol. 2: *Appeasement and Rearmament* (Toronto, 1981), 190–9; Hugh Francis Pullen, 'The Royal Canadian Navy between the Wars,' in James Boutilier, ed., *The RCN in Retrospect, 1910–1968* (Vancouver, 1982), 70–5.
16 Stacey, *Arms, Men and Governments*, 3–4; W.A.B. Douglas, *The Official History of the Royal Canadian Air Force*, vol. 2: *The Creation of a National Air Force* (Toronto, 1986), 125–35.
17 C.P. Stacey, *Canada and the Age of Conflict: A History of Canadian External Policies*, vol. 2: *1921–1945* (Toronto, 1981), 175–99.
18 Douglas, *Air Force*, 193–247.
19 Bennett's decision was based on two primary motives: McNaughton had become a political liability at National Defence, and Bennett wanted to replace H.H. Tory.
20 John Swettenham, *McNaughton*, vol. 1, 321.
21 Eggleston, *National Research*, 98–115.
22 Cited in Swettenham, *McNaughton*, vol. 1, 326.
23 Swettenham, *McNaughton*, vol. 1, 321–8.
24 National Archives of Canada (NAC), Papers of J.T. Henderson, vol. 4, file 4, Henderson to McNaughton, 9 April 1938.
25 The Royal Canadian Navy had not forgiven McNaughton for his 1933 proposal, when the Treasury Board had slashed defence spending by over $3.6 million, that 'since the RCN was too small for any useful role, it should be scrapped in order to preserve the Royal Canadian Air Force and the army.' David Zimmerman, *The Great Naval Battle of Ottawa* (Toronto, 1989), 7.
26 Granatstein, *The Generals*, 8–9.

27 Ibid., 33.
28 Ibid., 16–17.
29 Zimmerman, *Naval Battle*, 10, 24.
30 NAC, General H.D.G. Crerar Papers, vol. 10, Crerar to Pope, 14 May 1936.
31 Ibid., Crerar to Torr, 13 January 1936, 9 September 1936.
32 Ibid. Crerar to Torr, 13 April 1937; Report of Visit to Washington, 18–21 October 1937.
33 Ibid., Coulter to Crerar, 24 June 1938.
34 Ibid., Crerar to Major General R.H. Haining, Director of Military Operations and Intelligence, The War Office, 12 March 1937.
35 The Air Ministry's Committee for the Scientific Study of Air Defence (CSSAD) had been created in 1934. It was succeeded by the Committee for the Scientific Survey of Air Offence and the Air Defence Research Sub-Committee of the Committee of Imperial Defence.
36 Clark, *Tizard*, 122–38.
37 NAC, Papers of General A.G.L. McNaughton, Vol. 116, George Drew File: *Financial Post*, 26 March 1938; McNaughton to Drew, 8 April 1938. Drew was elected leader of the Ontario Conservative Party later in 1938.
38 Ibid., Drew to McNaughton, 7 May 1938; ibid., McNaughton to Drew, 16 May 1938. Because of the political uproar, the King government was forced to appoint a royal commission under the Hon. Henry Hague Davis, a judge of the Supreme Court of Canada, to investigate Drew's charges of wrongdoing. Although Davis's subsequent report exonerated the government, federal authorities soon established a Defence Purchasing Board that set forth an elaborate system of how companies could bid for military contracts, and limited their profit to 5 per cent per annum of the average amount of capital invested. Stacey, *Arms, Men and Governments*, 101–2.
39 NAC, Records of the National Research Council (NRC), 85/86, vol. 10, file 4C-9-21, McNaughton to Lieutenant-Colonel LaFleche, 25 September 1935.
40 Only HS mustard gas had been used in the First World War; HT mustard gas, developed during the interwar years, was more lethal and more stable for use. National Security Archives (NSA) Washington, CW files. CAB 4/24, YN 05546, Report, 8 July 1936; WO 32/3663, War Office Report, 7 August 1936.
41 NAC, NRC, 87/88, 104, vol. 69, file 32-1-12, #1, McNaughton to Major General E.C. Ashton, Chief of the General Staff, 23 December 1937.
42 John Bryden, *Deadly Allies: Canada's Secret War, 1937–1947* (Toronto, 1989), 21; G.B. Carter, *Porton Down: 75 Years of Chemical and Biological Research* (London, 1992), 39.
43 Bliss, *Banting*, 262–4.

44 Eggleston, *National Research*, 145–52.
45 Gingras, *Physics*, 71–2.
46 A.B. McKillop, *Matters of Mind: The University in Ontario, 1791–1951* (Toronto, 1994); Frederick Gibson, *To Serve and Yet to Be Free: Queen's University*, vol. 2: *1917–1961* (Kingston/Montreal, 1983).
47 John K. Smith, 'World War II and the Transformation of the American Chemical Industry,' in E. Mendelsohn et al., eds, *Science, Technology and the Military*, vol. 12 (Boston, 1988), 307–21; Roger Geiger, *To Advance Knowledge: The Growth of American Research Universities, 1900–1940* (New York, 1986).
48 *Bibliobiography of Publishing Scientists in Ontario between 1914 and 1939*, ed. Philip Enros (Thornhill, 1985), 297.
49 Ibid., 400.
50 One of Parker's most outstanding students was (Sir) Charles Frederick Goodeve, who later became one of Britain's leading defence scientists during the Second World War. F.D. Richardson, 'Charles Frederick Goodeve, 21 February 1904–7 April 1980,' *Biographical Memoirs of Fellows of the Royal Society of London* (London, 1981), 307–49.
51 Spinks, who joined the department in 1930, was instrumental in recruiting Herzberg, who had been dismissed from his position in Nazi Germany despite his reputation as a brilliant molecular physicist. John Spinks, *Two Blades of Grass: An Autobiography* (Saskatoon, 1980), 15–31.
52 *Publishing Scientists*, 21, 25.
53 *World Who's Who in Science* (Chicago, 1968), 1827.
54 McGill University Archives (MUA), Otto Maass Papers, 'Presidential Address to the Canadian Institute of Chemistry, 5 May 1939, London, Ontario,' 6. Bain was honoured at the 1939 meetings.
55 Hermann Fischer obtained his Ph.D. in chemistry in Jena in 1922; he came to Toronto in 1937. Baer came the same year.
56 Flood, 'Otto Maass,' 183–204.
57 Ibid., 194.
58 King, *Steacie*, 8; MUA, Otto Maass Papers, T.W. Richards to Maass, 10 October 1919.
59 The first eight NRC presidents were: A.B. McCallum (1916– 21); R.F. Ruttan (1921); R.A. Ross (1921–2); F.D. Adams (1922–3); H.M. Tory (1923–35); A.G.L. McNaughton (1935–44); C.J. Mackenzie (Acting Pres. 1939–44; Pres. 1944–52); E.W.R. Steacie (1952–62). Eggleston, *National Research*, 458.
60 In March 1939 Maass turned down a position as the NRC's director of the Chemistry Division. Instead, the job went to his close friend and colleague E.W.R. Steacie. Stanley Frost, *McGill University*, vol. 2: *1895–1971* (Montreal/Kingston, 1984), 180.

61 In 1939 the CIC membership was 911. One of its former presidents was Monsignor Vachon, Chair of Chemistry at Laval.
62 MUA, Otto Maass Papers, 'Presidential Address to the Canadian Institute of Chemistry, May 5, 1939, at London, Ontario.'
63 Daniel Kevles, *The Physicists: The History of a Scientific Community in Modern America* (New York, 1979); Henry Etzkowitz, 'The Making of an Entrepreneurial University: The Traffic among MIT, Industry, and the Military, 1860–1960,' in E. Mendelsohn et al., *Science, Technology, and the Military*, vol. 12, 515–39; Paul Hoch, 'The Crystallization of a Strategic Alliance,' 87–115.
64 McGill attracted a high percentage of its M.Sc. graduates (40%) into its Ph.D. program; in contrast, only 23 percent of physics doctoral students at the University of Toronto had degrees from that institution. Gingras, *Physics*, 71.
65 McLennan supervised 25 out of the 27 Ph.D. students in physics at Toronto before his retirement in 1932; he was awarded a total of $25,000 in NRC grants – more than the entire McGill physics department. Ibid., 72.
66 *Publishing Scientists*, 72, 75, 140, 396.
67 By 1939 Toronto's physics department had four professors – E.F. Burton, J. Satterly, L. Gilchrist, H.A. McTaggart – and seven associate-assistant professors. *University of Toronto Calendar for 1939–40*, 25.
68 Leopold Infeld, *Why I Left Canada: Reflections on Science and Politics* (Montreal, 1978), 22–68.
69 With the creation of the endowed Chown Chair in 1919, Queen's was able to attract A.L. Hughes as research professor in physics, and then J.A. Gray in 1923. Gingras, *Physics*, 45, 76.
70 In 1910 H.L. Bronson (Yale Ph.D., 1904) became head of physics at Dalhousie; he had previously been Rutherford's collaborator at McGill. One of his outstanding students, George Hugh Henderson, joined the department in 1923. Ibid., 77.
71 The patron of the institute, its equipment, and professorships was William Macdonald, the millionaire founder of Imperial Tobacco. J.L. Heilbron, 'Physics at McGill in Rutherford's Time,' in Mario Bunge and William Shea, eds, *Rutherford and Physics at the Turn of the Century* (New York, 1979), 42–73.
72 B.W. Sargent, 'Recollections of the Cavendish Laboratory directed by Rutherford,' mimeo (Kingston, ON: Queen's University Nuclear Physics Laboratory n.d.), 1.
73 Heilbron, 'Physics at McGill,' 57–61.
74 Queen's Archives, J.A. Gray Papers, Rutherford to Gray, 8 May 1909.
75 A.L. Clark, *The First Fifty Years: A History of the Science Faculty of Queen's University, 1893–1943* (Kingston, 1943), 68.

76 Gray Papers, Bragg to Gray, 12 March 1924; Gibson, *Queen's University*, 121–5.
77 James Chadwick won his prize in 1935 for his discovery of the neutron; Patrick Blackett received his award in 1948, and John Cockcroft in 1951.
78 Gray Papers, Rose to Gray, 28 October 1927.
79 A.S. Eve, *Rutherford: Being the Life and Letters of the Rt. Hon. Lord Rutherford, O.M.* (Cambridge, 1939), 283; W.E.K. Middleton, *Physics at the National Research Council of Canada, 1929–1952* (Waterloo, ON, 1979), 8–9.
80 Lawrence obtained two degrees at Dalhousie before going to Cambridge. Rose did two degrees at Queen's under J.A. Gray before he went to Cavendish. John T. Henderson, a McGill student of A.S. Eve and J.S. Foster, worked on atmospherics and radio under Edward Appleton. Middleton, *Physics*, 52, 57, 61.
81 McGill University Archives, Sir Arthur Currie Papers, Box 63, file 1110, Currie to Rutherford, 28 December 1932; Rutherford to Currie, 14 January 1932.
82 Jerry Thomas, 'John Stuart Foster, McGill University, and the Renascence of Nuclear Physics in Montreal, 1935–1950,' in *Historical Studies in the Physical Sciences*, vol. 14, pt 2 (1984), 360.
83 Foster and Lawrence had studied together at Yale University; they were also close friends. J.L. Heilbron and Robert W. Seidel, *Lawrence and His Laboratory: A History of the Lawrence Berkeley Laboratory*, vol. 1 (Berkeley, CA, 1989); B.W. Sargent, 'Nuclear Physics in Canada in the 1930s,' mimeograph (Kingston, ON: Queen's University, Nuclear Physics Laboratory, n.d.), 22.
84 Gingras, *Physics*, 74.
85 Thomas, 'Foster,' 363–6.
86 Interview with Harry Thode, 6 May 1982, Hamilton, Ontario.
87 In 1942 Thode was joined at McMaster by Harry Duckworth, who had also obtained his degree in atomic physics at the University of Chicago.
88 Zinn decided to accept a position at the City College, New York City, rather than take a job with the NRC. Gray Papers, Zinn to Gray, 23 March 1931; ibid., B.W. Boyle to Gray, 8 April 1931.
89 Interview with George Volkoff, 8 June 1938.
90 *World Who's Who in Science*, 783; W.B. Lewis, 'George Hugh Henderson,' *Royal Society, Obituary Notices* (London, 1950), 155–66.
91 The Canadian contingent at Cavendish included David Keys and A.S. Eve of McGill; Robert Boyle, Donald Rose, and George Lawrence of the NRC; John Gray, H.M. Cave, and B.W. Sargent of Queen's; and George Hugh Henderson of Dalhousie. Sargent, 'Recollections,' 1.
92 S.E.D. Shortt, ed., *Medicine in Canadian Society: Historical Perspectives* (Montreal/Kingston, 1981); Murray Barr, *A Century of Medicine at Western* (Lon-

don, ON, 1977); R.D. Defries, *The First Forty Years: Connaught Medical Research Laboratories* (Toronto, 1968).
93 Bliss, *Banting*, 126.
94 Ibid., 213.
95 Richard Kapp, 'Charles Best, the Canadian Red Cross Society, and Canada's First National Blood Donation Program,' *Canadian Bulletin of Medical History*, 12 (1) (1995), 28–42.
96 Bator and Rhodes, *Within Reach of Everyone*, 32–121.
97 *Publishing Scientists*, 406–7.
98 Wilder Penfield, *No Man Alone: A Neurosurgeon's Life* (Toronto, 1977); Jefferson Lewis, *Something Hidden: A Biography of Wilder Penfield* (Toronto, 1981).
99 Interview with R.G.E. Murray, London, Ontario, 25 January 1994.
100 Terrie Romano, 'The Associate Committee on Medical Research and the Second World War,' in Richard Jarrell and Yves Gingras, eds, *Building Canadian Science: The Role of the Natiional Research Council* (Ottawa, 1992), 71–87.
101 NRC, 88/89, vol. 5, 4-M-4-7, pt. 1, Report McNaughton to Council, 14 March 1938; ibid., ACMR organizational report, 19 April 1938.
102 National Research Council Archives (Ottawa), Frederick Banting Papers, Box 3, Dr. C.B. Stewart, Assistant Secretary ACMR to Dr. L.S. Klinck, president of the University of British Columbia, 10 November 1938.
103 Peter Kuznick, *Beyond the Laboratory: Scientists as Political Activists in 1930s America* (Chicago, 1987), 18–19.
104 'The A.S.W. – Twenty Years History,' *The Scientific Worker* (Autumn 1939), 68. See also Roy and Kay MacLeod, 'The Contradictions of Professionalism: Scientists, Trade Unionism and the First World War,' *Social Studies of Science* 9 (1979), 1–32.
105 Gary Werskey, *The Visible College: The Collective Biography of British Scientists and Socialists of the 1930s* (New York, 1978), 166, 215. Maurice Goldsmith, *Sage: A Life of J.D. Bernal* (London, 1980), 63.
106 Sir Bernard Lovell, 'Patrick Maynard Stuart Blackett, Baron Blackett of Chelsea,' *Biographical Memoirs of the Royal Society* (London, 1975), 1–107.
107 Werskey, *Visible College*, 44–60, 67–75, 82–170.
108 Bukharin was the leading theoretician of the Communist Party in Russia after the death of Lenin; he was also co-leader of the party with Stalin from 1925 to 1928. Goldsmith, *Sage*, 54–5.
109 Andrew Sinclair, *The Red and the Blue: Intelligence, Treason and Universities* (London, 1986), 51.
110 J.D. Bernal, *The Social Function of Science* (London, 1939), 412.
111 Kuznick, *Beyond the Laboratory*, 151.
112 Of the 485 papers presented, 170 were by Russian scientists. Ibid., 156–9.

113 University of Toronto Rare Book Room, Sir Frederick Banting Diary, 12 June–21 July 1935: 26 June 1935.
114 Ibid., 28 July, 7 August 1935.
115 Bliss, *Banting*, 228.
116 Roderick Stewart, *Bethune* (Markham, ON, 1980), 1–71.
117 Eugene Forsey and King Gordon, both founding members of the Canadian Commonwealth Federation (CCF), were examples of other Canadian intellectuals who were impressed by 'the pulsating life of the New Russia.' NAC, Eugene Forsey Papers, letter to his mother, 7 July 1932.
118 G.B. Reed, 'Socialized Medicine,' *Queen's Quarterly* (Autumn 1935), 295–308.
119 The CLPD was a branch of the World Committee for the Struggle against War, formed in October 1934 by a combination of communists and prominent civil libertarians. NAC, Frank and Libbie Park Papers, vol. 20, file 297, 'How to Organize a Local Branch of the CLPD,' 2 April 1938.
120 The story of the Mac-Paps and Norman Bethune have been told elsewhere. See Victor Hoar, *The Mackenzie–Papineau Battalion* (Toronto, 1969), 105–24; Stewart, *Bethune*, 85–114; Dorothy Livesay, *Right Hand, Left Hand* (Erin, ON, 1977), 242–68.
121 Under this act, passed by the House of Commons on 19 March 1937, it became a criminal offence for a Canadian citizen to enlist in the military service of a foreign power. Initially, it was used primarily to exclude members of the Mackenzie–Papineau Battalion from re-entering the country.
122 Edward came to McGill in 1934, and obtained his Ph.D. in 1941. His thesis supervisor was Raymond Boyer. Interview with Jack Edward, McGill University, 8 March 1994.
123 *Old McGill, 1925–1929*, 175.
124 Interview Jack Edward.
125 Interview with Gordon Lunan, 31 August 1990.
126 Interview Jack Edward.
127 *The Scientific Worker* (June 1936), 1, 54.
128 Until the Spanish Civil War the BAScW was inclined toward a pacifist outlook, although this had already begun to change by the time of the Brussels World Peace Conference of September 1936. *The Scientific Worker* (April 1937), 150–2.
129 Ibid., February 1939, 14–18.
130 Ibid., December 1938, 26–35.
131 United States, 83d Congress, House Committee on Un-American Activities, 1953, 'Communist Methods of Infiltration (Education).'
132 Kuznick, *Beyond the Laboratory*, 243.
133 Ibid., 232, 243, 258.

134 *Nature* (22 January 1938), 169.
135 *Nature* (13 August 1938).
136 Previous AAAS meetings in Canada were held at Montreal, 1857; Montreal, 1882; Toronto, 1899; Toronto, 1921. In 1938 AAAS membership was 19,307. *Science* (29 April 1938), 385; ibid., 20 May 1938, 445.
137 *Science* (29 July 1938), 87.
138 Participating Canadian scientists included chemists J.W.N. Spinks (Saskatchewan), W. Lash Miller (Toronto), and Harry Thode (McMaster); engineers T.R. Loundon (Toronto) and I.I. Sikorsky; physicists J.T. Henderson (NRC) and J.S. Foster (McGill); bacteriologists G.B. Reed (Queen's) and J.H. Craigie (Toronto); and psychologists Edward A. Bott. Ibid.
139 More of Canada's scientific 'stars' held forth at the June meetings of the Royal Society, including nine papers by Otto Maass and his McGill associates. Ibid., 24 June 1938, 580.
140 Lathe's paper was entitled 'World Natural Resources,' while Wallace spoke on 'The Changing Values of Science.' R.C. Wallace, a British-trained geologist, had become principal of Queen's in 1936 after serving for seven years as president of the University of Alberta. Gibson, *Queen's University*, vol. 2, 134–50.
141 'Chemistry and the Future,' *Science* (12 August 1938), 133–7.
142 Ibid., 'Physics and the Future' (5 August 1938), 115–21.
143 Kevles, *Physicists*, 287–301.
144 The AASW would itself experience a serious division over the issue of neutrality in June 1940. *New York Times* (15, 20, 30 June 1940).
145 Cited in Eggleston, *National Research*, 186

2 Building the Defence Science Alliance, 1940–1943

1 R.D. Cuff and J.L. Granatstein, *Ties That Bind: Canadian–American Relations in Wartime from the Great War to the Cold War* (Toronto, 1977), 101.
2 Colonel Stanley Dziuban, *Military Relations between the United States and Canada, 1939–1945* (Washington, 1959), 18–47.
3 Mel Thistle, *The Mackenzie–McNaughton Wartime Letters* (Toronto, 1975), 151.
4 In March 1941 Mackenzie briefly explored the possibility of making the NRC the wartime research branch of the Ministry of Munitions and Supply on the grounds that the NRC research activity 'was almost 100% war,' and that such an arrangement would solve its financial problems. Eggleston, *National Research*, 183–4.
5 The order-in council of 22 March 1943 stipulated that the NRC 'be classified as a war unit for priority in the purchase of equipment and supplies necessary for its war work.' Ibid.

6 Ibid., 109–22.
7 McNaughton did insist, however, that all such students participate in Canadian Officer Training Corps programs at their universities. NRC, vol. 46, file 17-15-9-2, #1, McNaughton to Dr. H.J. Cody, 16 September 1939.
8 University of Manitoba Archives (UMA), Box 57, file 23, Wallace to Sidney Smith, 18 October 1940. NAC, C.J. Mackenzie Papers, vol. 1, Mackenzie Diary, 30 October 1939.
9 Eggleston, *National Research*, 122.
10 By 1939 the NRC provided $30,000 for scholarships and another $150,000 for extramural research grants. Ibid., 119–25.
11 UMA, President's Papers, Box 57, file 5, Mackenzie to Sidney Smith, 23 June 1943.
12 Maass's wartime role as special assistant to Mackenzie did create some controversy among McGill's faculty. McGill University Archives, Cyril James Papers, vol. 93, James to Dean O'Neill, 5 September 1941; NRC, vol. 46, 17-15-9-2, #1, Maass to Mackenzie, 23 October 1943.
13 NRC, vol. 46, 17–15–9–2, McNaughton to Martin, 24 October 1939.
14 Forsey suggested that his talents could best be used 'in helping in the preservation and extension of democracy and liberty within Canada.' James Papers, Box 40, 170, Forsey to Keys, 19 September 1939.
15 Ibid., Tabulation of responses from McGill faculty.
16 Ibid., Hibbert to Keys, 18 September 1939.
17 Ibid., Maass to Martin, 10 October 1939.
18 Ibid., Foster to Keys, 10 October 1939.
19 Margaret Gowing, *Britain and Atomic Energy, 1939–1945* (London, 1964), 27–44; Mark Walker, *National Socialism and the Quest for Nuclear Power, 1939–1949* (Cambridge, 1989), 13–45.
20 James Papers, Box 39, file 167, Keys to Burton, 8 December 1939.
21 University of Toronto Archives (UTA), H.J. Cody Papers, vol. 45, E.A. Allcut, Mechanical Engineering, to Cody, 22 June 1940.
22 Ibid., Box 40, file 167, Bott to Martin, 13 January 1940.
23 Ibid., vol. 45, Report, 15 July 1940.
24 Public Archives of Ontario (PAO), H.J. Cody Papers, file 7, A.R. Gordon to Cody, 5 October 1940.
25 Forty-four corporations contributed, with Bronfman's Seagram Corporation, Eaton's, International Nickel Company, the Canadian Pacific Railway Company, and Consolidated Mining and Smelting Company providing 78 per cent of the total ($1,050,000). Most of the other corporate donors were mining companies. NRC Archives, Banting Papers, vol. 1, F.T. Rosser, The History of the Sir Frederick Banting Fund, 1 October 1960, 1–4.

26 Ibid., 6-8, 9. The sequence of events that resulted in the establishment of the WTSDC has been extensively documented in Eggleston, *National Research*, 162–4.
27 NRC, 88-89/046, vol. 48, file 4-59-14, S.P. Eagleson, Secretary WTSCD, to B.G. McIntyre, Department of Finance, 1 October 1940.
28 Mackenzie funded the secret biological warfare project (M-1000) out of this fund until National Defence assumed responsibility for this work in August 1942.
29 Rosser, *History of the Banting Fund*, 6–10.
30 UMA, President's Papers, Box 57, file 5, Mackenzie to the Presidents of Canadian Universities, 20 August 1940.
31 Ibid., University Comptroller to Unemployment Insurance Commission, 27 October 1943.
32 UTA, Cody Papers, vol. 46, King to Cody, 24 July 1940.
33 NRC, vol. 17, file 32-1-13, Mackenzie to Captain G.M. Hibbard, Chief of Naval Equipment and Supply, RCN, 15 May 1942.
34 NRC, vol. 17, file 32-1-13, Inventions Board, Report for April 1942.
35 Cook found this latter suggestion similar to the joke 'that Mussolini had designed a new type of truck, with only one speed forward and three speeds in reverse.' NRC, vol. 17, file 32-1-13, Cook to DesRosiers, 11 November 1942.
36 Ibid., Hibbard to Mackenzie 23 May 1942.
37 Most qualified francophone candidates already had good jobs and did not want to move to Ottawa. Ibid., S.J. Cook to Dean Armand Circe, École Polytechnique, University of Montreal, 2 January 1941; ibid., Circe to S.J. Cook, 15 January 1941.
38 Luc Chartrand, Raymond Duchesne, and Yves Gingras, *Histoire des sciences au Québec* (Montreal, 1987), 401–11.
39 One notable exception was Franco Rasetti, an émigré nuclear physicist who refused to participate in war work because of his pacifist convictions. The University of Montreal profited from the arrival of two French refugee atomic scientists, Henri Laugier and Louis Rapkine. Ibid.; Tizard Papers, 58, A.V. Hill to Brigadier Charles Lindemann, 28 February 1941.
40 'La participation des Canadiens français à la 2e Guerre mondiale: mythes et réalités," *Bulletin d'histoire politique* 3 (3–4) (1995), 230–8.
41 National Archives, Washington (NAW), State Department Records (842.20/150), 'Memorandum: Decisions Required if Military Assistance Is to Be Afforded to Canada in the Immediate Future,' 5 July 1940.
42 Ibid., Moffat to The Honourable Sumner Welles, Acting Secretary of State, 14 August 1940.

296   Notes to pages 51–4

43 NAC, Brooke Claxton Papers, vol. 96, 'Memorandum: Canada–US Defence Cooperation,' n.d.
44 NAW, State, Pierrepont Moffat to Secretary of State, 30 August 1940.
45 Ibid., Moffat to Sumner, 30 August 1940.
46 Moffat had quickly assured his superiors that Crerar's point of view did not find 'any favor' with the King government. Ibid., Moffat report, 12 October 1940, Welles to Moffat 23 October 1940.
47 Ibid., Welles to Secretary of State, 28 February 1941.
48 Ibid., Moffat to State, 25 April 1941.
49 David Zimmerman, *Top Secret Exchange: The Tizard Mission and the Scientific War* (Montreal/Kingston, 1996), 65–102. Although this study is generally in agreement with Zimmerman's overall thesis about Anglo-Canadian-American defence science arrangements, there are three important differences in emphasis. *The Science of War* pays much more attention to the important liaison role assumed by Brigadier Charles Lindemann during the spring of 1940; it regards the Conant Mission of February 1941 as a vital stage in Anglo-American cooperation; and it provides a more extensive account of the activities of the British Central Scientific Office.
50 William McGucken, *Scientists, Society, and State: The Social Relations of Science Movement in Great Britain, 1931–1947* (Columbus, OH, 1984), 161.
51 Hill's official position was a supernumerary Air Attaché technically on loan from the Royal Society. Imperial War Museum, Sir Henry Tizard Papers (HTP), file 58, Hill to Tizard, 4 December 1939; ibid., Tizard memo, 5 January 1940.
52 Churchill College Archives, A.V. Hill Papers (AVHP), 2/7, Millikan to Hill, 27 May 1940; ibid., Cannon to Hill, 29 May 1940.
53 HTP, 58, Hill Report, 18 June 1940.
54 HTP, 251, cable Marquess of Lothian, 23 April 1940.
55 Tizard also asked for specific information about atomic bomb research, while still admitting that he remained 'completely sceptical about its possible military uses.' AVHP, 2/7, Tizard to Hill, 6 May 1940.
56 Imperial War Museum, Sir Henry Tizard Diary, 3 May 1940.
57 AVHP, 2/7, Tizard to Hill, 16 April 1940.
58 AVHP, 3/4, Hill to Sir Henry Dale, 9 May 1940.
59 AVHP, 2/7, Hill to Mackenzie, 20 May 1940.
60 AVHP, Mackenzie to Hill, 22 May 1940.
61 AVHP, 2/7, 'Proposal Re RDF Development in Canada,' 28 May 1940.
62 Ibid.
63 AVHP, Archibald Sinclair to Hill, July 1, 1940.
64 In May, Hill had stressed the advantages of appointing a full-time Air Min-

istry scientific officer in Ottawa. AVHP 2/7, Hill to C.J. Mackenzie, 20 May 1940; HTP, 228, Hill to Tizard, 3 July 1940.
65 HTP, Sinclair to Hill, 22 June 1940; AVHP, 2/1, Pye to Hill, 24 June 1940.
66 Tizard Diary, 'Note of Conclusions of meeting ... 9th July 1940.'
67 Lindemann's task was facilitated by the many valuable Washington contacts that Arthur Purvis, chairman of the British Purchasing Commission, had already made, including Henry Morgenthau, Jr, American Secretary of the Treasury (1934–45).
68 HTP, 233, Lindemann to Tizard, 9 July 1940.
69 Ibid., 31 July 1940.
70 According to Lindemann, one of the advantages of the Norden bombsight 'was its telescope attachment which gave enemy searchlights a very small chance of "blinding" the bomber.' Ibid. Lindemann to Tizard, 31 July 1940.
71 Lindemann also obtained valuable information from Air Commodore Pirie, head of the British Air Mission in Washington. Ibid.
72 During his Ottawa trip, Lindemann met with C.J. Mackenzie, Robert Boyle, J.T. Henderson, J.H. Parkin, and Frederick Banting; he was particularly impressed by Boyle because of his expertise in asdic and radar research and because he had been 'a pupil of Rutherford's.' HTP, 233, Lindemann to Tizard 9 July 1940.
73 Ibid.
74 University of Toronto Rare Book Room (UTRB), Frederick Banting War Diary, 6 August 1940.
75 Clark, *Tizard*, 257–8.
76 HTP, 253, Report, 8 August 1940.
77 Tizard was pleased with the performance of his military staff members, especially Colonel F.C. Wallace, who later served with the National Research Council. Tizard Diary, 13 August 1940.
78 Ibid., 19 August 1940.
79 HTP, 58, Fowler to Dr. E.V. Appleton, Secretary DSIR, 16 August 1940.
80 Ibid., 255, Tizard to Vincent Massey, Canadian High Commissioner, 22 July 1941.
81 NAC, Mackenzie King Diaries, transcript/typescript, microfiche, 16 August 1940.
82 King's diary entry for this date makes no reference to any such commitment, although he did leave for Washington to meet with President Roosevelt the next day. Mackenzie King Diaries, transcript/microfiche, 16, 17 July 1940.
83 C.J. Mackenzie Diary, 20 August 1940.
84 In contrast, Frederick Banting was not so impressed with Tizard: 'He has a

fine record for Science & Daring & Initiative. But he does not know very much about details than I do.' Cited in Bliss, *Banting*, 286.
85 Tizard Diary, 26 August 1940.
86 HTP, 58, Hill to Lindemann, 28 February 1940.
87 Tizard Diary, 2 September 1940. Atomic energy was another subject of discussion.
88 Some of these discussions occurred at military research establishments such as Fort Monmouth (Signals), Edgewood (Explosives and Chemical Warfare), Langley Field (Army Airforce), and Norfolk (Naval research). Other sessions were held at the industrial laboratories of Bell Telephone (New York), General Electric (Schenectady) and RCA (Camden, NJ), as well as at the major universities – MIT, Harvard, Columbia, Johns Hopkins, Chicago.
89 Tizard Diary, 14, 17, 27 September 1940.
90 Ibid., 12, 28 September 1940.
91 Ibid., 25, 26 September 1940.
92 HTP, 58, Report of Meeting, 14 October 1940.
93 Ironically, the Admiralty, which had been sceptical of Tizard's mission, seemed to have profited most from the venture; its representative reported that the interchange of information with the American Navy now 'was functioning well.' Ibid.
94 Ibid.
95 McGucken, *Scientists, Society, and State*, 171–2.
96 Tizard obtained much useful information from John Cockcroft, both before and after his return from North America in December 1940; he used this information in his briefs to Hankey's Committee. Tizard Diary, 17 December 1940.
97 Clark, *Tizard*, 271.
98 HTP, 58, Hill to Brigadier Lindemann, 28 February 1941.
99 Although Tizard would remain influential, given his membership on the Aircraft Supply Council, the Air Council, and the Radio Policy Committee, he was clearly frustrated by his uncertain status. In the spring of 1941 he threatened to resign from his position as scientific adviser to Lord Beaverbrook, Minister of Aircraft Production; in September 1942 he withdrew from any direct role in the British war effort. Clark, *Tizard*, 270–333. Between September 1942 and November 1944, he temporarily withdrew from wartime service.
100 Bliss, *Banting*, 305–8.
101 Mackenzie Diary, 23, 24 February, 25 April, 1 May 1941.
102 Fowler returned to England in April 1941; his successor as Air Ministry scientific liaison officer was Sir William Lawrence Bragg. Ibid., 15, 16 April 1941.

103 Ibid., 18 January 1941.
104 NAW, Papers of the Office of Scientific Research and Development (OSRD), Box 129, Appendices of Summary of Report of OSRD Activities in the European Theatre during the Period March 1941 through July 1945. OSRD, Box 129, Bush to Hull, Nov. 20, 1940.
105 Clark, *Tizard*, 210–30.
106 Harvard University Archives, Nathan Pusey Library, James B. Conant Papers, Mission to England, 141, notes, etc. In addition to being president of Harvard, Conant was chairman of Division B of the NDRC and an active member of the Committee to Defend America by Aiding the Allies. James Hershberg, *James B. Conant: Harvard to Hiroshima and the Making of the Nuclear Age* (New York, 1993), 132–5.
107 Ironically, Tizard was not part of the British negotiating team, a slight that he conveyed to Conant when the two men met informally: 'The P.M. [Churchill] who sent me there [the United States], had not time to see me since my return, and had not even acknowledged a preliminary report that I sent him.' Tizard Diary, 8 March 1941.
108 OSRD, Box 131, Conant Mission, Wilson to Bush, 16 March 1941.
109 Ibid., Conant to Bush, 30 March 1941.
110 British expertise in operational research was another area Conant felt should be exploited. Ibid., Wilson to Bush, 16 March 1941.
111 OSRD, Box 131, Conant Mission, Anglo-American Scientific Co-Operation on Defence Matters, notes of Meeting held on 6th March, 1941.
112 Webster had worked with Cockcroft at the Cavendish Laboratory before the war; he followed his mentor to the Ministry of Supply and then to the Tizard mission, where he was responsible for assembling the reports. Imperial War Museum, Henry Tizard Papers, 256, Draft BCSO History.
113 HTP, 58, Hill to Lindemann, 28 February 1941.
114 In July 1941, Conant, through Tizard's efforts, was elected as a foreign member of the Royal Society, an honour that Conant regarded as 'symbolic of the close connection between the scientific workers on the two sides of the Atlantic.' HTP, 256, Conant to Tizard, 24 July 1941.
115 HTP, 255, Wilson to Webster, 20 August 1941; ibid., Webster, joint letter, 23 August, 1941.
116 HTP, 256, Draft BCSO History.
117 Roy MacLeod, '"All for Each and Each for All": Reflections on Anglo-American and Commonwealth Scientific Cooperation, 1940–1945,' *Albion* 26 (1994), 26: 79–112.
118 Ibid., 108; HTP, 255, 'Memorandum of Procedure for Cooperation between the British Central Scientific Office and Representatives of Dominion Scientific Effort for War Purposes, W.L. Webster, 30 June 1941.'

119 By 1943 over 500 Australians were working in Washington on a variety of military and scientific agencies. McLeod, 'All for Each,' 106–12.
120 In July 1942 Webster resigned from the BCSO and was hired to direct the NRC Associate Committee on Explosives and Ballistics. In 1943 he joined the British atomic liaison team in Washington. Ibid., 108.
121 HTP, 260, Tizard to Sir Edward Appleton, 11 December 1941.
122 Ibid., 57, Geoffrey Hill to Tizard, 23 July 1942.
123 Both Bragg and Thomson took great interest in determining how Canadian scientific talent could be utilized for British research projects. HTP, 57, A.V. Hill to Tizard, 23 July 1941; Mackenzie Diary, 11 February 1942.
124 HTP, 260, Mackenzie to Tizard, 20 August 1942.
125 The Australians had established their London office some months earlier; New Zealand and South Africa would soon follow. MacLeod, 'All for Each,' 96–112.
126 In 1942 Howlett was replaced by A.G. Shenstone, a prominent physicist who had previously represented the NRC in Washington. Eggleston, *National Research*, 177, 211–51.
127 Mackenzie Diary, 7, 15 January 1943.
128 Ibid., 18 January 1943; ibid., 26 July 1943.
129 Ibid., 28 November 1942.
130 Phillipson, 'International Research,' 20–5.
131 NAW, OSRD Papers, Box 9, Policy and Procedures: Interchange of Technical Information with Canada, memorandum of 25 October 1940.
132 Irwin Stewart, *Organizing Scientific Research for War: The Administrative History of the Office of Scientific Research and Development* (New York, 1948), 168; OSRD, Box 32, NRC Carroll Wilson to A.G. Shenstone, 28 May 1941.
133 The presidential order of February 1943 further improved the status of Canadian scientists working in the United States. OSRD, Box 9, Technical Exchange, Caryl Haskins to Mr. Cleveland Norcross, 27 March 1942.
134 Ibid., Report of Visit to Canada, 10–11 September 1941; ibid. file 32, NRC Haskins to Mackenzie, 20 November 1942.
135 Ibid., Box 32, NRC file, Manson to Louise Paddock, Administrative Aide, OSRD, 23 November 1943.
136 In 1942 the Canadian General Staff created a special staff unit 'to follow technical developments affecting weapons' in the United States. OSRD, 32, William Eaton, Technical Aide, OSRD to Major Hammerlee, 20 December 1942.
137 Ibid., NRC file, Julian Smith to Haskins, 29 July 1942.
138 Ibid., Webster to Haskins, 14 November 1942.

139 Ibid., 32, Shenstone to Carroll Wilson, 31 January 1942; ibid., 9, Melville Eastham, Radiation Lab to Caryl Haskins, 23 June 1942.
140 Mackenzie Diary, 24 April 1943; OSRD, 32, Mackenzie to Bush 11 June 1941; ibid., Bush to Mackenzie 12 June 1941.
141 NAW, Joint Chiefs of Staff Collection (JCS), file 6-24-42 #2, memorandum, JCS, 9 November 1945; Donald Avery, 'Allied Scientific Cooperation and Soviet Espionage in Canada, 1941–45,' *Intelligence and National Security* 8 (3) (July 1993), 104–7.
142 Major General H.F.G. Letson preceded Pope as the Army's Military Attaché to the Canadian legation in Washington from 19 August 1940 to February 1942.
143 Maurice Pope, *Soldiers and Politicians: The Memoirs of Lt. Gen. Maurice A. Pope* (Toronto, 1962), 184.
144 NAC, C.D. Howe Papers, vol. 55, Arnold Heeney, secretary of the Privy Council to Howe, 18 June 1942.
145 Although Pope carried the burden of the liaison work, he was assisted by Commodore Victor Brodeur and Air Commodore George Walsh on issues primarily affecting either the RCN or the RCAF. Pope, *Soldiers*, 184–5.
146 In September 1939 Pope, as a Colonel, had been appointed Director of Military Operations and Intelligence; in 1940 he became senior staff officer to Major General Harry Crerar in Canadian Military Headquarters in London. In December 1941 he was promoted to Major General and became Vice-Chief of the General Staff before becoming head of the Canadian Military Mission. Granatstein, *Generals*, 207–13.
147 The British government also had 1800 employees in the Supply Mission and another 1000 at the Embassy. Alex Danchev, *Establishing the Anglo-American Alliance: The Second World War Diaries of Brigadier Vivian Dykes* (London, 1990), 11.
148 Pope, *Soldiers*, 185, 190.
149 Danchev, *Alliance*, 1–9.
150 Pope, *Soldiers*, 193.
151 Ibid., 195.
152 NAW, JCS, file 3-31-42, Pope letter, cited in British Joint Staff Mission in Washington.
153 Ibid., Admiral William Leahy, Chief JCS, to Pope, 19 August 1942.
154 Pope was subsequently invited to attend those meetings of the U.S. Joint Intelligence Committee when Canadian interests were being discussed. Ibid., Major General Strong to Pope, 12 September 1942; ibid., Pope to Strong, 18 September 1942.
155 The American Navy, and especially Admiral Ernest King, Chief of Naval

Operations, strongly favoured the Pacific theatre. Meetings of the CCOS often ended in bitter confrontations between King and his allies on one side, and the U.S. Army and the British on the other. Danchev, *Alliance*, 7–15.
156 NAC, Diary of Maurice Pope (Pope Diary), 31 March 1943.
157 Ibid., 'The Quebec Conference, August 1943.'
158 Pope was opposed to the suggestion of having a separate Joint Staff Mission in London; he was overruled, and a Mission was established. Ibid., 8 March 1944; 10 June 1944.
159 OSRD, Box 129, Summary of Report on OSRD Activities in the European Theatre during the Period March 1941 through July 1945, submitted by Bennett Archambault, July, 1945; Stewart, *Organizing Scientific Research*, 172.
160 It was not until September 1944 that the NRC decided to move into the new BCSO offices; this did not mean that C.J. Mackenzie supported the call for 'a single approach ... made to U.S. authorities on behalf of the British Commonwealth.' MacLeod, 'All for Each,' 98.

3 Radar Research and Allied Cooperation, 1940–1945

1 Mel Thistle, ed., *The Mackenzie–McNaughton Wartime Letters* (Toronto, 1975), 48–9.
2 Ibid.
3 Eggleston, *National Research*, 173–4.
4 NAW, CCS 334, Joint Committee on New Weapons and Equipment Vol. 208, file 5–12–42, Minutes of Meeting, 12 May 1942; 2 August 1945.
5 McGucken, *Scientists, Society, and State*, 214–23.
6 Ibid., 224.
7 Churchill College Archives, A.V. Hill Papers (AVHP), vol. 57, Hill to Tizard, 13 March 1942.
8 AVHP, 'Memorandum: A Joint Technical Staff, 20 April 1942.'
9 House of Lords, *Debates*, 26 May 1942, 862.
10 AVHP, Hill to Lord Hankey, 7 May 1942.
11 *The Times*, London, 1 July 1942.
12 McGucken, *Scientists*, 241–59.
13 Ibid., 309.
14 In May–June 1943 the U.S. Special Radar Mission headed by Karl Compton visited England, at the invitation of the British Radio Board, and subsequently reported to the JCNW. NAW, CCS 334, JCNW, Box 208, 5-12-42, Minutes of Meeting of 17 June 1943; 3 November 1943.
15 NAW, CCS 385.2, 4-2-44, 'Technical Information Concerning the Existing

Means of Achieving Tactical Deception,' 4 April 1944. See also Robert Buderi, *The Invention That Changed the World* (New York, 1996), 192–4.
16 CCS 334, Box 208, 5-12-43, JCNW Meeting, 20 October 1943.
17 John Burchard, ed., *Rockets, Guns and Targets* (Boston, 1948), 147–216.
18 CCS 334, Box 208, 5-12-43, JCNW Meeting, 14 February 1943.
19 CSS 334, 2-8-44, Combined Chiefs of Staff, 'Proposed Early British Operational Use and Downgrading of Infra-Red Devices, 15 April 1944.'
20 E.G. Bowen, *Radar Days* (Bristol, 1987), 117–18.
21 Ibid., 121.
22 The most important of these was the British Mark VII and Mark VIII, both of which were installed in Beaufighters and Mosquitos in 1942, and the U.S. SCR 720 (Mark X), which became the principal set for the Army Air Force and the RAF during the duration of the war. Ibid., 131.
23 Ibid., 101, 107.
24 Ibid., 109–15.
25 Henry Guerlac, *Radar in World War II* (Philadelphia, 1987), 396.
26 Ibid., 473–86.
27 Daniel Kevles, *The Physicists: The History of a Scientific Community in Modern America* (New York, 1979), 306; Guerlac, *Radar,* 497–505, 507–10, 535.
28 Eggleston, *National Research,* 147.
29 NAC, J.T. Henderson Papers, vol. 2, file 30, Boyle to Henderson, 25 May 1939; ibid., file 31, Henderson to Appleton, 19 June 1939.
30 Ibid., vol. 4, file 15, Henderson to McNaughton, 10 May 1939; Report on Radio Projects, 18 October 1940.
31 Ibid.
32 Ibid, vol. 1, file 57, NRC, Radio Section, Progress Report, June, 1939 to 1 January, 1942; See also Zimmerman, *The Great Naval Battle of Ottawa,* 18–20.
33 Eggleston, *National Research,* 150–1.
34 Henderson Papers, vol. 4, file 2, Boyle to Sanders, 5 October 1939.
35 Imperial War Museum, Tizard Diary, 23 January 1940.
36 Ibid., 3 May 1940.
37 Ibid.
38 Harvey Sapolsky, *Science and the Navy: The History of the Office of Naval Research* (Princeton, 1990), 10–18.
39 Bowen, *Radar Days,* 160.
40 Ibid., 171.
41 By 1945 the Rad Lab had a staff of about 4000, including 500 physicists. Kevles, *Physicists,* 307.
42 Bowen, *Radar Days,* 176.
43 Ibid., 182.

44 The meeting also instructed John Cockcroft and Colonel Wallace to consult with the Radio Branch to ensure that the Canadian GL Mark III met the specifications of the British Chiefs of Staff. J. Middleton, *Radar Development in Canada: The Radio Branch of the National Research Council of Canada, 1939–1946* (Waterloo, ON), 19–22.
45 Imperial War Museum, Sir Henry Tizard Papers (HTP), 57, Report of 14 October 1940.
46 Middleton, *Radar Development*, 70–5.
47 At Dunkirk, Wallace was responsible for the destruction of the GL Mark I sets, which he apparently lined up on the beach and blew up before he himself escaped on a destroyer. Guerlac, *Radar in World War II*, 170.
48 Henderson Papers, vol. 1, file 56, memorandum, 6 September 1940.
49 Ibid., Report of 18 October 1940.
50 The Radio Branch was so strapped for equipment that in November 1940 Henderson asked General McNaughton if they might use his personal lathes and power tools 'for the duration of the war.' Ibid., Henderson to McNaughton, 15 November 1940.
51 Ibid., vol. 5, McKinley to Henderson, 26 December 1940.
52 Ibid.
53 Tizard Diary, 9 July 1941.
54 Cited in Thistle, *Wartime Letters*, 91–2.
55 NRC, vol. 178, file 3-25-4-31, Bush to Mackenzie 30 October 1940; ibid., Mackenzie to Bush 4 November 1940.
56 Ibid., Report of the NDRC Visit of 28 November 1940.
57 NRC, 25-4-34, Mitchell to Colonel Wallace, 14 December 1940.
58 Ibid., vol. 155, file 25-4-34, included in report from Brigadier H.F.G. Letson to Lieut. Colonel H.K. Taber, Director of Signals, DND, 5 December 1940.
59 Bell Labs concentrated on magnetron development; G.E. made the magnets, Sperry the scanning gear, and RCA the pulse modulator and power supply. Henderson Papers, vol. 4, file 2, Henderson to Boyle, 15 December 1940.
60 Ibid., Henderson report on trip to Boston, 16–17 December 1940.
61 Ibid., Mackenzie to Henderson, 15 January 1941.
62 Ibid., 25-4-31, DuBridge to Pitt, 24 January 1941.
63 Ibid., Sanders to Boyle, 7 February 1941.
64 Robert Buderi, *The Invention That Changed the World* (New York, 1996), 116, 125.
65 Ibid., 142–70.
66 NRC, vol. 155, 25-4-34 Colonel Mitchell to Mackenzie, 11 February 1941.
67 Ibid., Smith to Boyle, 7 January 1941.

68 NRC, vol. 178, 17-15-9-2, Mackenzie to Professor Young, Faculty of Engineering, University of Toronto, 18 December 1940.
69 Ibid., Mackenzie to Presidents of Canadian Universities, 20 August 1940.
70 Shenstone's colleague, the British-born Hugh Taylor, a prominent physical chemist, was also recruited for NRC liaison work. Eggleston, *National Research*, 177.
71 NRC, file 17–15–9–2, Smith to Boyle, 6 December 1940.
72 Thistle, ed., *Wartime Letters*, 96.
73 Queen's University Archives, J.A. Gray Papers, 'Offer of Canadian Government to British Admiralty,' n.d.
74 Henderson claimed that because the British Empire Air Training Scheme had just been launched, officials of the Department of National Defence 'were afraid to start any other kind of training.' University of Toronto Archives (UTA), President Cody Papers, vol. 49, Burton File, J.T. Henderson, Secretary, Radio Research Committee, 13 July 1940.
75 Six of the original twenty radar graduates were from Toronto as were all of the second batch of eleven. Ibid., 'The Story of Radio Training at the Department of Physics ... from April 1940 to July 1st 1942.'
76 Ibid., vol. 29, 'Memorandum re Training of Students for the Navy and the Air Force in Fundamentals of Scientific Methods of Anti-Submarine Warfare and Advanced Radar in the Department of Physics, University of Toronto,' 27 June 1941.
77 Mackenzie to McNaughton, April 9, 1940; cited Thistle, ed., *Wartime Letters*, 40.
78 The writer, H.G. Burchell, also mentioned that two of his friends were working as radar technicians on the battleships *King George V* and *Duke of York*). Cody Papers, vol. 49, Burton File, Burchell to Dr. J.M. Anderson, 15 January 1941.
79 Ibid., J.O. Cossette, Naval Secretary, Department of National Defence, Naval Service to Dr. E.F. Burton, 5 February 1941.
80 A total of ninety-one radar operators were sent; almost all became commissioned officers in either the RN or the RCN. In addition, 330 naval ratings were trained for antisubmarine work at Toronto between 1940 and 1942 in a three-month course designed by Burton and Commander Pressy. Ibid.
81 NAC, R.C. Dearle Papers, Vol. 3, McNaughton to Dearle, 30 September 1940.
82 Gray Papers, Gray to Boyle, 29 May 1940.
83 Ibid., Fowler to Gray, 20 August 1940.
84 Gray served as a member of the NRC Sub-Committee on Physics Problems and the NRC liaison subcommittee with Research Enterprises. He also took great pride in the fact that sixteen of his physics research assistants were

carrying out important war research. Gray Papers, Burton to Gray, 24 November 1943; ibid., Gray to Principal Wallace, 12 March 1943; ibid., Henderson to Gray, 16 May 1944.
85 Cody Papers, 49, Burton to Cody, 15 August 1940.
86 Ibid., 49, Burton to C.E. Higginbottom, Secretary, Board of Governors, 10 June 1941.
87 Ibid., 49, R. Common to Cody, 2 January 1941.
88 *Financial Post*, 7 January 1941; *Toronto Star*, 6 January 1941.
89 Cody Papers, vol. 50, Hunter to Cody, 22 January 1942.
90 Between 1940 and 1943 the foundation provided $25,000 to assist geophysical research in the physics department, with much of the money being used to pay the salaries of two physicists, to purchase equipment worth over $10,000 and to maintain the special laboratory at 49 St George Street. Ibid., vol. 60, Burton Report, 1943.
91 Engineering was also hard hit, losing key people such as T.R. Loudon, professor of applied mechanics. Ibid., 58, Burton to Cody, 3 September 1942; ibid., 61, Mackenzie to Cody, 3 January 1943; ibid., 62, Cody to James Duncan, 1 September 1944.
92 Ibid., 50, Cody to Mackenzie, 16 April, 7 June 1941.
93 Ibid., Mackenzie to Cody, 9 June 1941.
94 NRC, 87/88/104, vol. 69, file 32-60-1, Stanley to Ralston, 19 June 1940; Mackenzie to Stanley, 21 June 1940.
95 Ibid., Johnstone to McNaughton, 5 September 1939.
96 W.B. Lewis, 'George Hugh Henderson,' in *Obituary Notices of the Royal Society, 1950* (London, 1950), 155–66.
97 NRC, 87/88, 104, vol. 69, file 32-60-1, Mackenzie to H.W. Armstrong, Dept. of Finance, 16 August 1943.
98 Dearle was also able to utilize the services of Elizabeth Laird, who had just retired as head of the department of physics at Holyoke College, and had taken up residence in London. Ibid., Dearle to Boyle, 8 May, 26 October 1940.
99 Dearle Papers, vol. 3; Dearle to Boyle, 29 January 1940.
100 Dearle was particularly annoyed when his requests for an NRC research grant, and for Klystron oscillators from the U.S.-based Sperry Gyroscope Company, were rejected. Ibid., Dearle to Mackenzie, 2 November 1940; ibid., Mackenzie to Dearle, 6 November 1940.
101 Dearle Papers, Mackenzie to Dearle, 13 June 1941.
102 Ibid., Dearle to Mackenzie, 18 June 1941.
103 Ibid., Mackenzie to Dearle, 20 June 1941.
104 Ibid., vol. 3, mimeographed statement, undated, of research at UWO during the war.

105 Mackenzie Diary, 'Notes of Visit to University of Western Ontario,' January 1942.
106 Middleton, *Radar Development*, 40.
107 McGill Archives, Cyril James Papers, Vol. 94, Department of Physics Files, Foster to Shenstone, 2 August 1941.
108 Ibid., James to Mackenzie, 19 March 1941; ibid., Mackenzie to James 22 March 1941.
109 Ibid., Foster to Norman Shaw, 26 September 1941.
110 Ibid., Watson to James, 16, 19 November 1941; ibid., Shaw to James, 7 July 1941.
111 In July 1942 the NRC provided some assistance to McGill's beleaguered physics department by buying out some of the teaching time of N. Shaw, W.H. Watson, and F.R. Terroux. Ibid., Shaw to Colonel F.C. Wallace, 30 July 1942.
112 NRC, 88/89, S-25-4-31, Mackenzie to E.L. Bowles.
113 Ibid., Mackenzie to Foster, 20 October; ibid., Mackenzie to Bowles, 20 October 1941; ibid., Mackenzie to DuBridge, 28 October 1941.
114 Ibid., DuBridge to Mackenzie, 22 June 1942.
115 Ibid., DuBridge to Mackenzie, 20 October 1941.
116 Ibid., Wallace to Bowles, 2 October 1941.
117 In January 1942 Louis Ridenour of the NDRC visited Ottawa to discuss with NRC and Service officials the various radio sets in design and in operation. Ibid., DuBridge to Mackenzie, 27 September 1941; ibid., Colonel Wallace to E.L. Bowles, 5 January 1942.
118 Ibid., Henderson to Bowles, 24 October 1941.
119 NAW, OSRD Records, Series 1, Subject file, Radio Board, 'Notice to Representatives of the Allied Governments,' 19 August 1942.
120 Ibid.; See also Clark, *Tizard*, 321–31.
121 HTP, 325, Tizard to Air Chief Marshal Sir Charles Portal, 10 September 1942.
122 Buderi, *Invention*, 187–9.
123 The ERSA operated under the jurisdiction of the U.S. Defence Supplies Corporation of the War Production Board. OSRD, Series 1, file Electronics and Radio Cooperation with Canada, Ray Willis, Director of Radio and Radar Division, OSRD, to Senator Chapman Revercomb, 14 July 1943.
124 Ibid., Lieutenant R. Story, Washington Office, M & S to J.R. Pernice, Radio & Radar Division, WPB, 4 June 1943.
125 Ibid., Timmons to R.J. Woodrow, ERSA, 14 August 1943.
126 Zimmerman, *Naval Battle*, 55.
127 Ibid., 11.

128 Marc Milner, 'The Implications of Technological Backwardness: The Canadian Navy, 1939–1945,' *The Canadian Defence Quarterly* 19 (3) (Winter 1989), 46–52.
129 Zimmerman, *Naval Battle*, 26.
130 Bush's difficult relationship with the United States Navy, and especially with Admiral Ernest J. King, has recently been analysed by G. Pascal Zachary, *Endless Frontier: Vannevar Bush, Engineer of the American Century* (New York, 1997), 118–43.
131 Middleton, *Radar Development*, 47.
132 NRC, S-25-4-31, MIT, Henderson to Boyle 18 January 1941.
133 The 286 operated at 214 megacycles and a wavelength of 1.4 metres, while the CSC used 'a wave length of 1.5 metres [200 MHz].' Because of delays in getting British components, and production problems at REL, it was not until December 1942 that CSC/SWIC sets were available for RCN operational use. Zimmerman, *Naval Battle*, 38–45.
134 Michael Hadley, *U-Boats against Canada: German Submarines in Canadian Waters* (Toronto, 1985), 144–95.
135 Similar problems emerged in the saga of the 3 cm Type 268 (RX\F) 'probably the most distinctive Canadian contribution to the field of radar." Middleton, *Radar Development*, 54.
136 Hadley, *U-Boats*, 211, 246, 251.
137 Zimmerman, *Naval Battle*, 30–4, 58, 84, 126.
138 The Squid was a considerable improvement over the Hedgehog because it was linked with a narrow-beamed 147B asdic set that followed the target to within twenty feet, and guided the weapon. It became operational in September 1943.
139 Milner, 'Canadian Navy,' 47; Marc Milner, *North Atlantic Run: The RCN and the Battle for the Convoys* (Toronto, 1985).
140 Henderson Papers, vol. 1, file 57, 'Radio Section Progress Report, June 1939 to 1st January, 1942.'
141 Milner, *North Atlantic Run*, 76.
142 The IFF apparatus distinguished one aircraft from another by means of '"transponders" that swept through a band of frequencies from 157 to 187 MHz and returned a coded signal when the interrogating signal from the radar station was received. Frequencies of 165 and 171 MHz were assigned to GL-III C.' Middleton, *Radar Development*, 78.
143 Colonel Phillips claimed that by the end of the war REL had been forced to implement a total of 2740 engineering changes with an added cost of $1,396,904. Ibid., 43.
144 Hackbusch, an executive with Stromberg-Carlson Company, was appointed

manager of the radio side of REL in October 1940 by Colonel W.E. Phillips. Ibid., 42.
145 Buderi, *Invention*, 131–4.
146 Eventually 350 GL Mark IIIC sets were produced for the Canadian Armed Forces, and other sets were used by Australia, South Africa, and the Soviet Union. Middleton, *Radar Development*, 79, 84.
147 Several other promising NRC projects failed to materialize before the end of hostilities, namely the Counter-Bombardment Radar (CB), and the Microwave Zone Position Indicator. Ibid., 86–93.
148 Buderi, *Invention*, 154.
149 Ibid., 154.
150 W.A.B. Douglas, *The Official History of the Royal Canadian Air Force*, vol. 2: *The Creation of a National Air Force* (Toronto, 1986), 393–8, 536.
151 Ibid., 393–8.
152 Ibid., 100–1.
153 NRC, S-25-4-31, MIT, McKinley to Colonel Wallace, 18 January 1943; ibid., F.H. Sanders to DuBridge, 12 February 1943; ibid., Sanders to John Trump, Secretary, Radiation Laboratory, 25 May 1943.
154 Middleton, *Radar Development*, 104–10.
155 NRC, S-25-4-31, W.J. Henderson to C.B. Collins, MIT, 20 September 1944.
156 NAW, OSRD, Box 7, 'Elements of Loran,' April 1944.
157 Ibid., Commander L.M. Harding to Radiation Laboratory, 28 December 1942; ibid., Bush to Rear Admiral J.A. Furer, 12 February 1943.
158 Ibid., Compton to Bush, 13 August 1943; ibid., Compton to Rear Admiral Joseph Redman, 13 August 1943.
159 Buderi, *Invention*, 185–225.
160 NRC, S-25-4-31, MIT, F.H. Sanders to Dr. C.W. Giddings, 12 October 1943.
161 Nightingale worked for Bell Telephone and Shawinigan Water and Power companies in Quebec during the 1930s and served with the RCAF between 1941 and March 1945. 'The King vs. Matt Simon Nightingale, Before His Worship Magistrate Glenn Strike, K.C., Ottawa, 4 April 1946: "Testimony of George Maurice Fawcett, Director of Electronics Research and Development, RCAF."'
162 Stacey, *Arms, Men and Governments*, 485–528; Eggleston, *National Research*, 236–40.
163 The NRC continued to function as the science arm of the Navy, particularly in the area of asdic and acoustical research, until July 1945, when they parted company. Zimmerman, 'The Canadian Navy and the National Research Council, 1939–1945,' *Canadian Historical Review* 69 (2) (June 1988), 203–26.

164 Ibid., 221.
165 Eggleston, *National Research*, 137–8.
166 Mackenzie Diary, 18 December 1943.
167 Kevles, *Physicists*, 107–8

4 Weapons Systems: Proximity Fuses and RDX

1 Stacey, *Arms, Men and Governments*, 485–9.
2 Eggleston, *National Research*, 180–223.
3 NRC, vol. 288, file 1-1-115.
4 The division had ten sections: physical chemistry; organic chemistry, colloid and plastics; electrochemical; rubber; textiles; paints; corrosion; refractory and metallurgy.
5 The Aeronautical Labs included the Wind Tunnels, the Hydrodynamic Model Testing Basin, the Aircraft Instrument Laboratory, the Engine Laboratory, and the Gasoline and Oil Lab. In addition, there was a Mechanical section, a Structures Laboratory, and a Fire Hazard Testing Laboratory. Ibid.
6 Wilfrid Eggleston, *Scientists at War* (Toronto, 1950).
7 Mackenzie also agreed to the creation of the Special Committee on Applied Mathematics, with Toronto's John Synge as chair, despite his reservations about using 'mathematical geniuses' in NRC projects. Mackenzie Diary, 23 September, 13 November 1942.
8 While Mackenzie found some professional military officers difficult, he had excellent relations with Air Vice-Marshal Stedman, head of the RCAF Aeronautical Engineering Section, whom he appointed NRC liaison officer to the RCAF in 1942. Mackenzie Diary, 16 January, 14 March 1942.
9 Ibid., 17 February, 29 June 1942.
10 Ibid., 21 July 1942.
11 'Patrick Maynard Blackett, 18 November 1897–13 July 1974,' *Biographical Memoirs* (1975), 56–58.
12 NAW, Report of OSRD London Mission Activities in the Field of Proximity Fuzes for Shells, Bombs and Rockets (Washington, 1945).
13 Ralph Baldwin, *The Deadly Fuze: The Secret Weapon of World War II* (San Rafael, CA, 1980), 33.
14 Imperial War Museum, Henry Tizard Papers, Diary, 8 February 1941.
15 NAW, OSRD Records, Series 1, RG 227, Box 4, F/G, Gilbert Hoover, NRL, memorandum, 17 August 1940.
16 Ibid., Memorandum by R.C. Tolman of conference with members of the British Scientific Mission, 17 September 1940.
17 Baldwin, *Deadly Fuze*, 36.

18 In April 1942 D.B. Langmuir was appointed special NDRC adviser in England to provide liaison on proximity fuse work, as well as on radar countermeasures.
19 OSRD, Series 13, Box 37, Tolman to Vannevar Bush, 19 May 1941.
20 Baldwin, *Deadly Fuze*, 80.
21 Tuve also developed a fine working relationship with the engineers in the electronic laboratories of Hytron, Raytheon, Bell, and Rogers Majestic of Toronto. NRC, 90-91/151, vol. 17, 35-7-10-0, Pitt to Mackenzie, 16 July 1941.
22 Cited in Eggleston, *National Research*, 183.
23 OSRD, Series 13, Box 37, Tolman to Dr. Alexander Ellett, Chairman, Section E, NDRC, 2 February 1942.
24 Ibid., Bush to Dr. Hunsaker, 12 August 1941.
25 Ibid., Tuve to Conant, 11 February 1942.
26 In 1941 it was decided that most audio-fuses would use 'radio waves at about one hundred to one thirty megacycles, or wave lengths close to three meters.' Baldwin, *Deadly Fuze*, 143–68.
27 Ibid., 135, 150.
28 Ibid., 135.
29 OSRD, Series 13, Box 37, Tuve to Tolman, 23 February 1942.
30 Ibid., Bush to Members of NRDC, 10 March 1942; ibid., Adams to Bush 16 March 1942.
31 Baldwin, *Deadly Fuze*, 145–6.
32 Ibid., 151.
33 Guy Hartcup and T.E. Allibone, *Cockcroft and the Atom* (Bristol, 1984), 109–12.
34 NRC, MIT file, S-25-6-31, McKinley to Pitt, 18 February 1941.
35 NRC, 90/91, 35-7-10-0, Report W.L. Bragg to Mackenzie, 5 August 1941.
36 Ibid., Mackenzie to E.F. Burton, 6 August 1941.
37 The Army Technical Development Board agreed to provide $50,000 to cover expenses incurred between 1 April 1942 and 20 June 1942. Ibid., Brigadier K. Stuart to Mackenzie, 9 August 1941; Mackenzie to Victor Sifton, 16 June 1942.
38 Ibid., Report, Arnold Pitt, 27 February 1942.
39 Ibid., Maass to Mackenzie, 13 June 1942.
40 Ibid., Inspector of Naval Ordnance to Colonel Wallace, 29 October 1942.
41 Ibid., Pitt to Mackenzie 26 October 1942.
42 Interview with C.C. Gotlieb, University of Toronto, 20 May 1994.
43 Although Gotlieb did not consider Burton a great scientist, he admired his judgment and his interest in bright students. Ibid.

44 C.C. Gotlieb, P.E. Pashler, and M. Rubinoff, 'A Radio Method of Studying the Yaw of Shells,' reprint from *Canadian Journal of Research* 26 (Toronto, May 1948), 167–98.
45 Ibid.
46 Public Record Office (PRO), Cabinet Documents (CAB), 122/364, Memorandum by U.S. Chiefs of Staff to the Combined Chiefs of Staff, 6 August 1943.
47 Ibid., Chiefs of Staff to Joint Staff Mission, Washington, 11 August 1943.
48 NAW, CCS-334, Joint Committee on New Weapons, vol. 208, file 5-12-42, Minutes of the JNW Meetings, 24 February, 15 August 1944.
49 Ibid., Memorandum of the JCS, 7 February 1944; ibid., Chief of Staff Committee, British War Cabinet, 12 February 1944.
50 Ibid., Report by the Sub-Committee on the Allocation of Active Air Defences, British War Cabinet, 1 May 1944.
51 OSRD, Series 13, Box 37, Report, 16.
52 There was also fear the Germans might pass the technology to the Japanese, 'who would use it against U.S. B-29 attacks.' CAB, 122/364, Joint Staff Mission to Chiefs of Staff, 17 March 1944.
53 Ibid. JSM memorandum, 1 June 1944.
54 OSRD, Series 13, Box 37, Report of OSRD London Mission Activities in the Fields of Proximity Fuzes for Bombs and Rockets and Toss Bombing. (Washington, 1945), 19–21.
55 Ibid.
56 NAW, CCS-334, JCNW, vol. 28, file 5-12-44, Memorandum by the U.S. Chiefs of Staff, 20 October 1944.
57 Goering denied, however, that any proximity fuse information was passed on to the Japanese: 'We never gave them anything unless it was in production.' NAW, Carl Spaatz Papers, vol. 134, 'Interrogation of Reich Marshal Hermann Goering, 10 May 1945.'
58 OSRD, Series 13, Box 37, Report of OSRD London Mission Activities in the Fields of Proximity Fuzes for Bombs and Rockets and Toss Bombing (Washington, 1945), 19–21.
59 *Technical Manual, U.S. War Department, Military Explosives* (Washington, 29 August 1940).
60 Ibid.
61 The Luftwaffe had been the first to use aluminized explosives during the 1940–1 Blitz. The British took in the lead in developing plastic explosives, which were extensively used in commando raids and in sabotage behind enemy lines. NAW, OSRD, Papers, 'Report of ASTRIDE London Mission Activities in the Field of Explosives, Chemical Warfare and Chemistry

(Divisions 8, 9, 10, 11 NDRC) ... During the Period March 1941 through July 1945,' 15–17.
62 NDRC chemical work was divided among four divisions: explosives, chemical warfare agents, defensive CW equipment, flame-throwers/incendiaries. W.A. Noyes, ed., *Chemistry: A History of the Chemistry Components of the National Defence Research Committee, 1940–1946* (Boston, 1948), 36–124.
63 Ibid.
64 Ibid., 15–19.
65 James Richardson Donald, *Reminiscences of a Pioneer Canadian Chemical Engineer, 1890–1952*, ed. Robert V.V. Nicholls and Mario Onyszchuk (Department of Chemistry, McGill University, 1989), 65.
66 CIL was jointly owned by Imperial Chemicals Industries (ICI) and Du Pont; it was also the only Canadian manufacturer of explosives. Ibid., 66, 98.
67 Ibid., 68.
68 McGill Archives, J.R. Donald Papers, 'Report of the Technical Sub-Committee on Chemicals & Explosives (Canadian Section), September 1945.
69 Ibid. The five American members of the Sub-Committee represented the Army, Navy, the Chemical Warfare Services, and the Bureau of the War Production Board. The Canadian members were drawn from the Army, the Allied War Supplies Corporation, the National Research Council, Department of Munitions and Supply, and J.R. Donald, who was Chairman of the Canadian section of the Technical Sub-Committee.
70 In keeping with the close Anglo-Canadian cooperation in explosives, Harold Poole of the British Ministry of Supply was appointed director of the small pilot plant at Valcartier 'where propellants and explosives could be made to test methods of manufacture on an economical scale.' Goodspeed, *DRB*, 117.
71 Doug Downing, 'The RDX Research Program at U of T in World War II,' unpublished manuscript in possession of author. The chemical name for RDX was cyclotrimethylennetrinitramine, or cyclonite. Since the substance was a hard, crystalline, and rather sensitive material, it was necessary to desensitize the RDX crystals. OSRD, Series 1, box 13, RDX file, Report Frank Whitmore to U.S. Ordnance, 20 December 1941.
72 By 1940 there were three methods of making Woolwich RDX sufficiently desensitized for use in shells and bombs: a fluid mixture of 60% RDX–40% TNT; a combination of 91% RDX–9% beeswax; and 80% RDX–12% oil mixture for use in plastic explosives. OSRD, Series 1, 'RDX Subject File,' U.S. War Department, Office of the Chief of Ordnance, 20 December 1941.
73 Interview with John Edward, McGill University, 8 March 1994.
74 NRC, vol. 106, file 4-C-9-28, #1, Maass to Mackenzie, 12 August 1941.

75 Scheissler came to McGill in 1939 to work on his M.Sc. in chemistry; in 1944 he obtained his Ph.D. in physical chemistry from the University of Pennsylvania. John Edward, 'Wartime Research on RDX: A False Hypothesis Is Better Than No Hypothesis,' *Journal of Chemical Education* 64 (July 1987), 600.
76 Bachmann was fortunate in having John Sheehan (later famous for the synthesis of penicillin) as his graduate assistant. Ibid.
77 The Bachmann process combined the old nitrolysis technique with the Ross process through 'the addition of more ammonium nitrate and of acetic anhydride to the nitrolysis mixture.' This meant reducing the wastage of large amounts of nitric acid and formaldehyde, which 'were in critical supply,' by substituting acetic anhydride, which the United States possessed in abundance. Noyes, *Chemistry*, 37.
78 Edward interview.
79 Wright's team discovered that RDX produced by the Bachmann process contained small amounts of another explosive – nitramine, or HMX (high melting explosive). It too was used as an explosive, although not on the scale of RDX. Downing, 'RDX Program.'
80 Donald, *Reminiscences*, 132.
81 OSRD, Series 1, 'RDX,' Dr. Frank Whitmore, University of Pennsylvania to Vannevar Bush, 29 April 1942.
82 NAC, C.D. Howe Papers, vol. 40, file S-9-6, #2, Howe to Donald, 30 April 1941.
83 Ibid., Harris to Sir Clive Baillieu, Director-General, British Purchasing Commission, Washington, 7 July 1941; ibid., Baillieu to Hopkins, 5 September 1941.
84 Mackenzie Diary, 3 September 1942.
85 Howe Papers, vol. 40, S-9-16, Linstead to Ross, 8 July 1941; ibid., Donald to Howe, 14 July 1941.
86 Howe Papers, vol. 40, S-9-16, Howe to Donald, 21 July 1941.
87 Ibid., Donald to Howe, 17 July 1941; Donald, *Reminiscences*, 114.
88 Howe Papers, vol. 40, S-9-16, Howe to Donald, 21 July 1941.
89 NRC, vol. 106, file 4-C-9-28, #2, Wright to Maass, 8 August 1941; ibid., Maass to Wright, 7 August 1941.
90 Ibid., Wright to Maass, 20 August 1941.
91 James Phinney Baxter, III, *Scientists against Time* (Boston, 1948).
92 Howe Papers, vol. 40, S-9-16, Donald to Howe, 4 October 1941. Interview with Robert Nicholls, 4 March 1994.
93 By December 1941 the joint RDX Committee had decided that the Ross-Scheissler process had too many problems to be viable for mass production.

Ibid., Donald to Howe, 4 October 1941; OSRD, 32, NRC, Chadwell to Conant, 6 November 1941.
94 NRC, vol. 106, 4-C-9-28, Donald to Howe, 4 October 1941.
95 Ibid., Howe to Donald, 6 October 1941.
96 OSRD, 32, NRC, Adams to Conant, 6 November 1941; ibid., Chadwell to Conant, 7 November 1941; ibid., Conant to Harris, 14 November 1941.
97 Ibid., Harris Chadwell, NDRC, to Maass, 4 September 1941; Maass to Ralph Connor, Secretary RDX Committee (U.S.), 1 October 1941.
98 Ibid., 32, NRC, Mackenzie to Bush, 2 September 1941; ibid., Maass to Ralph Connor, Secretary RDX Committee, NRDC, 1 October 1941.
99 OSRD, Series 1, Box 13, RDX file: Minutes of the meeting of the RDX Committee, University of Michigan, 28 December 1941; OSRD, 32, NRC, Bush to Mackenzie, 30 November 1942.
100 Howe Papers, vol. 40, S-9-16, Donald to Howe, 28 October 1941.
101 NRC, vol. 106, 4-C-9-28, Gordon to Maass, 30 October 1941.
102 NAC, Cambron Papers, vol. 4, Cambron to S.J. Cook, 29 April 1943.
103 Mackenzie was increasingly relying on the advice of Ned Steacie, head of the NRC Chemistry Division, since Maass was busy with chemical warfare matters. Mackenzie Diary, 2 October 1942.
104 Ibid., 11 November 1942.
105 Mackenzie noted that about half of the munitions workers at Valcartier were women. Ibid., 16 November 1942.
106 Thistle, *Wartime Letters*, 122–3.
107 Cambron Papers, vol. 4, Cambron to S.J. Cook, 29 April 1943.
108 Ibid., R. McIntosh to Wright, 24 March 1942.
109 Ibid.
110 John Edward, 'The Scientific Career of Raymond Boyer: A Tribute,' July 1993, Montreal.
111 McGill Archives, Robert Nicholls Papers, Nicholls to Wright, 16 November 1942; Blomquist to Nicholls, 24 November 1942; Nicholls to Walter McCrone, 26 December 1942.
112 Ibid., McCrone to Nicholls, 8 March 1943.
113 G.K. Batchelor, 'Geoffrey Ingram Taylor, 7 March 1886–27 June 1975,' *Biographical Memoirs of the Royal Society of London* (1976), 596–608.
114 Howe Papers, vol. 40, S-9-16, 'Memorandum re Visit to United Kingdom June 29th 1942–August 9 1942.'
115 Ibid., Donald to Howe, 14 January 1943, 'Memorandum re Necessity for Expansion of RDX Production.'
116 Ibid; OSRD, Series 1, RDX, Ralph Connor to Roger Adams, 16 February 1942.

117 By 30 April 1942, the pilot plant had cost NDRC $350,978.80 and was producing about 2000 pounds of RDX a day. Ibid., Connor to Adams, 30 April 1942; ibid., Connor to Conant, 4 May 1942.
118 NAW, Records of the Joint Chiefs of Staff, CCS 471.6 (12-11-42), Report to the Combined Munitions Assignment Board, 11 December 1942.
119 Ibid., Connor to Boyer, 20 August 1943.
120 Eggleston, *Scientists at War*, 80–1; Baxter, *Scientists against Time*, 47–89, 255–60.
121 OSRD, 32, NRC, Webster to Haskins, 11 January 1943; ibid., Bush to Mackenzie, 30 November 1942.
122 Ibid., Haskins to Webster, 20 January 1943.
123 Bush was less successful in ensuring that RDX data and material reached its destination once it crossed the Canada–U.S. border; on one occasion 'six microscope slides of RDX samples' got lost in the customs loop. Noyes, *Chemistry*, 18, 38.
124 NAC, Canadian Mutual Aid Board, 7-M3-14, Karl Fraser to Hume Wrong, 8 September 1945.
125 Downing, 'RDX Program.'
126 In 1943, Boyer and Nicholls became actively involved with the Polymer Subcommittee of the Associate Committee on Explosives; their research was particularly concerned with explosive polymers and plastic explosives. Nicholls Papers, series of reports for 1943–4.
127 NRC, 90-91/151, Box 24, 4-E-5-9, Wright to Webster, 28 October 1942.
128 Ibid., Wright to Webster, 7 December 1942.
129 Noyes, *Chemistry*, 47.
130 Cited in Thistle, *Wartime Letters*, 113.
131 NRC, Box 24, 4-E-5-9, Report, Brigadier G.B. Howard, Chairman, Inspection Board, 25 November 1942.
132 ORSD, 32, NRC, Webster to Haskins, Jan. 9, 1943; Haskins to Webster, Jan. 28, 1943.
133 Noyes, *Chemistry*, 26, 38.
134 Cambron Papers, Mackenzie to Major J.E. Hahn, 24 September 1943; ibid, Mackenzie to M.G.O., 16 December 1943.
135 Given this new commitment to propellants, the NRC asked Woolwich to send a consultant to Valcartier to advise 'during the initial operations of the project.' W.J. Poole arrived in May 1942. Cambron Papers, Cambron to O.W. Stickland, British Supply Mission, Washington, 18 March 1943.
136 UTA, Cody Papers, vol. 57, Cody to W.L. Webster, NRC, 14 January 1943.
137 Mackenzie Diary, 15 January 1943.
138 As early as 1943 the U.S. Navy had requested 7½ tons of DINA for firing

trials. Cambron Papers, vol. 4, Report of the Army Technical Development Board, Status of Development Projects, 31 May 1945; Noyes, *Chemistry*, 48.
139 By 1945 the ACE had spent $60,321 on DINA.
140 Cambron and Steacie of the NRC were also involved in the planning of this survey. NRC, 90-91/151, vol. 24, file S-4-E-5-9, Boyer to Cambron, 14 June 1945.
141 Ibid., Boyer Report to C.J. Mackenzie, 21 August 1945.
142 Ibid.
143 Goodspeed, *DRB*, 119. See also H.P. Tardif, *Recollections of CARD/DREV, 1945–1995* (Valcartier, 1995), 1–96.
144 Cambron Papers, vol. 4, ATDB, Status of Development Program, 31 May 1945. Ibid., vol. 12, file 4-E-4-7, Proceedings of the Reorganization Meeting of the Associate Committee on Explosives and Ballistics, 10 May 1946.
145 The committee included George Wright, Adrien Cambron, and D.C. Rose, the new Chief Superintendent of Valcartier. Ibid.
146 In 1945 Wright was awarded the U.S. Medal of Freedom for his war work. *Biobibliography of Publishing Scientists in Ontario between 1914 and 1939*, comp. Philip Enros, (Thornhill, ON, 1985); *World Who's Who in Science* (Chicago, 1968), 1827.
147 UTA, Letterbooks of Professor George Wright, 15 January 1941–October 1942. NRC, vol. 106, file 4-C-9-28, #1, Wright to Maass, 8 August 1941; ibid., 20 August 1941.
148 Noyes, *Chemistry*, 28.
149 James Eayrs, *In Defence of Canada: Peacemaking and Deterrence* (Toronto, 1972), 161–4

5 Chemical Warfare Planning, 1939–1945

1 Nature, (21 May 1938), 893; ibid. (11 March 1939), 408; ibid. (6 May 1939), 741.
2 Jeffrey Legro, *Cooperation under Fire: Anglo-German Restraint during World War II* (Ithaca, NY, 1995), 144–202.
3 Edward Spiers, *Chemical Warfare* (London, 1986), 62, 68; Robin Ranger, *The Canadian Contribution to the Control of Chemical and Biological Warfare* (Toronto, 1976), 23–87.
4 C.B. Carter, *Porton Down: 75 Years of Chemical and Biological Research* (London, 1992), 30–55.
5 NAC, NRC, 87/88, 104, vol. 69, file 32-1-12, #1, Rabinowitch to McNaughton, 2 October 1939.

6 Rabinowitch was Chief of the Division of Metabolism at the Montreal General Hospital and Associate Professor of Medicine at McGill.
7 NRC, 87/88, 104, vol. 69, Stewart, 'Memorandum'; ibid., McNaughton to the Deputy Minister, DND, 6 October 1939.
8 Ibid., Banting to Mellanby, 21 October 1939.
9 Ibid., Banting to C.J. Mackenzie, 12 December 1939.
10 Ibid., Mackenzie to Rabinowitch, 27 February 1940.
11 Rabinowitch had tested 400 soldiers of the Canadian First Division stationed at Aldershot, England. Ibid., Rabinowitch to General J.C. Murchie, 1 March 1940.
12 British chemical warfare specialists such as R. Kingan were frustrated by the British Army's unwillingness to prepare for chemical warfare and predicted that 'when it [gas] is used there will be a mad rush to get things done.' NRC, Kingan to Rabinowitch, 28 March 1940.
13 Sir John Dill, memorandum: 'The Use of Gas in Home Defence,' 15 June 1940, cited in Robert Harris and Jeremy Paxman, *A Higher Form of Killing: The Secret Story of Gas and Germ Warfare* (London, 1982), 109–11.
14 Ibid.
15 Legro, *Cooperation under Fire*, 144–77.
16 Sweetenham, *McNaughton*, vol. 2, 134–5.
17 Mackenzie Diary, 20 June 1941.
18 NAC, DND files on chemical warfare and biological warfare (microfilm reels C-5001–5019), HQS, 4354-11-1 'Resume of the Status of Chemical Warfare,' prepared in the Technical Division, OC-CWS, 4 September 1940.
19 Ibid.
20 Harris and Paxman, *Higher Form of Killing*, 53–61.
21 Headquarters DND Records (HQS), 4354-11-1, vol. 2, 'Report of Canadian Mission,' 16 October 1940; Mackenzie Diary, 20 October 1940.
22 Ibid., Colonel Morrison to Can/Military, London, 16 October 1940.
23 NAW, OSRD, Box 9, 'Exchange Canada,' Carroll Wilson to Conant, 23 October 1940.
24 NRC, 4-C9-7, F.G. Green, Secretary, to Colonel D.E. Dewar, 24 October 1940.
25 HQS, 4354-3-1, E.A. Flood, 'Memorandum: Production of Gases in Canada,' 5 November 1940.
26 Ibid., Flood to Ned Steacie, 12 September 1940.
27 NRC, A. Cambron Papers, vol. 1, H.M. Barrett, Research Associate, Department of Physiological Hygiene to Dr. A. Cambron, NRC, 18 October 1940; ibid., Cambron to Barrett, 15 November 1940.
28 HQS, 4354-6-5-1, Lt.-Colonel Goforth Report, 7 November 1941.

29 At McGill, research on 'Z' was afforded 'exceptional' security protection. Ibid., Maass Report, 5 February 1942.
30 HQS 4354, McNaughton to General Crerar, 25 September 1940.
31 Mackenzie Diary, 14 July, 7 August 1941.
32 HQS, 4354, Morrison telegram to Can/Military, 9 December 1940.
33 HQS, 4354-1, Morrison memo, 9 November 1940.
34 NRC, 4-C-9-19, Victor Sifton to General Crerar, 9 December.
35 Ibid., Summary of Meeting Re Field Research Chemical Warfare in Canada held at Canada House, London, 1600 hrs., 17 December 1940.
36 Under PC1/6687 of 2 July 1941, almost all aspects of the joint Anglo-Canadian chemical warfare operation were clarified. In 1941–2 the British were expected to pay one-half of the basic costs of Suffield, or $2,028,425, while the Canadian government was to pay $2,497,548. NAC, J.L. Ralston Papers, vol. 396, file 54, Hankinson to N. Robertson, 1 March 1941.
37 Ibid., Ralston to Hon. C.G. Power, 31 March 1941.
38 On 9 April 1941, by order-in-council PC 2508, arrangements were made to secure the land from the province of Alberta on a ninety-nine-year lease; under PC 4458 of 28 June 20, the privately owned lands 'were expropriated.' HQS 4354-2, Hankinson to Robertson, 22 March 1941.
39 Mackenzie Diary, 27 March 1941.
40 HQS, 4354-2, Power to Ralston, 26 March 1941.
41 Ibid., Morrison memorandum to Sifton, 19 May 1941.
42 Ibid., order-in-council, 2 July 1941.
43 Ibid., Colonel H.S. DesRosiers, Acting Deputy Minister DND, to Sir Edward Beatty, 24 May 1941.
44 HQS, 4354-0-1, Report, 3 March 1942.
45 HQS 4354-6, Colonel W.D. Lambert, to DesRosiers, 8 August 1941.
46 Under the pay schedule, the chief superintendent received $8000, the superintendent of research $6000, and a Grade I scientist $4500. HQS, 4354-2, Colonel Morrison, memorandum, 8 May 1941.
47 Ibid., Maass to E.L. Davies, 19 September 1941.
48 NRC, 4-C-9-19, Mackenzie to Hon. J.A. Mackinnon, Chairman, Committee of the Privy Council on Scientific and Industrial Research, 2 June 1941; Directorate of History, Department of National Defence (DHist), 745.043, D3, 'Minutes of the Second Meeting of the Inter-Service Board,' 2 July 1941.
49 NRC, 4-C-9-19, Maass to Andy Gordon, 2 June 1941.
50 In January 1942, for example, 38 McGill students were given exemptions because of their 'special research work.' HQS 4354-5-4-1, Brigadier General, E. de B. Panet, DOC, M.D. #4, to Secretary, DND, 11 January 1942.

51 NRC, Banting Papers, vol. 1, F.T. Rosser, 'The History of the Sir Frederick Banting Fund,' 1 October 1960.
52 In the summer of 1942 Maass decided to leave supervision of these research grants to his McGill colleague R.L. McIntosh and to H.M. Barrett of Suffield, who distributed funds to the western Canadian universities. Ibid., Flood to McIntosh, 24 July 1942; Mcintosh to Major H.W. Bishop, 22 July 1942.
53 OSRD, Box 32, NRC file, C. Haskins, 'Memorandum of conversation with Dr. Chadwell (technical aide of Division B, NDRC), and W.L. Webster, 3 June 1941.'
54 McGill Archives, J.R. Donald Papers, Joint War Production Committee of Canada and the United States, Report of Technical Sub-Committee, 31 August, 1945.
55 Eggleston, *Scientists at War*, 105.
56 NRC, vol. 106, 4-C-9-19, Report of Dr. Watson, Suffield, to Maass, 10 May 1941; Dhist, Chemical Warfare Files, Report of the meeting of the Inter-Service Board, 2 July 1941. Animals were normally used at Suffield to measure the toxicity of various war gases.
57 Mackenzie Diary, 28 March 1942.
58 HQS 4354-2, Colonel Morrison to Secretary, Chemical Warfare Committee, NRC, 28 April 1941.
59 HQS 4354-3-1, Maass to Director of Technical Research, 15 January 1942; ibid., Maass, Report, 27 May 1942.
60 By 1943 the order for battle dress was expanded to 100,000; Canadian troops overseas were also issued American-type steel helmets, which were treated for gas protection. HQS 4354-5-2, Report, 19 February 1943.
61 HQS, 4354-1-8, Flood to Deputy Minister, Pensions and National Health, Ottawa, 29 December 1941; NAC, Privy Council Papers (PCO), vol. 10, 4C-9-22, Watson to Davies, 8 December 1941.
62 Harris and Paxman, *Higher Form of Killing*, 117–18.
63 British experts concluded that Japan would probably use chemical weapons in the near future. HQS, 4354-6-5-1, Minutes of the fifty-first meeting of the British Sub-Committee on Chemical Warfare, 23 February 1942.
64 Ibid., Stuart to R.L. Ralston, 2 February 1942.
65 HQS, 4354-6-5-1, Lieutenant-General Stuart, CGS, to Minister of Defence, 13 February 1942.
66 HQS, 4354-1-8, 'Minutes of the Meeting held 20 January 1942,' ibid., 'Minutes of Meeting on Chemical Warfare on 9 February 1942.'
67 C.P. Stacey, *Official History of the Canadian Army in the Second World War,* vol. 1: *Six Years of War. The Army in Canada, Britain and the Pacific* (Ottawa,

1966), 170–95; Ted Ferguson, *Desperate Siege: The Battle of Hong Kong* (Toronto, 1980), 128–45.
68 Peter Ward, 'British Columbia and the Japanese Evacuation,' *Canadian Historical Review* 57 (3) (September 1976), 289–309.
69 HQS, 4354-6-5-1, Report of Kenneth Stuart to Ralston, 7 February 1942.
70 Ibid., 'Inter-Service Board: Recommendations on Canadian Chemical Warfare Policy for Consideration of the Chiefs of Staff, 21–22 March 1942.'
71 Frederic Brown, *Chemical Warfare: A Study in Restraints* (Princeton, 1968), 207–30.
72 In early January 1941, U.S. Secretary of War Henry Stimson asked the War Plans Division of the General Staff for a full report of the existing U.S. capabilities in the event of a gas attack. HQS, 4354-6-5-1, Major General Kenneth Stuart to R.L. Ralston, 13 February 1942. Leo Brophy and George Fisher, *The Chemical Warfare Service: Organizing for War* (Washington, 1959), 49–51.
73 NAC, PCO, Series 18, vol. 38, D-19-A-2, Major General J.C. Murchie, Vice Chief of the General Staff, to J.L. Ralston, 29 June 1942.
74 HQS, 4354-8-1, Arnold Heeney to Murchie, 10 July 1942.
75 Ibid., Stuart to Secretary, Chiefs of Staff Committee, 21 July 1942.
76 Ibid., Memorandum to War Committee of Cabinet, 7 June 1942.
77 DHist, file HD, 745.043D, Fifteenth Meeting of the Inter-Service Board, 1 July 1942.
78 HQS, 4354-6-5-1, Major General Kenneth Stuart to R.L. Ralston, 13 February 1942. Brophy and Fisher, *Chemical Warfare Service*, 49–51.
79 HQS 4354-1-8, Maass memorandum, 21 August 1942.
80 NRC, 4-C-9-19, #2, Davies to Maass, 11 November 1942.
81 OSRD, Series 1, Chemical Warfare, Conant to Pratt, 16 July 1942.
82 Ibid., Conant to Major General Porter, 2 August 1942.
83 Ibid.
84 Ibid., Flood to Conant, 12 August 1942; ibid., Conant to Flood, 21 August 1942.
85 Ibid., Conant to Maass, 16 July 1942; ibid., Maass to Conant, 3 September 1942.
86 Ibid., Maass to Conant, 18 September 1942.
87 Ibid., Conant to Flood, 3 October 1942.
88 OSRD, Box 10, CW file, Conant to Bush, 26 May 1942.
89 The Technical Section received over 1000 CW reports a month from British and American sources. HQS, 4354-2-2, J.W. Young, Master General of Ordnance to Adjutant General Letson, 6 February 1943.
90 Lt.-Colonel W.R. Sawyer had the responsibility of keeping Canadian Mili-

tary Headquarters in England informed about chemical warfare developments. Ibid., Memorandum, Maass to Young, 9 February 1943.
91 Ibid., Major General Murchie, CGS, Memorandum, 18 July 1942.
92 HQS, 4354-2-1, Major J. Morris to E.A. Flood, 2 October 1942.
93 HQS, 452-7-3, N.R. Anderson, Air Vice-Marshal to Air Officers Commanding Eastern Air Command (Halifax) and Western Air Command (Victoria), 24 December 1942.
94 DHist, 745.043D, Minutes, ISB, 7 December 1942.
95 Ibid., Maass to J.P. Pettigrew, M & S, 18 May 1942.
96 PRO, CAB 122/1323, Joint Staff Mission to Chiefs of Staff, 22 July 1943.
97 Ibid., Memorandum by the Canadian Joint Staff Mission, 9 September 1943.
98 Ibid., Brigadier H. Redman to Combined Chiefs of Staff, 29 September 1943; Pope, *Soldiers and Politicians*, 182–208.
99 DHist, 745.043D, Minutes of the Inter-Service Board, 29 March, 22 July 1943.
100 Ibid., Minutes ISB, 22 July 1943.
101 HQS, 4354-32-1, Minutes ISB, 4 June, 28 September 1943.
102 The Canadian Navy was also interested in further exploring the advantages of using smoke screens to protect Canadian harbours from enemy planes or ships. HQS, 4354-6-2, Group Captain T.J. Desmond to Maass, 29 September 1942.
103 Maass regarded toxic smoke as having special battlefield application, and he personally supervised the work of Leo Yaffe, who obtained his M.Sc. degree in 1942 by studying 'the aerosol filtration problems connected with toxic smoke ... to develop a filter for protection against gas molecules.' Interview with Leo Yaffe, McGill University, 16 June 1984.
104 Eggleston, *Scientists*, 110.
105 The only problem in carrying out field tests was that Suffield did not have sufficient numbers of smoke generators, mortars, and bombs. Once again, an urgent call was made to Edgewood Arsenals for assistance. HQS 4354-32-1, 'Report of Smoke Weapons in Canada, February 2, 1943.'
106 HQS, 4354-24-3, Report on Smoke Weapons in Canada, 2 February 1942; ibid., Major Morris to Suffield, 17 August 1944.
107 Ibid., Colonel H.R. Lynn to Canadian Army Staff, Washington, 15 November 1943.
108 Maass had no difficulty gaining approval from the Army Directorate of Engineering Development (DED) for taking responsibility for flame research. Ibid., Maass to British PDWD, 22 December 1943.
109 By 1942 Edgewood scientists and engineers were able to increase the range of U.S. flame-throwers from 30 yards to an operational 150 yards, as well as to increase the duration and heat of the flame by mixing gasoline with vari-

ous 'metallic soap thickening agents known as Napalm.' Ibid., Major Wharton to Captain R.B. Harvey, 12 February 1944.
110 By 1943 emphasis on the use of flame-throwers and incendiary bombs against Japanese troop concentrations was reflected in Suffield research projects. HQS 4354-24-3-10, Colonel Goforth to Maass, 9 June 1943.
111 Ibid., Memoranda for General Staff, 17 June 1943.
112 HQS, 4354-24-3, Minutes of the Sub-Committee, 18–19 July 1944.
113 Field tests with mustard gas in the Australian experimental station at Proserpine, North Queensland, confirmed the laboratory tests. D.P. Mellor, *Australia in the War of 1939–45: The Role of Science and Industry* (Canberra, 1958), 372–80.
114 OSRD, Box 10, CW file, Conant to Bush, 17 September 1943.
115 HQS, 4354-24-6-1, Maass to J.W. Young (MGO), 11 December 1943. See also Brophy and Fisher, *Chemical Warfare Service*, 290.
116 HQS, 4354-32-1, Brunskill to Maass, 6 March 1944.
117 DHist, 740, 'Minutes of the twenty-eighth meeting C Inter-Service Board, 8 February 1944.
118 Ibid., Minutes of the ISB, 13 January 1944. Emphasis in original.
119 PRO, CAB 122/1323, Prime Minister's Personal Minute, 28 April 1944.
120 Ibid., Churchill to General Ismay, 21 May 1944.
121 Ibid., Report of 28 April 1944 to the Prime Minister from the Secretary, British Chiefs of Staff Committee.
122 The BCOS also did not feel that Churchill should repeat his May 1942 warning on the grounds that it 'would suggest to the German High Command that we are apprehensive of the effects that gas would have against our invasion, and thus might influence them to use gas rather than to deter them from doing so.' Ibid.
123 CAB 122/1323, BCOS Report, 24 April 1944.
124 HQS, 4354-23-7, Report to War Office, 8 February 1944; ibid., Major J.L. Blaisdell, RCAMC, Report of interview with Mr Davidson-Pratt, 7 January 1944.
125 NAW, George Merck Papers, Box 186, British Intelligence Objectives Sub-Committee Final Report no. 44, July 1945.
126 The German High Command had also made plans to shift its production priority to the even more deadly sarin, which was projected for 500 tons a month by March 1945, as well as to explore the potential of soman, which 'was found to be twice as toxic as Sarin.' Ibid.
127 The Germans also carried out unsuccessful trials of charging the warhead of the V-1 rocket with phosgene rather than high explosives. Ibid., Report no. 9; report compiled between 17 July and 2 August 1945.

128 Cited in Harris and Paxman, *Higher Form of Killing*, 127–9.
129 Churchill also received strong protests from General Eisenhower over any planned use of chemical or biological weapons. CAB 122/1323, Minutes of 251st meeting of Chiefs of Staff, 28 July 1944.
130 John McCloy, Assistant Secretary of War to Stimson, 29 May 1945, cited in John Ellis van Courtland Moon, 'Project SPHINX: The Question of the Use of Gas in the Planned Invasion of Japan, '*The Journal of Strategic Studies* 12 (3) (September 1989), 304. See also John Dower, *War without Mercy: Race and Power in the Pacific War* (New York, 1986), 121–230.
131 In the battle for Okinawa the United States suffered 39,000 casualties, with 12,500 killed, while the Japanese had almost 120,000 casualties, of whom 109,629 were killed. Moon, 'Project SPHINX,' 306–7.
132 HQS, 4354-24-1, F.F. Lowe to Major D.C. Turner, 20 June 1944.
133 HQS, 4354-20-14, 'Canadian Chemical Warfare Inter-Service Board Cooperation with Inter-Departmental Policy Committee, Research and Development, War against Japan, 6 July 1944.
134 HQS, 8641-9, vol. 7, 'Minutes of a meeting ... held in the MGO conference room, Army Building, Ottawa, at 0945, on Monday 8 January 45.'
135 HQS, 4354-32-1, Major General Brunskill to Otto Maass, 6 March 1944.
136 Brown, *Chemical Warfare*, 290.
137 Legro, *Cooperation under Fire*, 144–202.
138 Spiers, *Chemical Warfare*, 74–7, 88.
139 Although the JCS seriously considered the proposal, in the end they concluded that such an ultimatum might prod the Nazis into 'the use of gas and possibly other inhumane methods of warfare with a resulting great loss of life, among non-combatants as well as military personnel.' NAW, CCS 385.3, file 9-16-44, Peter Bergson to JCS, 16 September 1944; ibid., Admiral William Leahy to Hebrew Committee of National Liberation, 4 October 1944.
140 Otto Maass Papers, Porter to Maass, 10 November 1945.
141 E.A. Flood, 'Otto Maass, 1890–1961,' *Biographical Memoirs of Fellows of the Royal Society of London*, vol. 8 (1962), 183–95.

6 Canadian Biological and Toxin Warfare Research: Development and Planning, 1939–1945

1 John Bryden's recent study *Deadly Allies: Canada's Secret War, 1937–1947* (Toronto, 1989) provides a popular account of Canada's involvement in biological warfare during the Second World War.
2 Protocol for the Prohibition of the Use in War of Asphyxiating, Poisonous or

Other Gases, and of Bacteriological Methods of Warfare, signed at Geneva, 17 June 1925; entered into force 8 February 1928, cited in Susan Wright, ed., *Preventing a Biological Arms Race* (Cambridge, MA, 1990), 369.
3 Martin Hugh Jones, 'Wickham Steed and German Biological Warfare Research,' *Intelligence and National Security* 7 (4) (1992), 379–402.
4 In this instance, the source was an unnamed German Jewish scientist who claimed that the German military were preparing both anthrax and typhus bombs. *Toronto Star*, 27 March 1935.
5 Banting placed the agents into three categories: (1) water-borne – typhoid, cholera, bacillary dysentery, and amoebic dysentery; (2) airborne – spinal meningitis, virulent haemolytic streptococcus, influenza; and (3) Insect-borne – bubonic plague, yellow fever, malaria, sleeping sickness. Banting felt that the worst danger of all was 'presented by virus diseases which could be distributed absorbed on dust.' National Archives of Canada (NAC), National Research Council Papers (NRC), 87/88, 104, vol. 69, McNaughton memorandum, 17 September 1937; ibid., Banting to McNaughton, 16 September 1937.
6 Ibid., memorandum 17 September 1937.
7 It does not appear that Banting was aware of the biological warfare research being conducted by special units of the Japanese Army in northern Manchuria under the direction of Colonel Shiro Ishii. Peter Williams and David Wallace, *Unit 731: The Japanese Army's Secret of Secrets* (London, 1989), 4–14; Sheldon Harris, *Factories of Death: Japanese Biological Warfare 1932–45 and the American Cover-up* (New York, 1994), 1–112.
8 McNaughton also conferred with Minister of Defence Ian Mackenzie, who expressed 'great interest' in the subject, and later raised the matter at the September meeting of the Committee of the Privy Council on Scientific and Industrial Research. NRC, vol. 69, McNaughton to Banting, 27 September 1937; ibid., McNaughton to Lieutenant-Colonel L.R. LaFleche, Deputy Minister of National Defence, 1 October 1937.
9 *The Times*, 14 October 1937.
10 NRC, vol. 69, N.E. Gibbons to Dr Newton (NRC), 23 September 1937.
11 Banting had previously received assurances from University of Toronto President Cody that his top-secret biological research could be conducted at the university. Ibid., Banting to McNaughton, 16 September 1938; University of Toronto Rare Book Room, War Diary of Frederick Banting (Diary) Box 20, Hunter to Banting, 14 February 1938.
12 NRC/87-88/104, vol. 69, file 36-5-0-0, Banting to McNaughton, 16 September 1938; ibid., L.R. LaFleche to McNaughton, 21 September 1938; ibid., McNaughton to Banting, 24 September 1938.

13 These reports included the 1938 British policy document 'Proposals for an Emergency Bacteriological Service to Operate in War.' Stephen Roskill, *Hankey: Man of Secrets*, vol. 3: *1931–1963* (London: Collins, 1974), 321–4.
14 NRC/87-88/104, vol. 69, 36-5-0-0, Report of Conversation with Sir Edward Mellanby in the office of Dr Wodehouse, Deputy Minister, Department of Pensions and Health, 29 September 1938; ibid., McNaughton to Major General LaFleche, 3 October 1938.
15 Ibid., LaFleche to McNaughton, 24 October 1938; ibid., Skelton to LaFleche, 22 November 1938.
16 Ibid., Extracts from Eighteenth Annual Report of the Chemical Defence Research Department, 1938.
17 Michael Bliss, *Banting: A Biography* (Toronto, 1984), 254–97.
18 Ibid., Rabinowitch to McNaughton, 2 October 1939.
19 Banting was particularly impressed by the work of Drs Topley and Heartley, experts in tetanus and gas gangrene research, and of Sir Patrick Laidlaw of Cambridge. NRC, vol. 69, Banting to C.B. Stewart, NRC, 4 December 1939; Banting Diary, 30 November 1939.
20 Banting Diary, 9 December 1939; Roskill, *Hankey*, vol. 3, 321.
21 Banting Diary, 3, 8 January 1940.
22 Ibid., 11, 12 January 1940.
23 Bliss, *Banting*, 268.
24 Banting Diary, 11, 12 February 1940.
25 Ibid., 17 May 1940.
26 NRC, 87/88/104, vol. 69, Banting to Mackenzie, 24 June 1940; ibid., Mackenzie to Banting, 25 June 1940; Banting Diary, 27 June 1940.
27 NRC, 87-88/104, vol. 69, H. Desrosiers to Secretary NRC, 12 July 1940.
28 Banting Diary, 3, 31 December 1940.
29 In October 1940, for instance, Banting had informed Sir Edward Mellanby and Air Vice-Marshal H.E. Whittingham of the British Air Ministry about the successful experiments Craigie had been carrying out 'in meeting a demand for typhus vaccine.' Ibid., 31 December 1940; NRC, vol. 69, H.E. Whittingham Air Vice-Marshal to C.J. Mackenzie, 13 November 1940.
30 Mitchell highlighted rinderpest because it was one agent 'which might be employed by the enemy and against which we had at present no defence.' NAC, E.G.D. Murray Papers, vol. 29, Mitchell to Murray, 9 June 1942.
31 Murray had come to McGill from Cambridge in 1930 and founded the Department of Bacteriology and Immunology.
32 NRC, vol. 69, Murray to Banting, 27 June 1940.
33 One of the more interesting projects at McGill was the study of the 'synthetic oestrogen, stilboestrol,' which was claimed to create temporary steril-

ity 'following oral treatment.' The McGill team speculated on its possible use 'as a water poison' to produce widespread sterility in German cattle, thereby enhancing the effectiveness of the allied blockade. NRC, vol. 69, J.B. Collip to Mackenzie, 21 November 1940.

34 Valdee had originally contacted his friend R.D. Defries, director of the Connaught Laboratories, who, in turn, informed Banting that Valdee and other members of the National Institute were 'keenly interested in doing everything possible to help in the war.' NRC, vol. 69, Valdee to Defries, 23 December 1940; ibid., Defries to Banting, 28 December 1940; ibid., Banting to Valdee, 30 December 1940; Valdee to Banting, 9 January 1941.

35 Barton Bernstein, 'Origins of the U.S. Biological Warfare Program,' in Susan Wright, ed., *Preventing a Biological Arms Race*, 9–25. National Archives Washington (NAW), RG 165, Entry 488, Files of George W. Merck, Special Consultant, 1942–1946 (Merck Papers), Box 187, E.B. Fred File, Fred to George Merck, 16 August 1943.

36 NAW, Records of the Joint Chiefs of Staff, 1948–1950, 385.2, file 12-17-43, 'Report to the Secretary of War by Mr. George Merck, Special Consultant for Biological Warfare,' November 1945. The report was publicly released 3 January 1946.

37 NRC, vol. 69, extract of letter sent by H.R. Elkins, United China Relief, to Joe Morris, Foreign News Editor, United Press, February 1942; ibid., cited in *The New Republic*, 9 March 1942. The seriousness of the Japanese biological warfare threat was confirmed in a March 1942 report by the U.S. Office of Naval Intelligence.

38 G.P. Gladstone et al., 'Paul Gordon Fildes, 1882–1971,' *Biographical Memoirs of the Royal Society of London* (London, 1973), 317–41; PRO, CAB 120/782, 58921, Hankey to Lord Ismay, GCOS, 6 December 1941.

39 Ibid., Lord Ismay to Churchill, 3 January 1942.

40 The Hankey Report was also sent to J.B. Collip, chairman of the NRC Associate Committee on Medical Research, along with eleven specialized studies. These included 'Experiments on Air-Borne Infection,' 'Experiments on Methods of Spread of Bacteria,' 'Studies on Air-Borne Virus Infection,' 'The Performance of Sprays from Aircraft considered from the point of view of Bacteriological Warfare,' 'The Use of Anthrax against Cattle in Warfare,' and 'Swine Fever.' NRC, vol. 69, A. Landsborough-Thomson, British Medical Research Council, to Collip, 16 December 1941.

41 Other members of the Special Committee were Drs J.W. Swaine, A.G. Lochhead, E.A. Watson, and Charles Mitchell of the Department of Agriculture; Lt.-Colonel A.C. Rankin of the Department of Defence; G.D.W. Cameron, Hygiene Laboratory, Department of Pensions and National Health; G.B.

Reed, Department of Bacteriology, Queen's University; E. Frederick Smith, Department of Bacteriology, McGill; and University of Toronto scientists Donald Fraser, Department of Hygiene and Preventative Medicine, and P.H. Greey, Department of Bacteriology. NRC, vol. 69, Collip to Murray, 16 November 1941.
42 Ibid., Mackenzie to Murray, 19 December 1941.
43 The Committee also discussed the possibility that cities such as Vancouver, with its questionable sanitary standards and large rat population, would be vulnerable to an outbreak of bubonic plague. NAC, R.L. Ralston Papers, vol. 30, 'Recommendations to the Department of Defence of Canada by the Committee of Project M-1000,' included in letter from Collip to Colonel A.A. Mageem, Assistant Minister of National Defence, 29 December 1941.
44 The usual sources of rinderpest vaccine were the Veterinary Research Institute, Muktesar, U.P., India; or the Veterinary Research Laboratory, Onderstepoort, Transvaal, South Africa. Ibid.
45 Ibid.
46 Murray Papers, vol. 26, 'Notes,' 9 September 1940.
47 One of these projects was the distribution of the mosquito *Aegis aegypti*, long regarded as the major vector for yellow fever and malaria. Murray Papers, vol. 26, Notes on Washington Visit, 2 January 1942; NRC, vol. 69, Murray to Colonel James Simmons, 5 January 1942.
48 NRC, vol. 69, Minutes of the First Meeting of Sub-Committee on Project M-1000.
49 Murray Papers, vol. 29, Murray to Fred, 26 January 1942.
50 Ibid., Murray to Fred, 24 January 1942; ibid., Fred to Murray 27 January 1942.
51 NRC, vol. 69, Minutes of the Meeting on Project M-1000, 28 January 1942.
52 Reed described some of his own experiments with botulinum toxin with special reference to the persistence of its toxicity in natural waters and the fact 'that chlorination does not appear to destroy the toxin.' Ibid.
53 In February the NRC received a secret British report of German biological warfare tests showing 'that toxins can be carried by bomb shells without being destroyed ... The toxins of Anthrax are ... of particular efficiency.' NRC, vol. 69, Enclosed in letter from H.C. Bazett to Mackenzie, 4 February 1942.
54 Another topic that would receive greater Allied attention was the use of herbicides. Dr. Sherman of WBC suggested, for example, 'that the destruction of the rice crop would cripple the Japanese just as destruction of the potato crop would the Germans.' NRC, vol. 69, Minutes of the Meeting on Project M-1000, 28 January 1942.

55 Murray Papers, vol. 29, Murray to Fred, 30 January 1942.
56 E.A. Watson and Charles Mitchell had serious reservations about Grosse Île, claiming that 'four miles of water did not afford sufficient barrier against the spread of Rinderpest.' Ibid.
57 Ralston Papers, vol. 30, Ralston to the Hon. James Gardiner, 10 March 1942; ibid., General Murchie to Ralston, 6 March 1942.
58 Murray Papers, vol. 29, Murray to Fred, 30 January 1942.
59 On 24 July, Dean Fred, General J. Kelser, and Colonel Defandorf (CWS) visited Grosse Île and were entertained by Murray and Maass. Ibid., Fred to Murray, 31 July 1942; ibid. Kelser to Murray, 3 August 1942; ibid., Murray to Kelser, 5 August 1942.
60 NRC, vol. 69, Second Report of the Committee M-1000 on Biological Warfare, 12 June 1942.
61 The Canadian commissioners were Murray (chair), C.A. Mitchell, J. Craigie, and G.B. Reed; the U.S. members were General R.A. Kelser (chair), E.B. Reed, R.E. Dyer (NIH), and H.W. Schoening (Dept. Agriculture).
62 It was not, however, until 30 December 1942, by order-in-council PC 69/11742 that the Canadian government gave its official authorization. NAC, HQS, 4354-20-1, Review of Canadian Chemical Warfare Policy.
63 The other members of the Commission were Otto Maass (*ex officio*), Guilford Reed, James Craigie, Charles Mitchell, and J.B. Collip. It was designed as a secret subcommittee of the Directorate of Chemical Warfare and Smoke.
64 Bernstein, 'Origins of the U.S. Biological Warfare Program,' 11–15.
65 Rexmond C. Cochrane, *History of the Chemical Warfare Service in World War II (1 July 1940–15 August 1945*, vol. 2: *Biological Warfare Research in the United States* (Historical Section, Plans, Training and Intelligence Division, Office of the Chief, Chemical Corps, November 1947) (hereafter *Biological Warfare Research*).
66 NAW, RG 165, Records of the War Department: General and Special Staffs: New Developments Division (Merck Papers), Box 183, Liaison with U.K. file, Merck memorandum for Harvey Bundy, 10 September 1942; Cochrane, *Biological Warfare Research*.
67 Murray was quite impressed with Merck, as was C.J. Mackenzie, who described him as '6'5" tall, beautifully proportioned and an extraordinarily fine man. He is many times a millionaire, is a personal friend of the president's.' Mackenzie Diary, 26 October 1942.
68 Merck was accompanied to Ottawa by his special assistant, writer John Marquand, General Raymond Kelser, and Colonel Willard Peake, CWS. Secretary of War Stimson provided Merck with special authorization to cross the border without interference by American or Canadian customs

officials. Merck Papers, Box 187, Canada File, Stimson to Ralston, 21 October 1942.
69 HQS, 4354-1-23, Murray to Maass, 19 October 1942.
70 Ibid.
71 While Maass was in Britain in August 1942, he had been asked by Fildes whether Canada could support the British anthrax project. Murray Papers, vol. 29, Maass to Murray, 21 August 1942; Mackenzie Diary, 24 November 1942.
72 HQS, 4354-8-4-1 Murray to Maass, 2 December 1942; ibid., 4354-33-13-3, Reed to Maass, 1 December 1942.
73 GIN was an abbreviation for the Grosse Île (N) Anthrax project; GIR stood for the Rinderpest (R) research project. Ibid., Murray to Maass, 8 December 1942.
74 HQS, 4354-33-13-3, Murray to General Kelser, 27 January 1943; Reed to Fildes 3 March 1943.
75 Ibid., Murray to Maass, August 30, 1943.
76 In February 1943 a member of the C-I Committee was sent to Porton to become acquainted with the 'N' charging machine. HQS, 4354-27-1-1, General Young to Ralston, 16 February 1943.
77 In December 1942 Murray had informed Maass that he had discussed possible 'N' field tests with A.E. Cameron of the Department of Health and Welfare, who agreed to the proposal that there be double fencing around the Suffield test area, with a safety buffer of at least five square miles.' Ibid., Murray to Maass, 12 December 1942.
78 Murray's major contact at Suffield was Major J.C. Paterson. HQS, 4354-1-23, Murray to Maass, 23 July 1943; ibid., 4354-33-13-3, Major Bishop to Paterson, 23 April 1943; ibid., Murray to Reed, 18 June 1943.
79 In August 1943 the Inter-Service Chemical Warfare Board approved the decision by the Special Weapons Division 'to implement a Most Secret project involving biological work' and agreed that Suffield could be used to test these weapons. DHist, 745.043D, Minutes ISB.
80 HQS, 4354-33-13-5, Murray to Maass, 26 December 1942.
81 HQS, 4354-33-13-3, Murray to Maass, 8 December 1942.
82 HQS, 4354-33-13-5, Shope to Murray, 18 July 1943.
83 Ibid., Shope to Maass, 21 October 1943.
84 HQS, 4354-1-23, Murray to Maass, 7 August 1943.
85 HQS, 4354-33-13-3, Murray to Major E.C. Duthie, 12 July 1944; ibid., 31 July 1944.
86 Ibid., Murray to Duthie, 14 August 1944.
87 By August 1944, Grosse Île had produced 'a stock' of 500 litres of anthrax

with 'an average spore count of 4 × 10/10 ... equivalent to some 2,000 Type F 4-lb bombs.' Merck Papers, Box 186, 'Minutes of the meeting of the C-I Committee together with representatives of the U.K. and Suffield, held in Ottawa 8 January 1945.'
88 In January 1946 the U.S.–Canadian Commission agreed to turn most of its rinderpest stockpile over to the United Nations Relief and Rehabilitation Agency 'for use in China.' HQS, 4354-1-23, Reed to Maass, 14 January 1946; Robert Paterson, U.S. Secretary of War, to Douglas Abbott, Minister of Finance, 8 May 1946. NAW, State/War/Navy Coordinating Committee, Lot 57, Box 74, file 15 M, Fred to Dr Richard Tolman, 26 May 1946.
89 HQS, 4354-1-23, Murray to Colonel Chittick, 23 December 1943.
90 Cochrane, *Biological Warfare Research*.
91 It was estimated the concentrated toxin would only require 0.75 gamma (one-thousandth part of a milligram) to kill a 75 kg man.
92 The usual process of 'bot tox' refinement involved nutrition, harvesting, storage, and preparation for dissemination, as a liquid, dry powder, or crystalline toxin. The latter process (crystallization) was particularly important. Cochrane, *Biological Warfare Research*.
93 In July 1944 Murray informed Colonel Paterson that Reed's 'grinding of X' was going well and that '10 lbs of mud will be treated as quickly as possible.' Reed's major problem was how 'to get the mass particle size down to 10 microns or less.' HQS 4354-1-23, Murray to Paterson [Davies's assistant] 31 July 1944.
94 HQS, 4354-33-13-2, Murray to Davies 14 August 1944. See also Merck Papers, Box 186, Minutes of the Meeting of the C-I Committee Together with representatives of the U.K., and Suffield, Ottawa, 8 January 1945.
95 Ibid., Davies to Fildes, 25 November 1944 (telegram).
96 Merck Papers, Box 186, 'Minutes of CI meeting ... Ottawa 8 January 1945.'
97 HQS 4354-1-23, Fildes to Davies 7 December 1944; ibid., Maass to Davies 5 April 1945.
98 Ibid., Davies to Maass, 5 April 1945.
99 Field tests at Suffield had used 'virulent Strain 19 of Brucella abortus in a 'Type F 4-lb bomb charged slurry.' Merck Papers, Box 186, 'Minutes of CI meeting ... Ottawa January 8, 1945.'
100 The 'shot-gun shell,' or SS, was a Camp Detrick innovation that worked on the principle 'that the smaller the bomb, the larger the burst-cloud relative to the amount of the filling.' Cochrane, *Biological Warfare Research*.
101 Ibid.
102 HQS, 4354-33-16-2, Fothergill to Maass, 15 February 1945; ibid., Maass to Fothergill, 20 February 1945.

103 Ibid.; Barton Bernstein, 'America's Biological Warfare Program in the Second World War,' *Journal of Strategic Studies* 11 (3) (September 1988), 308–10.
104 By January 1945 American and British military planners were hopeful that the Vigo plant would be able to produce one million anthrax bombs, sufficient to obliterate six German cities; this estimate soon proved unrealistic. John Moon, 'United States Biological Warfare Planning and Preparedness: The Dilemmas of Policy,' in Erhard Geissler and John Ellis van Courtland Moon, eds, *Biological and Toxin Weapons: Research, Development, and Use from the Middle Ages to 1945* (Stockholm International Peace Research Institute, forthcoming).
105 Merck Papers, 187, Marquand File, Marquand to Harvey Bundy, 31 August 1943; ibid., William Sarles to Merck, 20 April 1944.
106 Paul Fildes was the committee's technical director. Merck Papers, Box 187, Colonel Paget to Sarles, 16 November 1944.
107 In carrying out its assessment of enemy intentions, ISSCBW was instructed to maintain close contact with the Joint Intelligence Sub-Committee. Ibid.
108 Bryden, *Deadly Allies*, 121.
109 Merck Papers, Box 187, Sarles File, Sarles to Merck, 30 September 1944.
110 Ibid., Sarles to Merck, 7 August 1944.
111 Maass had been recommended for membership by General Porter of the CWS. Ibid., Box 186, Sarles to Merck, 20 November 1944.
112 In terms of military precedent, Merck was advised that there was 'nothing in the Combined Chiefs of Staff documents relating to Canadian liaison.' Ibid., Note for Lord Stamp, 17 November 1944; ibid., Sarles to Merck, 20 November 1944.
113 Stimson immediately approved Merck's recommendation. An embarrassed Colonel Paget was forced to declare 'that he had no right to object to the Canadians being invited to the USBWC and *that his previous expressed objections were unofficial.*' (Emphasis in the original.) Ibid., Merck to Stimson, 2 December 1944.
114 Merck Papers, Box 186, 'Minutes of the Meeting of the C-I Committee Together with representatives of the U.K., and Suffield, Ottawa, 8 January 1945.'
115 HQS, 4354-33-15, Lt.-Col. H.N. Worthley, CWS Special Projects to Murray, 27 November 1944; Murray to Worthley, 27 February 1945.
116 HQS, 4354-33-17, Paget to Maass, 5 October 1944.
117 Ibid., Col. Goforth, memorandum to Chief General Staff, 16 October 1944.
118 Murray also suspected that SES superintendent E.L. Davies had divided loyalties, and that he gave priority to British rather than Canadian projects. Ibid., Murray to Davies, 24 November 1944.

119 Murray also charged Davies with using unqualified scientific personnel in dealing with dangerous agents. Ibid., Murray to Maass, 13 December 1944.
120 Merck Papers, vol. 186, Fildes to Colonel J.H. Defandorf, 7 December 1944.
121 NAW, Joint Committee on New Weapons (JCNW), Box 19, G. Edward Buxton, Acting Head of OSS, to JCS, 17 December 1943.
122 Ibid.
123 NAW, Records of the United States Joint Chiefs of Staff, (CC) file 385.2, 12-17-43, Report by the Joint Committee on New Weapons and Equipment, 20 February 1944.
124 HQS, 4354-33-13-6, Young to Ralston, 17 December 1944.
125 HQS, 4354-33-13-6, Report of Murray of 28 December 1944 Conference at U.S. Army Surgeon General's Office, Washington.
126 Jonathan King and Harlee Strauss, 'The Hazards of Defensive Biological Warfare Programs,' in Susan Wright, ed., *Preventing a Biological Arms Race*, 120–32.
127 Cochrane, *Biological Warfare Research*, 274–5.
128 HQS, 4354-33-13-6, Murray to Maass, 20 March 1944.
129 Ibid., Maass to C.J. Mackenzie, 15 February 1944; ibid., Murray to Maass, 22 March 1944.
130 Ibid., Reed to Colonel Morris, 23 March 1944.
131 Ibid., Murray to Maass, 6 April 1944.
132 Ralston Papers, vol. 30, Stuart to Murchie, 19 May 1944.
133 Ibid., Murchie to Stuart, 25 May 1944.
134 Adolph Hitler opposed using biological weapons until the Allied bombing of Dresden in February 1945. Erhard Geissler, 'Biological Warfare Activities in Germany, 1923–1945,' in Geissler and Moon, eds., *Biological and Toxin Weapons*.
135 Ralston Papers, vol. 30, Murchie to Stuart, 14 June 1944.
136 Cochrane, *Biological Warfare Research*, 280.
137 Ralston Papers, vol. 30, Stuart to Murchie, 8 June 1944.
138 HQS, 4354-33-19, Minutes of the sub-committee of the CWISB, 20 January 1945.
139 Harris and Paxman, *Higher Form of Killing*, 131–4.
140 Cochrane, *Biological Warfare Research*, 437–82.
141 In October 1944, General William Porter asked James Conant for assistance in developing an anthrax spore munition that could be dispersed 'in clouds of powder or liquid droplets ... from airplanes in precision bombing operations. OSRD, 227, General Porter to Conant, 16 October 1944.
142 Merck Papers, vol. 186, Lt.-Col. Woolpert, memorandum to William Sarles, 25 October 1944.

143 The rice pathogens were brown spot of rice (code named E) and rice blast (R). Cochrane, *Biological Warfare Research*, 395–400.
144 Merck Papers, vol. 186, Report J. Davidson-Pratt, 1 March 1945; ibid., Minutes of the Inter-Services Committee on Chemical Warfare, 13 February 1945.
145 DND, Directorate of History, 74/714, Report, Directorate of Intelligence, R.C.A.F. headquarters, 14 February 1945.
146 NAW, OSRD, Box 7, War Department Report, Military Intelligence Division, 1 September 1945.
147 Ibid., Lt.-Colonel Arvo Thompson, CWS to Chief, Special Projects Division. CWS, 1 January 1945.
148 Ibid., Carroll Wilson, to Alexander King, BCSO, 26 April 1945.
149 By February 1945 twenty-five balloons had been recovered, four of them in Canada, including one as far east as Moose Jaw, Saskatchewan. Ibid., Report War Department, 12 February 1945.
150 HQS, 4354-32-1, Maass to General Munchie, 23 February 1945.
151 Ibid., Minutes of the Thirty Fifth Meeting of the Canadian Chemical Warfare Inter-Service Board, 1 March 1945.
152 OSRD, file 227, Bush to George Merck, 2 July 1945; ibid., General W.A. Borden to Bush, 11 July 1945; ibid., William Sarles to Bush, 12 July 1945.
153 It is estimated that thousands of Chinese civilians and prisoners of war were used in a series of brutal experiments carried out by Japanese medical scientists, including being deliberately exposed to bubonic plague, cholera, and typhus pathogens. They all died. See Sheldon Harris, *Factories of Death*, 50–146.
154 Since the relevant Japanese records were destroyed, one is left with these denials. Michael Unsworth, '"A Threat Well within the Capabilities," Japanese Balloon Bombs and Biological Warfare,' paper presented at the American Military Institute Annual Meeting, Bethesda, Maryland, 5 April 1986.
155 Peter Williams and David Wallace, *Unit 731*, x–xxviii, 75. See also Charles G. Roland, 'Allied POWs, Japanese Captors and the Geneva Convention,' *War & Society* 9 (2) (October 1991), 83–101.
156 On 3 October 1946 Maass advised Guilford Reed about the Joint Intelligence Report on Canada's postwar defence that was being prepared for Cabinet and asked if he could submit 'an appreciation of the possible use of B.W. agents in war, to be incorporated in this joint report.' HQS 4354-29-13-1.
157 The British delegation of Air Marshal Bottomley, Major General Brunskill, and Davidson-Pratt noted that the United States was 'developing two or more weapons based on non-epidemic and two or more epidemic.' Among

the latter category, they were most impressed by American efforts to mass produce brucellosis (brucella suis) which had a toxicity 'about 10,000 times that of "N" [anthrax]' for use in four-pound bombs. Merck Papers, vol. 186, Minutes of the ISSCBW meeting, 1 February 1945.
158 Merck Papers, vol. 186, Meeting ISSCBW, 1 February 1945.
159 Ibid., Meeting of ISSCBW, 11 June 1945.
160 Maass reported that Section E at Suffield was available for future Porton tests as long as the C-I Committee considered them 'safe from an epidemic point of view.' Ibid.
161 On 25 July 1945 Murray submitted his resignation as chairman of the C-I Committee to Otto Maass. Predictably, he recommended Guilford Reed as his successor. HQS, 4354-33-17, Murray to Maass, 25 July 1945.
162 In July 1945 the British Medical Research Council declared that its role was 'to promote by research, knowledge conducive to the saving of human life, whereas the primary object of BW is destruction.' Merck Papers, vol. 186, Minutes of the ISSCBW meeting, 25 July 1945.
163 According to David Henderson, a future director of biological warfare research at Porton, biological weapons research was essential so 'that no Nation would dare initiate it, however desperate its situation might be.' Ibid., Minutes of the ISSCBW meeting, 25 July 1945.

7 Atomic Research: The Montreal Laboratory, 1942–1946

1 Cited in Eggleston, *National Research*, 314.
2 Kevles, *The Physicists*, 326; Spencer Weart, *Nuclear Fear: A History of Images* (Cambridge, MA, 1988), 97–132.
3 Robert Bothwell, *Nucleus: The History of Atomic Energy of Canada Limited* (Toronto, 1988), 107, 127, 286, 414.
4 Margaret Gowing, *Britain and Atomic Energy, 1939–1945* (London, 1965); Bertrand Goldschmidt, *Atomic Rivals* (New Brunswick, NJ, 1990); Spencer Weart, *Scientists in Power* (Cambridge, 1979).
5 Report by M.A.U.D. Committee on the Use of Uranium for a Bomb,' cited in Gowing, *Britain and Atomic Energy*, 394–436.
6 The estimated cost of a plant that could produce 1 kg of uranium 235 a day, 'or 3 bombs per month,' was estimated at $5,000,000. Ibid., 77.
7 Ibid., 96.
8 Anderson had been trained as a physical chemist and had worked on the chemistry of uranium. Ibid., 107.
9 Akers was the director of research at Imperial Chemical Industries; he was well regarded as a scientific administrator and had many powerful friends

in government and industry. *Biographical Memoirs of Fellows of the Royal Society*, vol. 1 (London, 1955), 1–4.
10 Members of the advisory committee were Lord Hankey, Lord Cherwell, Sir Edward Appleton, the Minister of Aircraft Production, and Sir Edward Dale of the Royal Society. Growing, *Atomic Energy*, 78–103.
11 Weart, *Scientists in Power*, 65–103.
12 Stewart Cockburn and David Ellyard, *Oliphant: The Life and Times of Mark Oliphant* (Adelaide, 1981), 108–9.
13 Ibid., 122–3.
14 In November 1941 Anderson told the OSRD representative in Britain that he was 'disturbed about the possibility of a leakage of information to the enemy' since the United States was not yet at war. Gowing, *Britain and Atomic Energy*, 94.
15 Ibid., 115–46; Martin Sherwin, *A World Destroyed: The Atomic Bomb and the Grand Alliance* (New York, 1977), 36–53.
16 Kevles, *Physicists*, 326–7.
17 NAW, OSRD, (Sherwin file), Bush to E.V. Murphree, V.P. of Standard Oil, 15 May 1942.
18 Gowing, *Britain and Atomic Energy*, 185–9; Bothwell, *Nucleus*, 20–43.
19 Cited in Ferenc Szasz, *British Scientists and the Manhattan Project: The Los Alamos Years* (New York, 1990), 9.
20 PRO, AB1/379, Lord Anderson to Malcolm Macdonald, 6 August 1942.
21 John Wheeler-Bennett, *John Anderson, Viscount Waverley* (New York, 1962), 291–4.
22 As a member of the Tizard mission, John Cockcroft had shown great interest in the uranium-graphite research of George Lawrence and Weldon Sargent. Mackenzie had been kept informed of major British developments in the nuclear field by G.P. Thomson. Christine King, *E.W.R. Steacie and Science in Canada* (Toronto, 1989), 79.
23 Bothwell, *Nucleus*, 20–88.
24 NAC, Papers of Clarence Decatur Howe (Howe Papers), vol. 13, file F30, 'Memorandum on proposed procedure for a cooperation project as between the Governments of the United Kingdom and Canada,' 26 September 1942; NRC, vol. 284, Mackenzie to Edward Appleton, 2 October 1942.
25 Ronald Clark, *Sir Edward Appleton* (Toronto, 1971), 125.
26 With the fall of France, Halban and his colleague Lew Kowarski carried the Paris laboratory's precious supply of heavy water to Cambridge University, where they conducted slow neutron heavy water research before coming to Canada. Weart, *Scientists in Power*, 66, 72, 167.
27 PRO, AB1/379, Halban to Perrin, 22 December 1942; Clark, *Appleton*, 126.

28 Ibid., Malcolm Macdonald to Akers, 25 September 1942.
29 Goldschmidt, *Atomic Rivals*, 160–80.
30 Goldschmidt's Chicago experience was also a great asset when he and Paneth set up the new chemistry laboratory, especially since he had 'made a complete list of all the apparatus and chemicals used in Seaborg's laboratory.' AB1/359, Akers to Perrin, 5 November 1942.
31 NRC, vol. 283, Halban to Mackenzie, 15 February 1943.
32 Ibid., Mackenzie to Halban, 19 February 1943. There is some debate why the Laval scientist was blacklisted. According to Mackenzie, there were concerns about Rasetti's 'political background,' while Luc Chartrand et al. in their book *Histoire des sciences au Québec* (Montreal, 1987), 409, claim that Rasetti refused to participate in war work because of his pacifist beliefs.
33 During the prolonged negotiations, Akers often became frustrated with Mackenzie's inability to 'understand how short in physicists the team is, and how much work has got to be done.' PRO, AB1/359, Akers to Perrin, 9 December 1942.
34 Among the British born, the most important were physicists Alan Nunn May, A.G. Maddock, and ICI engineers R.E. Newell and D.W. Ginns. PRO, AB1/50, Perrin to Canadian Immigration Branch, London, 10 November 1942.
35 In his diary he described William Akers, the project's director, as being 'very bright ... scientifically sound and a first class engineer.' Mackenzie Diary, 12 November 1942.
36 Ibid., 10 December 1942.
37 B.W. Sargent, Queen's; H.G. Thode, McMaster; and H.M. Cave and W. Henderson of the NRC had already been designated as excellent candidates for the Montreal Lab.
38 University of Manitoba Archives, President's Papers, Box 71, folder 15, Thomson to C.J. Mackenzie, 1 May 1942, 'Visits to Universities at Winnipeg, Saskatoon, Edmonton, Vancouver and Victoria, 18–26 May 1942'; Interview with Gerhard Herzberg, Ottawa, 31 August 1990; Lawrence Stokes, 'Canada and an Academic Refugee from Nazi Germany: The Case of Gerhard Herzberg,' *Canadian Historical Review* 57 (2) (June 1976), 150–70.
39 Interview George Volkoff, Vancouver, 8 June 1982.
40 NRC, vol. 283, Mackenzie to Professor Gordon Shrum, 13 November 1942.
41 Ibid., F.A. Paneth, 'Report on Visit to Dr. Thode,' 11 March 1943.
42 NRC. vol. 283, James Conant to C.D. Howe, 20 February 1943; ibid., Mackenzie to W.J. Bennett, Executive Assistant, Minister of Supply, 13 March 1943.

43 Ibid., Mackenzie to C.P. Gilmour, Chancellor, 19 March 1943; Interview with Harry Thode, Hamilton, Ontario, 15 May 1982.
44 Interview with J. Carson Mark, Los Alamos, 15 November 1993. J. Carson Mark, 'A Maverick View,' unpublished recollections, in possession of author.
45 Interview with Leo Yaffe, McGill University, 16 June 1984.
46 The key positions were as follows. Theoretical Physics: Head, G. Placzek. Physics: Head, P.V. Auger; Section Leaders, B. Pontecorvo and A.N. May. Chemistry: Head, F.A. Paneth; Section Leaders W.J. Arrol; B.L. Goldschmidt; J. Gueron. Engineering: Head, Mr. R.E. Newell; Administrative Secretary: J.F. Jackson. Instrument Making Shop: N.Q. Lawrence; Radio: H.F. Freundlich; Counter-making: N. Veall.
47 NRC, vol. 283, R.W. Boyle to George Lawrence, 2 March 1943.
48 Although Mackenzie had initially favoured having the laboratory in Ottawa, he eventually conceded to the arguments that the University of Montreal site was superior because of the availability of housing and the proximity to McGill's scientific resources. NAC, vol. 173, Lesslie Thomson memo, 18 November 1942; ibid., Mackenzie to Conant, 12 December 1942.
49 NRC, vol. 173, R.C. Nickle to Lesslie Thomson, 29 January 1943; ibid., Halban to Lesslie Thomson, 13 January 1943. Goldschmidt, *Atomic Rivals*, 241-6.
50 Yaffe interview; interview with George Lawrence, Deep River, Ontario, 20 November 1981.
51 Even his French colleague, Bernard Goldschmidt, admitted that Halban 'as a manager of his team ... had been a failure.' Goldschmidt, *Atomic Rivals*, 205-6.
52 PRO, AB1/379, Akers to Perrin, 23 July 23 1943.
53 Ibid., Akers to Perrin, 29 July 1943.
54 Ibid., Akers to Perrin, 23 July 1943.
55 This subject has been extensively discussed in Gowing, *Britain and Atomic Energy*; James Conant, *My Several Lives: Memoirs of a Social Inventor* (New York, 1970), 286-304; James Hershberg, *James B. Conant: Harvard to Hiroshima and the Making of the Nuclear Age* (New York, 1993), 154-94; and Sherwin, *A World Destroyed*, 77-9.
56 Sherwin, *A World Destroyed*, 77-9; Hershberg, *Conant*, 183-5.
57 NRC, vol. 284, Conant to Mackenzie, 2 January 1943; Conant, *My Several Lives*, 286-95.
58 Goldschmidt, *Atomic Rivals*, 180.
59 This offer did not extend to British and Canadian engineers, who were excluded from contact with the Chicago research teams. On the other hand,

Notes to pages 186–9   339

Conant did offer to release all of the Trail, B.C. heavy water for the use of Montreal Laboratory scientists. Ibid.; Bothwell, *Nucleus*, 31–3.
60 NAC, Mackenzie Diary, 29 December 1942.
61 Gowing, *Britain and Atomic Energy*, 154–64.
62 NAC, Mackenzie Diary, 7 January 1943.
63 Mackenzie's assessment of Sir John Anderson as an obstinate and naïve politician did not change after their short and nasty meeting in London in 1943. Ibid., 11 May 1943.
64 Ibid., 18 January 1943.
65 Ibid., 18 January 1943.
66 Ibid., 3 August 1943, 26 July 1943.
67 Goldschmidt, *Atomic Rivals*, 180.
68 PRO, AB1/379, Akers to Perrin, 29 July 1943.
69 In the spring of 1943 British authorities had carried out a feasibility study of what it would cost to construct a separation plant for uranium 235, a heavy water plant, and a large plutonium-producing reactor. The estimate was prohibitive: a tenfold increase in expenditures. Goldschmidt, *Atomic Rivals*, 192.
70 NAC, Mackenzie Diary, 3 August 1943, 26 July 1943.
71 Not all American nuclear scientists approved of Conant's aggressive American-first approach. Indeed, one of his most outspoken critics was Nobel laureate Harold Urey, who, in a June 1943 letter, denounced Conant's tactics for causing 'a delay of a year or more, in establishing the feasibility of a homogeneous heavy water pile.' NAW, Records Manhattan Engineer District (hereafter MED), vol. 201, Urey File, Urey to Conant, 21 June 1943.
72 Conant and Bush, however, had been concerned that Roosevelt might reverse American policy because of Churchill's sustained pressure for a renewal of cooperation. Conant, *My Several Lives*, 298–300.
73 NAW, Bush-Conant File Relating to the Development of the Atomic Bomb, 1940–1945, microfilm, roll 1, Conant to Bush, 6 August 1943.
74 The negotiations leading up to the Quebec Agreement are extremely complex and have been extensively analysed by Martin Sherwin, Margaret Gowing, Bertrand Goldschmidt, and others.
75 OSRD (Sherwin file), Bush to Conant, 2 September 1943.
76 NRC, vol. 273, Akers Report, 13 September 1943; OSRD (Sherwin file), Bush to Anderson, 6 August 1943.
77 NRC, vol. 284, Akers to Halban, 18 October 1943.
78 NAW, OSRD (Sherwin file), Bush to Roosevelt, 23 August 1943.
79 NAW, Bush-Conant File Relating to the Development of the Atomic Bomb, 1940–1945, Groves to Conant, 2 November 1943.

80 After the Quebec Conference, Mackenzie had little doubt that Groves dominated the American atomic program, for one fundamental reason: 'He is the man who is actually entrusted with the spending of hundreds of millions of dollars.' Mackenzie Diary, 10 September 1943.
81 NRC, vol. 284, I, Chadwick to Mackenzie, 31 December 1943.
82 Mackenzie Diary, 17 September 1943. See also Leslie Groves, *Now It Can Be Told: The Story of the Manhattan Project* (New York, 1962); William Lawren, *The General and the Bomb: A Biography of General Leslie R. Groves, Director of the Manhattan Project* (New York, 1988).
83 NRC, vol. 284, Chadwick to Mackenzie, 31 December 1943.
84 Andrew Brown, *The Neutron and the Bomb: A Biography of Sir James Chadwick* (New York, 1997), 195–298.
85 Howe Papers, vol. 14, file 32, 'Memorandum by Brigadier General L.R. Groves,' 10 December 1943.
86 Cockburn and Ellyard, *Oliphant*, 47–93.
87 Szasz, *British Scientists*, xix, 16–31.
88 On 21 May 1946, while helping in the preparations for the Bikini tests, Slotin was involved in an accident that exposed him to a fatal dose of radiation. *Bulletin of Atomic Scientists*, 1 June 1946.
89 NRC, vol. 283, Personnel File, Webster to Mackenzie, 27 February 1945.
90 Goldschmidt, *Atomic Rivals*, 190–210; PRO, AB1/379, Akers to Perrin, 8 December 1943.
91 This was certainly the advice of Mark Oliphant, who in April 1944 pressed Chadwick to wind up the Montreal show since it had 'no relation whatever to the war effort.' PRO, AB1/58, Chadwick to Appleton, 17, 20 April 1944.
92 Arthur Compton, *Atomic Quest: A Personal Narrative* (New York, 1958), 149–215.
93 NAC, Mackenzie Diary, 18–20 September 1943.
94 NRC, 284, Minutes of Meeting in Chicago on 8 January 1944 to Discuss Future Collaboration between the Chicago and Montreal Laboratories Engaged in Tube Alloy Research.
95 Goldschmidt, *Atomic Rivals*, 210–11.
96 PRO, AB1/58, Chadwick to Appleton, 5 February 1944.
97 PRO, AB1/379, Akers to Perrin, 8 December 1943.
98 PRO, AB1/58, Chadwick to Appleton, 17 April 1944.
99 King, *Steacie*, 86.
100 NAW, Bush–Conant File, #33A, Conant to Groves, 18 February 1944.
101 PRO, AB1/485, Webster to Chadwick, 23 March 1944.
102 Mackenzie Diary, 17 February 1944.

103 During the discussions, Chadwick became increasingly annoyed with Conant's anti-British arguments. According to Mackenzie, he even 'talked about withdrawing and going back to England if they [the CPC] did not agree to the Montreal proposals.' NAC, Mackenzie Diary, 17 February 1944; NRC, vol. 284, Chadwick to Mackenzie, 24 February 1944.
104 Mackenzie Diary, 17 February 1944; NRC, vol. 284, Chadwick to Mackenzie, 24 February 1944.
105 While these delicate negotiations were being conducted, the Canadian government nationalized the Eldorado Mining Company, a major supplier of uranium for both the Montreal Laboratory and the Manhattan Project. For a complete account of the ongoing dispute over Canadian uranium, see Robert Bothwell, *Eldorado: Canada's National Uranium Company* (Toronto, 1984), 107–54.
106 Mackenzie Diary, 9 June 1944.
107 Howe Papers, vol. 14, file 32, Résumé of Report of Subcommittee on Joint Development of a heavy water pile to the Combined Policy Committee, 10 April 1944.
108 NAW, MED, Canadian Liaison File, Minutes of meeting 8 June 1944 of the CPC sub-committee; NAW, OSRD, Box 63, Canadian Liaison, A Review of Liaison Activities between the Canadian and the United States Atomic Energy Projects, 20 February 1947.
109 Cockcroft welcomed the arrival of the two MED scientists, suggesting that if Groves wanted to send another ten U.S. physicists, he would be delighted. PRO, AB1/278, Cockcroft to Appleton, 8 May 1944; MED, Canadian Liaison File, minutes of meeting 8 June 1944 of the CPC sub-committee.
110 One of Benbow's tasks was to make arrangements for the RCMP to assign a permanent security officer to the Montreal Laboratory. MED, Box 21, 'Liaison with the Canadians,' memorandum Benbow to Groves, 13 July 1944.
111 Goldschmidt, *Atomic Rivals*, 209.
112 Since June 1944 there had been extensive collaboration between Montreal and Chicago on uranium 233, with the exchange of monthly reports and reciprocal visits. MED, Box 21, 'Liaison with the Canadians,' Report, 23 October 1944.
113 Ibid., Report, 5 August 1944, 18 November 1944.
114 Sherwin, *World Destroyed*, 108–11, 284–5.
115 Ibid., 111.
116 PRO, AB1/278, Cockcroft and Steacie to Mackenzie, 27 September 1944.
117 Cockcroft felt that Mackenzie was overreacting, in part because of fears that

the King government would be defeated in the June election, and that he would have to justify the Montreal project 'to a new Administration.' PRO, AB1/193, Cockcroft to Chadwick, 22 May 22; NAC, Mackenzie Diary, 17 May 1945.
118 Cockcroft also proposed that Britain would supply about ten high-quality scientists 'and that other Dominions might welcome the opportunity of contributing staff.' Ibid.
119 Although none of the British Los Alamos team were transferred to Chalk River, Rudolph Peierls and Otto Frisch visited the site on their way back to England. PRO, AB1/485, Chadwick to Peierls, 6 April 1945; Otto Frisch, *What Little I Remember* (Cambridge, 1969), 192–6.
120 PRO, AB1/485, Chadwick to Moon, 10 September 1945.
121 Gar Alperovitz, *The Decision to Use the Atomic Bomb, and the Architecture of an American Myth* (New York, 1995). Historians supporting Truman's decision include McGeorge Bundy, *Danger and Survival: Choices about the Bomb in the First Fifty Years* (New York, 1988), and Robert Donovan, *Conflict and Crisis: The Presidency of Harry S. Truman, 1945–1948* (New York, 1977). In addition to Alperovitz, the most strident critics of the decision have been Barton Bernstein, ed., *The Atomic Bomb: The Critical Issues* (Boston, 1976) and Sherwin, *A World Destroyed*.
122 Alperovitz, *Decision*, 370–3.
123 Mackenzie Diary, 5 July 1945.
124 Ibid., 30 July 1945.
125 Ibid., 1–12 August 1945; Howe Papers, vol. 13, file 30, Press Release by Honourable C.D. Howe.
126 In February 1945 British High Commissioner Malcolm Macdonald told King about the possibility that atomic bombs might be used against Japan. This message was reinforced a month later when King visited President Truman in Washington. J.W. Pickersgill and D.F. Forster, *The Mackenzie King Record*, vol. 3: *1945–46* (Toronto, 1968), 448.
127 James Eayrs, *In Defence of Canada: Peacemaking and Deterrence* (Toronto, 1972), 274–8.
128 Mackenzie King Diaries, transcript (microfiche), 6 August 1945.
129 Dhist, file 3-11-2-1, Colonel Goforth to General Letson, (memorandum of telephone call), 8 August 1945.
130 NRC, 4-H-6-9, W.G. Mills to Mackenzie, 21 September 1945.
131 NRC, 83/84, vol. 196, file 4-H-6-9, Norman Robertson to Mackenzie, 10 August 1945.
132 Howe Papers, 13/30, W.J. Bennett to Robertson, 11 August 1945.
133 Sir John Anderson was appointed chairman of the committee; other members were Patrick Blackett, James Chadwick, G.P. Thomson, and Sir Henry

Dale. Royal Society, London, Patrick Blackett Papers, D-184, Attlee to Blackett, 17 August 1945.
134 Ibid., File D-185. Memorandum to Sir James Chadwick from the Montreal Laboratory Scientists, 30 August 1945.
135 On 5 December 1945, in a widely publicized speech, C.D. Howe stated Canada's fundamental position on nuclear weapons: 'We have not manufactured atomic bombs, we have no intention of manufacturing atomic bombs.' Canada, House of Commons *Debates*, 5 December 1945, 2959.
136 Bothwell, *Nucleus*, 54–76.
137 In the fall of 1945 Groves encouraged the leading American nuclear scientists to visit Chalk River. These included Glenn Seaborg (Nov. 1945), Walter Zinn (Sept. 1945), Eugene Wiger (Oct. 1945), and J.A. Wheeler and Edward Teller (Sept. 1946). Ibid., Groves to Chadwick, 6 November 1944; OSRD, Reports, May 1945–July 1946, passim; Mackenzie Diary, 16 July 1945.
138 In 1945, the Americans sent Chalk River ten tons of uranium rods, a number of radioisotopes, additional irradiated slugs, radiation detecting devices, and various instruments. OSRD, Canadian Liaison, Report, 17 November 1945.
139 Goldschmidt, *Atomic Rivals*, 280. In September 1946 British physicist W.B. Lewis was appointed scientific director of Chalk River, a position he held for twenty-seven years. Ruth Fawcett, *Nuclear Pursuits* (Montreal/Kingston, 1994), 40–64.
140 Goldschmidt, *Atomic Rivals*, 282.
141 NAW, Bush-Conant Files, Mackenzie File, Mackenzie to Conant, 14 August 1945.
142 Ibid., Conant to Mackenzie, 24 August 1945.
143 Historian Robert Bothwell has endorsed some of the criticisms of Mackenzie in his article 'Weird Science: Scientific Refugees and the Montreal Laboratory,' in Norman Hillmer et al., eds, *On Guard for Thee: War, Ethnicity, and the Canadian State, 1939–1945* (Ottawa, Supply and Services Canada, 1988), 217–32. In contrast, almost all of the Canadian scientists who worked in the Montreal Laboratory gave Mackenzie high marks.
144 Interview with Harry Thode, Hamilton, 15 May 1982

8 Secrets, Security, and Spies, 1939–1945

1 H. Montgomery Hyde, *The Atom Bomb Spies* (London, 1980); Chapman Pincher, *Too Secret Too Long: The Great Betrayal of Britain's Crucial Secrets and the Cover-up* (London, 1984); John Sawatsky, *Men in the Shadows: the RCMP Security Service* (Toronto, 1980).
2 Reg Whitaker and Greg Kealey, eds, *R.C.M.P. Security Bulletins, The War Series, 1939–1941* (St. John's, 1989), 1–15.

3 NAC, Transcript of the evidence presented to the Royal Commission on Espionage (microfilm), 'Exhibit 58, H.Q. 124-1-25, Lieutenant Colonel N.O. Carr, memorandum, 7 January 1935; ibid., H.Q. 1233, circular letter, 2 February 1938.
4 *The Report of the Royal Commission to investigate the facts relating to and the circumstances surrounding the communication, by public officials and other persons in positions of trust, of secret and confidential information to agents of a foreign power* (Ottawa, 1946), 313. Hereafter RC, Espionage, *Report*.
5 RC, Espionage, *Report*, 505.
6 University of Chicago Archives, Federation of American Scientists Papers, Box 21, 4, Norman Veall to Dr Melba Phillips, 28 March 1946.
7 Paul Fussell, *Wartime: Understanding and Behaviour in the Second World War* (New York, 1989).
8 Ronald Stent, *A Bespattered Page? The Internment of His Majesty's 'Most Loyal Enemy Aliens'* (London 1980), 27–9, 47–56.
9 NAC, Mackenzie King Papers, C-239964, memorandum for prime minister, 22 May 1940; J.L. Granatstein, *A Man of Influence: Norman A. Robertson and Canadian Statecraft, 1929–1968* (Toronto, 1981), 80–92.
10 Although the large German-Canadian community found itself for the second time in the century, classified as enemy aliens only about 800 were arrested and interned. Italian Canadians had a higher ratio of internment. Robert Keyserlingk, 'Breaking the Nazi Plot: Canadian Government Attitudes towards German Canadians, 1939–1945,' in Norman Hillmer et al., eds, *On Guard for Thee: War, Ethnicity, and the Canadian State, 1939–1945* (Ottawa 1988), 53–70; Bruno Ramirez, 'Ethnicity on Trial: The Italians of Montreal and the Second World War,' ibid., 71–84.
11 The internment of Japanese Canadians is extensively discussed in Ken Adachi, *The Enemy That Never Was: A History of the Japanese Canadians* (Toronto, 1976), and Peter Ward, *White Canada Forever: Popular Attitudes and Public Policy toward Orientals in British Columbia* (Montreal 1978).
12 Stent, *Bespattered Page*, 1–70.
13 Fuchs spent about a year in a Canadian camp until he was repatriated to Britain. In 1943 he returned to North America as part of the British nuclear team at Los Alamos. During the next two years he gave the Russians much valuable information about the plutonium bomb. Hyde, *Atom Bomb Spies*, 93–116.
14 UTA, Cody Papers, vol. 62, Mendel to Cody, 22 November 1943.
15 Ibid., vol. 1048, Cody to T. Mulligan, 27 September 1940.
16 Cody felt it necessary to emphasize that there 'was no element of anti-

semitism' in connection with the Levine case. Ibid., Cody to Professor Norman Levinson, MIT, 6 January 1942.
17 *Canadian Forum* 21 (250) (November 1941), 245–7.
18 Cody Papers, vol. 1048, Halperin to Cody, 14 December 1941; ibid., Cody to Halperin, 17 December 1941.
19 Ibid., Levinson to Cody, 19 December 1941.
20 Ibid., Levine to Cody, 9 April 1942.
21 R. Douglas Francis, *Frank Underhill: Intellectual Provocateur* (Toronto, 1986), 114–29.
22 Stanley Frost, *McGill University*, vol. 2: *1895–1971* (Montreal, 1984), 200–3.
23 DND, 2159, 54-27-35-60-1, Church to Ralston, 25 March 1943; ibid., Ralston to Cody, 14 April 1944.
24 NRC, vol. 170, 32-1-54, Synge to W.L Webster, 13 January 1943; ibid., Webster to Synge, 27 January 1943; ibid., Mackenzie to Synge, 18 February, 1943.
25 NRC, vol. 69, Macklin to Colonel Walter James Brown, Executive Secretary, University of Western Ontario, 14 March 1942.
26 Interview with Leo Yaffe, 16 June 1984.
27 In 1939 the Intelligence Section of the RCMP was a six-man operation attached to the Criminal Investigation Branch; many of the officers moved between the two fields of investigation. C.W. Harvison, *The Horsemen* (Toronto, 1967), 93–4, 122–67, 205–7; Charles Rivett-Carnac, *Pursuit in the Wilderness* (Toronto, 1965), 301–20.
28 Larry Hannant, 'Inter-War Security Screening in Britain, the United States and Canada,' *Intelligence and National Security* 6 (4) (1991), 711–735, and *The Infernal Machine: Investigating the Loyalty of Canada's Citizens* (Toronto, 1995).
29 In the first years of the war, the RCMP apprehended some 500 members of various Nazi and fascist organizations, as well as over 130 members of the Communist movement. Reg Whitaker, 'Official Repression of Communism during World War II,' *Labour/Le Travail*, 17 (1986), 135–66.
30 CSIS Documents, folder 141, file 117-91-99, Report Supt. E.W. Bavin, 29 October 1940.
31 Christopher Andrew, *Secret Service: The Making of the British Intelligence Community* (London, 1985), 460. Richard Powers, *Secrecy and Power: The Life of J. Edgar Hoover* (New York, 1987), 255–6. Christopher Andrew and Oleg Gordievsky, *KGB: The Inside Story of Its Foreign Operations from Lenin to Gorbachev* (London, 1990), 254–7.
32 RC, Espionage, Transcript, 'Exhibit 64': H.Q. 171, vol. 2.
33 Ibid., 'Exhibit 64,': N.D.H.Q., General Young memorandum, 29 May 1944.
34 Mackenzie Diary, 27 January 1941; 17 June 1944; NRC, vol. 106, 'RDX file,' Linstead to Maass, 18 September 1941.

35 RC, Espionage, Transcript, Exhibit 64, Memorandum, Major A. Bebrisay, Directorate of Artillery, 31 August 1942.
36 Mackenzie Diary, 27 March 1941; NAC, HQS, 4354-2-0, Colonel C.P. Morrison, memorandum for Victor Sifton, 10 May 1941.
37 DND, Records of the Directorate of Chemical Warfare (on microfilm) file 4354-22-5, W.W. Stewart to Maass, 25 September 1942.
38 NRC, vol. 69, Fred to Murray, 21 April 1942; ibid., Murray to Fred, 22 April 1942.
39 Ralston Papers, vol. 30, 'Recommendations to the Department of Defence of Canada by the Committee on Project M-1000,' enclosed in letter, J.R. Collip to Colonel A.A. Magee, Executive Assistant to the Minister of National Defence, 29 December 1941; Murray Papers, 'Minutes of the M-1000 meeting, 19 December 1941.'
40 HQS 4354-2-1, Inspector E.H. Perlson, Assistant Intelligence Officer to Bishop, 21 July 1942.
41 Murray Papers, vol. 29, Murray 'Notes,' 22 February 1942.
42 In the spring of 1942, for example, 500 stencils of U.S. confidential reports were temporarily misplaced, causing much consternation in Washington and Ottawa. NRC, vol. 69, Edwin Fred to Murray, 21 April 1942.
43 RC, Espionage, Transcript, Exhibit, Dr D.W.R. McKinley to Supt. G.E. Rivett-Carnac, Intelligence Branch, RCMP, 3 April 1946.
44 Transcript, *The King vs. Matt Simon Nightingale*, 4 April 1946.
45 NAW, Joint Chiefs of Staff Collection (JCS), CGS 413.44, file 6-24-42 #2, memorandum, JCS, 9 November 1945.
46 Ibid., file 11-22-44, Roosevelt to Bush, 17 November 1944.
47 Ibid., CGS 385.2 (6-24-42); ibid., Memorandum JSC, 9 November 1945.
48 Martin Sherwin, *A World Destroyed: The Atomic Bomb and the Grand Alliance* (New York, 1977), 44–7.
49 Ibid., 85–6.
50 Richard Hewlett and Oscar Anderson Jr, *The New World, 1939–1945: A History of the United States Atomic Energy Commission* (University Park, 1962), 89–170.
51 Sherwin, *World Destroyed*, 59, 73.
52 Stanley Goldberg, 'Groves and the Scientists: Compartmentalization and the Building of the Bomb,' *Physics Today* (August) 1995), 42.
53 NAW, Records of the Manhattan Engineer District (hereafter MED), file, 'Liaison Montreal,' minutes of meeting, 8 June 1944 of CPC subcommittee.
54 Ibid., Report, 20 January 1945; ibid., Report, 17 February 1945; NRC, vol. 284, Cockcroft to Mackenzie, 4 December 1944.
55 Sargent did, however, indicate that most of the Montreal scientists met with

the famous Danish physicist 'one evening at Dr. Cockcroft's house.' Queen's University Archives, J.A. Gray Papers, Sargent to Gray, 24 March 1945. For a concise account of why U.S. officials suspected Bohr, see Sherwin, *World Destroyed*, 106–20.
56 MED, 201, Chadwick File, Groves: Notes on meeting with Dr J. Chadwick, 20 January, 1945.
57 Ibid.
58 PRO, AB1/193, Cockcroft to Chadwick, 16 July 1945; ibid., Chadwick to Cockcroft, 24 July 1945.
59 Groves claimed that Joliot-Curie was a dangerous Communist who would convey atomic secrets to the Soviet Union. MED, Box 21, file, Liaison with the Canadians, 'Memorandum on Security Problems Raised by Employment of French Scientists in Canadian NRX Project,' 20 January 1945; ibid., Mackenzie to Groves, 26 December 1944.
60 Ibid., 'The Substance of Recommendations Which Will Be Made by Major General Groves," 20 January 1945; ibid., Memorandum on Security Problems Raised by Employment of French Scientists in Canadian NRX Project, 20 January 1945.
61 PRO, CAB 122/937, War Cabinet, Allied Supplies Executive Report, 30 May 1944; Foreign Office (FO) 371/36928, Notes on the Working of the Anglo-Soviet Agreement, 28 June 1943.
62 A.J.P. Taylor, *Beaverbrook* (London, 1972), 495; Gabriel Gorodetsky, *Stafford Cripps' Mission to Moscow, 1940–42* (Cambridge, 1984).
63 Imperial War Museum, Tizard Papers, file 320, memorandum to the War Cabinet Chiefs of Staff Committee, 1 June 1942; ibid., file 390, Minutes of the fourth meeting of the Moscow Mission held at the Ministry of Supply, 20 May 1943.
64 Ibid., file 390, Tizard to Oliver Lyttleton, Minister of Aircraft Production, 29 January 1943; ibid., Tizard to Sir Andrew Duncan, Minister of Supply, 19 March 1943; CAB 122/104, Meeting of the Combined Chiefs of Staff, 10 March 1943, Note by the representative of the British Chiefs of Staff.
65 Both the Chief of Army Intelligence, General George Strong, and J. Edgar Hoover, Director of the FBI, suspected a secret exchange system with the Soviets. Ibid., telegram Viscount Halifax to Foreign Office, 7 August 1941; E.H. Beardsley, 'Secrets between Friends: Applied Science Exchange between the Western Allies and the Soviet Union during World War II,' *Social Studies of Science* 7 (1977), 447–73.
66 Tizard Papers, 390, memorandum on agreement between the British and Soviet governments on the exchange of technical information, 23 July 1943.
67 Beardsley, 'Secrets between Friends,' 454–65.

348   Notes to pages 215–17

68 PRO, CAB 122/104 British Joint Intelligence to American Joint Intelligence Committee, Washington, 20 March 1943.
69 PRO, CAB 122 936, Report of the Allied Supply Executive to War Cabinet, 30 October 1943.
70 Ibid., Report of the Firebrace Committee, 17 November 1943.
71 PRO, CAB 122/936, Report of the Allied Supply Executive, 17 November 1943; Eggleston, *National Research*, 186–250.
72 In 1942 Dana Wilgress was sent to the temporary Soviet capital of Kuibyshev as envoy extraordinary and minister plenipotentiary; in 1943, he moved to Moscow, where he remained as Canadian ambassador until 1947. Denis Smith, *Diplomacy of Fear: Canada and the Cold War, 1941–1948* (Toronto, 1988).
73 Gusev was being groomed for more important duties. In August 1943 he was appointed ambassador to Great Britain; his successor, George Zaroubin, arrived in Ottawa on 26 May 1944. RC, Espionage, *Report*, Exhibit 543.
74 NAC, Canadian Mutual Aid Board Records (CMAB); Aloysius Balawyder, 'Canada in the Uneasy War Alliance,' in Aloysius Balawyder, ed., *Canadian Soviet Relations, 1939–1980* (Oakville, 1981), 1–14; Donald Page, 'Getting to Know the Russians, 1943–1948,' ibid., 15–35.
75 John Kolasky, *The Shattered Illusion: The History of Ukrainian Pro-Communist Organizations in Canada* (Toronto, 1974), 27–50; Robert Bothwell and Jack Granatstein, *The Gouzenko Transcripts: The Evidence Presented to the Kellock-Tascherean Royal Commission of 1946* (Ottawa, 1982), 15.
76 The director of military operations and planning, Colonel John Jenkins, later informed the Royal Commission that most of Zabotin's official requests were so reasonable that they 'had all been quickly approved.' RC, Espionage, *Report*, 621–2; Exhibit 444, DND Directive, 23 December 1943, Colonel Jenkins for Chief of Staff.
77 Ibid., Exhibit 444, DND Directive, 23 December 1943, Colonel Jenkins for Chief of Staff.
78 CMAB, 8–23–2, J.R. Donald to K.C. Fraser, 21 July 1943; ibid., British Supply Mission (Washington) to British High Commissioner, Ottawa, 4 August 1943.
79 Ibid., file 7-C-1, Minutes of a special meeting of the Committee on RDF, 30 June 1943.
80 Ibid., file 7-1, Karl Fraser to Howe, 23 August 1943.
81 Ibid., Howe to Fraser, 6 September 1943.
82 In September 1943 the Soviets made their most audacious request: 'fifty tons of uranium salts and a small quantity of uranium metal.' Ibid., 7-M3-14, Fraser to Hume Wrong, 8 September 1945.

83 PRO, CAB, 122/936, memorandum of the British members of the Combined Sub-Committee on Technical Information, 24 January 1944.
84 CMAB, file 7-1, J.R. Donald to G.K. Shields, DM, Department of Munitions and Supply, 7 August 1944; ibid., G.R. Heasman to Shawinigan Chemical Ltd., 19 July 1944.
85 Ibid., J.R. Donald to G.K. Shields, DM, Department of Munitions and Supply, 7 August 1944; ibid., G.R. Heasman to Shawinigan Chemical Ltd., 19 July 1944.
86 Ibid., 7-M3-14, Fraser to Hume Wrong, 8 September 1945.
87 Testimony of Kenneth Cheethan, cited at Raymond Boyer's trial in March 1947; *Ottawa Citizen*, 27 March 1947. Another witness, who had worked at the Shawinigan plant, claimed 'that during 1943 he shipped hundreds of thousands of pounds of the explosive (RDX) to Russia under the Mutual Aid plan.' *Montreal Gazette*, 6 December 1947.
88 In Washington, the Soviet ambassador Maxim Litvinoff had already gained a reputation for hosting lavish parties and cultivating alliances with rich and influential Americans. By the fall of 1943, however, Stalin decided that Litvinoff was 'too pro-American' and had him replaced by Andrei Gromyko, a dour embassy functionary. David Brinkley, *Washington Goes to War: The Extraordinary Story of the Transformation of a City and a Nation* (New York, 1988), 153, 156.
89 H.S. Ferns, *Reading from Left to Right: One Man's Political History* (Toronto, 1983), 184.
90 *Montreal Star*, 28 November 1947.
91 June Callwood, *Emma: The True Story of Canada's Unlikely Spy* (Toronto, 1984), 94–5, 102.
92 John Sawatsky, *Gouzenko: The Untold Story* (Toronto, 1984), 289.
93 Colonel Motinov's primary responsibility was coordinating the work of Alan Nunn May. Colonel Sokolov's major contact was Raymond Boyer; Sokolov and his wife also recruited Emma Woikin, a cipher clerk in the Department of External Affairs. RC, Espionage, *Report*, 118, 268, 356.
94 RC, Espionage, *Report*, 123–59. Lunan interview.
95 Department of External Affairs (DEA), vol. 2620, file 7-1-5-9. RCMP interview with Gousekno [sic], Memo I & II, n.d.
96 By the 1950s the NKVD had evolved into the KGB, an elaborate secret police and intelligence organization, which exerted a powerful influence over Soviet foreign policy until the 1991 breakup of the USSR. Studies analysing the evolution of the GRU and its wartime relationship with the NKVD include Andrew & Gordievsky, *KGB*, and Victor Suvorov, *Soviet Military Intelligence* (London, 1984).

97 RC, Espionage, *Report*, 57–77; Bothwell and Granatstein, *Gouzenko Transcripts*, 187–232, 245–322.
98 Bothwell and Granatstein, *Gouzenko Transcripts*, 43–112.
99 RC, Espionage, *Report*, 74–8.
100 Ferns, *Reading from Left to Right*, 193.
101 Cited in Callwood, *Emma*, 107.
102 Sawatsky, *Gouzenko*, 295.
103 RC, Espionage, *Report*, 44.
104 DEA, 2620, 50242–4D, Arnold Smith, 'Summary of Certain Points in the Report,' 4 July 1946.
105 Interview with Gordon Lunan, 31 August 1990, Ottawa; RC, Espionage, *Report*, 57–70, 123–61.
106 RC, Espionage, Transcript, Mazerall Testimony, 929–40; Smith Testimony, 2658–78.
107 Mazerall was an electrical engineering graduate from the University of New Brunswick; Smith was a physics specialist from McGill. Both joined the Radio Branch in 1942. RC, Espionage, *Report*, 130–45.
108 Halperin, a Ph.D. in mathematics from Yale, taught at Queen's University before becoming a major in the Army's Directorate of Artillery; Shugar, a Ph.D. in physics from McGill, joined the Royal Canadian Navy in 1943 as an electrical sub-lieutenant. Ibid.
109 Squadron-Leader Nightingale, a McGill-trained electrical engineer, joined the RCAF directorate of signals and telephone transmission engineering. Ibid., 261–79.
110 Ibid., 450–1.
111 The Committee for Allied Victory, formed in 1942, was one of the first Canadian pro-Soviet wartime organizations. It actively campaigned for a second front in Europe in order to assist the beleaguered Red Army.
112 During 1942 there was also a major campaign for the release of the Communist leaders who had been interred in 1940 under the Defence of Canada Regulations. Public Archives of Ontario (PAO), Workers Education Association Papers (WEA), Box 51, file 51/170, 'circular letter, Canadian Civil Liberties Association, Toronto, 16 April 1942; Norman Penner, *The Canadian Left: A Critical Analysis* (Toronto, 1977), 119, 140–2.
113 *Globe and Mail*, 23 June 1943.
114 Donald Avery, 'Canadian Communism and Popular Front Organizations,' paper presented at the meeting of the Canadian Political Science Association, June 1983, Vancouver.
115 Similar activities were carried out in the United States by the National Council of American Soviet Friendship. Between 1942 and 1945 it had

branches in twenty-eight U.S. cities. Ralph Levering, *American Opinion and the Russian Alliance, 1939–1945* (Chapel Hill, NC, 1975), 100–2.
116 NAC, Hazen Sise Papers, vol. 5, *Council Bulletin*, May 1944; ibid., July 1944.
117 WEA, 11/5, Minutes of the NCCSF meeting, 6 December 1943.
118 Ibid.
119 Significantly, many of those associated with the NCCSF Scientific Committee would also assume an important role in the creation of the Canadian Association of Scientific Workers.
120 Wilder Penfield, *The British-American-Canadian Surgical Mission to the U.S.S.R.* (Ottawa: NRC, 1943), 1–2.
121 Roy and Kay MacLeod, 'The Contradictions of Professionalism: Scientists, Trade Unionism and the First World War,' *Social Studies of Science* 9 (1979), 1–32; McGucken, *Scientists, Society, and State*, 60–184.
122 Donald Avery, 'Atomic Scientific Cooperation and Rivalry: The Anglo-Canadian Montreal Laboratory and the Manhattan Project, 1943–1946,' *War in History* 2 (3) (1995), 274–305.
123 RC, Espionage, Transcript, 4557, Testimony of Norman Veall.
124 Scattered minutes of the Montreal and Ottawa CAScW chapters, as well as copies of the Montreal and Ottawa *Bulletins* and the *Canadian Scientist*, are located in the Hazen Sise collection at the National Archives of Canada.
125 Sise Papers, vol. 33, 'Formation and Draft Program of the Canadian Association of Scientific Workers,' December 1944.
126 According to Veall, Boyer's selection as president came because of his reputation as a social activist and his prestige as an outstanding chemist. RC, Espionage, Transcript, 4596.
127 The CAScW also recruited members through a series of lectures and workshops involving high-profile guest speakers such as British radar scientist Robert Watson-Watt and U.S. nuclear physicist Hans Bethe. *Canadian Scientist*, January 1945.
128 PRO, CAB/939, Chiefs of Staff Committee, 4 January 1945.
129 Beardsley, 'Secrets between Friends,' 455–65.
130 Churchill College Archives, J.D. Cockcroft Papers, file 11/6, Invitation from Soviet Ambassador F. Gousev to Cockcroft, 19 May 1945.
131 Ibid., 11/6 Report of the Conference by BMH Tripp, 12 August 1945. Because of his key role in Britain's atomic program, Cockcroft declined the invitation to attend.
132 See the memoirs of Sir Nevill Mott, *A Life in Science* (London, 1986), for his account of the controversy.
133 A.V. Hill Papers, file 4/77, joint letter to the President of the Royal Society, 14 June 1945, signed by Bernal, Blackett, Mott, Norrish, and Dirac.

134 Ibid., E.A. Milne to Hill 14 June 1945; Hill to Milne 18 June 1945.
135 Cockcroft Papers, 11/6, Tripp Report, 12 August 1945.
136 Another version of the Moscow meetings was provided by Irving Langmuir, who was a member of the U.S. delegation. MED, Box 99, file 8, Lt. Parrish to General Groves, Interview with Dr. Langmuir re: 220th anniversary, Russian Academic of Science, 13–14 July 1945.
137 Selye represented the Royal Society of Canada; he briefly describes his experiences in his autobiography *The Stress of My Life: A Scientist's Memoirs* (Toronto, 1979), 85.
138 *The Canadian Scientist*, January 1945; ibid., December 1945.
139 *Bulletin, Montreal Branch of the Canadian Association of Scientific Workers*, January–February 1946.
140 University of Chicago Library, Papers of the Atomic Scientists of Chicago, *Newsletter of the American Association of Scientific Workers*, May–June 1945; Statement Goals, AAScW, June 1945.
141 Loren Butler, 'Robert Mulliken and the Politics of Science and Scientists,' *Historical Studies in the Physical and Biological Sciences*, vol. 25, part 1 (1994), 31–6.
142 Kevles, *Physicists*, 351, 382; Matt Price, 'Roots of Dissent: The Chicago Met Lab and the Origins of the Franck Report,' *Isis* 86 (2) (June 1995), 222–44.
143 Alan Nunn May had previously been a member of the BASW national executive. *Canadian Scientist*, November 1945.
144 John Simpson, representative of the FAS, was even more effusive in his praise about the conference. *Canadian Scientist*, May 1946; FAS, Box 21, #4, Simpson, Report on a Recent Visit to England and Paris, March 1946.
145 In July 1946 Joliot-Curie was elected president of the WFSW, with Bernal and Semenov of the USSR as joint vice-presidents. The CAScW was represented, but held no executive positions. *Canadian Scientist*, October 1946.
146 As part of their public education campaign, the CAScW asked Canadian physicists George Volkoff and Phil Wallace to prepare a report on the dangers of a nuclear arms race. *Canadian Scientist*, December 1945.
147 The CAScW also repeated the warnings of leading British atomic scientists 'that any industrial nation could have atomic bombs within five years, and any determined industrial nation within two.' Hazen Sise Papers, vol. 33, 'Statement on Atomic Energy ... 19 November 1945.'
148 University of Chicago Library, Papers of the Federation of American Scientists (FAS), Box 21, file 4, CAScW program, December 1945.
149 The first CAScW national convention was held in Ottawa on 27 May 1945. *Canadian Scientist*, January 1946; *Bulletin of the Montreal Branch*, February 1945.
150 The executive of the BASA included some of Britain's most outstanding

atomic scientists: N.F. Mott (president); R.E. Peierls (vice-president executive); with P.M.S. Blackett, J.D. Cockcroft, Lord Cherwell, and M.L. Oliphant as general vice-presidents. FAS, 21/3, Report of H.S.W. Massey, 23 February 1946; *Newsletter*, February 1947.
151 *Canadian Scientist*, April 1946; ibid., December 1945.
152 RC, Transcripts, Exhibits, Veall to Mrs. L. Kislova, VOKS, Moscow, 22 February 1945; Veall to I. Volenko, Third Secretary, USSR Embassy, 16 May 1945.
153 Vannevar Bush's claim that the Russians did not know 'this fuse development was going on' is clearly erroneous, since Colonel Zabotin was asking about it in early 1945. Vannevar Bush, *Pieces of the Action* (New York, 1970), 109.
154 DND, HQS, 4354-26-1-1, Report, July 1945.
155 In contrast, Inspector Harvison claimed that the RCMP Intelligence Branch had been aware for some years 'that the USSR's Embassy in Ottawa was engaged in espionage and in recruiting espionage agents.' But no proof is provided. Harvison, *The Horseman*, 149; Sawatsky, *Gouzenko*, 70.
156 CSIS Documents, folder 141, file 117-91-99, RCMP superintendent C.W. Rivett-Carnac, memorandum for file, 10 July 1945.
157 William Lanouette, *Genius in the Shadows: A Biography of Leo Szilard, the Man behind the Bomb* (Chicago, 1992), 305-13; Cochrane, *Biological Warfare Research*, 479.

9 Scientists, National Security, and the Cold War

1 Kathleen Willsher, the deputy registrar in the British High Commission in Ottawa, was among those arrested on 15 February. Although his name was also on the list, Communist MP Fred Rose was not apprehended until 20 March. Bothwell and Granatstein, *Gouzenko Transcripts*, 5-13; Sawatsky, *Gouzenko*, 78-9.
2 DEA, 8531-A 40, Hume Wrong to Tommy Stone, 2 August 1946.
3 RC, Espionage, *Report*, 27 June 1946, 637.
4 George de B. Robinson, a University of Toronto mathematician who had played an important role in Canada's code-breaking work during the war, was asked by the RCMP to assess the coded material which Gouzenko had brought with him. DEA, 2620, file 7-1-5-9, Robinson to Norman Robertson, 15 March 1946.
5 John Bryden, *Best-Kept Secret: Canadian Secret Intelligence from the Second World War to th Cold War* (Toronto, 1993), 275.
6 RC, Espionage, *Report*, 641.
7 Ibid., 640.

8 Not only had Gouzenko's GRU superiors criticized the quality of his cipher work, Vitali Pavlov, head of NKVD operations in Ottawa, might have forwarded his own negative assessment. Andrew and Gordievsky, *KGB*, 125–200, 305; Sawatsky, *Gouzenko*, 6, 20.
9 Reg Whitaker and Gary Marcuse, *Cold War Canada: The Making of a National Insecurity State, 1945–1957* (Toronto, 1994), 30–5.
10 There is evidence that the RCMP had contingency plans to seize the embassy documents if Gouzenko attempted suicide. Ibid., 30–1.
11 Both High Commissioner Malcolm Macdonald and William Stephenson, head of the Washington-based British Control Commission, were initially involved; they were soon joined by Peter Dwyer and Roger Hollis of British intelligence. Sawatsky, *Gouzenko*, 65–80; Granatstein, *A Man of Influence*, 168–82; David Stafford, *Camp X: Canada's School for Secret Agents, 1941–45* (Toronto, 1986), 257–69.
12 On 12 September 1945, J. Edgar Hoover wrote President Truman about Soviet espionage activities, particularly with regard to the atomic bomb; and the fact that the Soviets had 'spies in many of the Canadian government departments and are attempting to recruit others in order to have complete coverage of Canadian government activity.' NAW, Conant-Bush Correspondence, vol. 3B, Hoover to Matthew Connelly, secretary to the president, 18 September 1945.
13 Whitaker and Marcuse, *Cold War Canada*, 34–6.
14 Gregg Herken, *The Winning Weapon: The Atomic Bomb in the Cold War, 1945–1950* (New York, 1980), 34–58.
15 Sawatsky, *Gouzenko*, 69.
16 According to Tommy Stone, Canadian chargé d'affaires in Washington, J. Edgar Hoover was Drew Pearson's source of information: 'Hoover wanted to force the issue.' DEA, 2620, 7-1-5-9, Stone to Robertson, 19 February 1946.
17 In January 1946 Gordon Lunan had been sent to Canada House in London for public relations work; on 15 February he was recalled on the pretext that his services were immediately required in Ottawa. DEA, vol. 2620, file 50242-40, memo for Norman Robertson, 13 January 1946.
18 Whitaker and Marcuse, *Cold War Canada*, 56–63; Gordon Lunan, *The Making of a Spy: A Political Odyssey* (Montreal, 1995), 174–9.
19 DEA, 2620, 7-1-9, Canadian Embassy, Press Analysis Section, 18 February 1946.
20 Ibid., Massey to Department of External Affairs (DEA), 22 February 1946; ibid., John Holmes to DEA, 23 February 1946.
21 Ibid., John W. Holmes to Hume Wrong, 23 February 1946.

22 On 19 February Robert Ford, the chargé d'affaires, reported that Mackenzie King's statement of 15 February had been cited in *Izvestia* 'without comment.' Ibid., Ford to DEA.
23 Ibid., chargé d'affaires to SSEA, telegrams, 20, 22 February 1946.
24 Mackenzie King's proposal that he go to Moscow and meet with Stalin was scuttled by ambassador Dana Wilgress, who suggested that King would be snubbed by the Soviet dictator and that this 'would be a studied insult to Canada.' DEA, 2620, 7-1-5-9, Wilgress to Robertson 15 April 1946.
25 Ibid., report chargé d'affaires to DEA, telegram, 22, 24, February 1946. Dana Wilgress returned to Moscow during the first week of March. Among his many delicate tasks was negotiating the removal of the three Soviet embassy personnel named in the RC *Report* as 'spy masters.' Ibid., Wilgress to Louis St Laurent, 28 March 1946; ibid., 2619, 8531-40C #2, memorandum, 12 July 1946.
26 Ibid., Malcolm Macdonald to King, 23 February; DEA, 8531-40C, Robertson to SSEA, 12 October 1946; Denis Smith, *Diplomacy of Fear: Canada and the Cold War, 1941–1948* (Toronto, 1988), 133.
27 After May's arrest, pressure mounted on the British government to make it clear 'that no more than a small fraction of the total of the secret United States and British information available with respect to atomic energy' could have been turned over to the Soviets. DEA, 2020, 8531-40C, Acting High Commissioner to SSEA, telegram, 20 March 1946.
28 Ibid., 2081, AR 13/13, Pearson to Robertson, 4 March 1946.
29 King strongly endorsed Churchill's speech, as did Lester B. Pearson, who read an earlier draft while visiting Churchill in Washington. Ibid., Pearson to Robertson, 4 March 1946.
30 Bothwell and Granatstein, *Gouzenko Transcripts*, 156–332.
31 Patrick Boyer, *A Passion for Justice: The Legacy of James Chalmers McRuer* (Toronto, 1994), 192–5.
32 Lunan described himself as a reluctant spy. 'I tried in my naïve way to improve the situation; it was an illegal act, motivated by idealism.' Interview with Gordon Lunan, 31 August 1990, Ottawa.
33 Durnford Smith in his testimony before the commission claimed that the microwave zone position indicator set 'was at one time considered for sale to the Soviet government,' and that 'certain Russian officers were taken through the set.' RC, Espionage, *Report*, 298.
34 NAC, John G. Diefenbaker Papers, 1940–1956 Series, Shugar to Mackenzie King, 10 March 1946; Shugar to Diefenbaker, 12 April 1946; RC, Espionage, *Report*, 671–3.
35 The prosecution's case was also based on the allegations of University of

Toronto chemist F.E. Beamish that Shugar's actions in his laboratory were suspicious. Ibid., 281–318; RC, Espionage, Transcript (microfilm), Exhibit 508, 5033–5.
36 Bothwell and Granatstein, *Gouzenko Transcripts*, 316.
37 Boyer, Lunan, and Mazerall had all cooperated with the commission, and their testimony was used against them during their trials.
38 RC, Espionage, *Report*, 616.
39 DEA, 2620, 7-1-5-9, Robertson to Wrong, 29 May 1946, telegram.
40 RC, Espionage, *Report*, 70–1.
41 Cited in Whitaker and Marcuse, *Cold War Canada*, 85; CSIS Documents, folder 141, file 117-91-99, Security Screening NRC and CAScW, Report RCMP Toronto Special Section, 17 January 1946.
42 In September 1945 the U.S. Army had sponsored the May-Johnson bill, which became the focal point of congressional debate for the next ten months. For its part, the American scientists' movement mobilized public and political support behind its campaign, and was successful in having the issue referred to the special Senate Committee on Atomic Energy, chaired by Senator Brien McMahon. Herken, *The Winning Weapon*, 115–30.
43 While Veall claimed he was not a member of the Communist Part, he admitted being associated with the Young Communist League while a youth in Great Britain, and that his political sympathies 'would closely correspond' with Communism. RC, Espionage, *Report*, 513, 519; RC, Espionage, Transcripts, 4557–4603.
44 Veall also sent copies of his report to Melber Phillips secretary of the AScW. University of Chicago Archives, Papers of the Federation of American Scientists (FAS), Box 21, file 4, Veall to Higinbotham, 16 February 1946.
45 Alice Kimbell Smith, *A Peril and a Hope: The Scientists' Movement in America, 1945–47* (Chicago, 1965), 63–122.
46 FAS, Box 21, file 4, Higinbotham to Veall, 21 February 1946.
47 On 18 February Lester B. Pearson, Canadian ambassador in Washington, reported that General Groves had charged that FAS lobbying efforts had 'lost more security on the bomb ... than during the entire wartime development period.' DEA, vol. 2619, 6226–40, vol. 2, Pearson to Robertson, 18 February 1946.
48 The CAScW brief referred to C.D. Howe's statement that he was unaware 'of any person engaged on the [atomic] project having been arrested.' FAS, Box 21, file 4, The Spy Scare and Atomic Energy, 20 February 1946.
49 Ibid., Veall to Higinbotham, 4 March 1946; Higinbotham to Veall, 6 March 1946.
50 Ibid., M. Phillips to Veall, 28 March 1946.

51 FAS, vol. 13, #1, 'Summary of Published Facts Concerning Canadian Spy Case,' 12 April 1946; ibid., Veall to Phillips, 15 April 1946.
52 Ibid., Veall to Phillips, 28 March 1946.
53 The commission grilled Veall on his association with Alan Nunn May in England and in Montreal. Reference was made to a GRU document that discussed whether he posed a threat to May's status at the Montreal Laboratory because Veall was 'pretty well known in the laboratory as a "Red."' In the end the commission found no evidence to prove that Veall had divulged any secrets, probably because 'the Russians designedly did not ask him.' Ibid.; RC, Espionage, *Report*, 505, 526.
54 FAS, Box 21, #4, The CAScW and the Canadian Espionage Case, 3 April 1946. The Australian Association of Scientific Workers also protested against the arrest of CAScW members 'for the alleged disclosure of secret information.' CSIS Documents, folder 141, file 117-91-99, T.C. Davis, High Commissioner for Canada, Canberra, to DEA, 15 April 1946.
55 FAS, Box 21, #4, Veall to Phillips, 5 May 1946.
56 Ibid., Veall to Phillips, 12 May 1946.
57 Furry had studied with Robert Oppenheimer before joining Harvard's physics department in 1937, the year he became an active member of the American Association of Scientific Workers, along with his brother-in-law Israel Halperin. Furry would become a target of the House Un-American Activities Committee during the late 1940s. Harvard Archives, Pusey Library, Wendell Furry Papers, vol. 3, Notes on the Canadian Spy Case.
58 Edward Condon, Victor Weisskopf, Louis Ridenour, and Hans Bethe were all invited. Ibid., Phillips to Taylor, 2 May 1946; ibid., Phillips to Veall, 15 May 1946.
59 NAW, State Department Records, 861.20242; Transcript of *Rex v. Alan Nunn May*, Old Bailey, London, 1 May 1946. Enclosed in report, W.J. Gallman, 22 May 1946.
60 British atomic scientists compared the British legislation very unfavourably with the U.S. McMahon Bill. FAS, Box 21/3, Burhop to James Simpson, 6 May 1946.
61 *Canadian Scientist*, 23 October 1946.
62 Ibid. The RCMP carefully monitored the CAScW conference and its activities across the country. On 11 June 1946 the Winnipeg detachment informed RCMP headquarters that the CAScW was 'making no progress whatever' in their city, and that some scientists from the University of Manitoba had left the organization because of 'the pro-Russian attitude of the executive.' CSIS Documents, folder 141, file 117-91-99.
63 A more sobering news item of that day was the death of Louis Slotin, a

Winnipeg-born physicist, of radiation poisoning at Los Alamos. *Globe and Mail*, 31 May 1946.
64 FAS, 13/1, J.H. Rush to W. Thomson, 23 October 1946.
65 Ibid., Higinbotham to A.S. Bishop, 24 May 1946.
66 Mackenzie Diary, 8 September 1945, 15 February 1946.
67 Ibid., 12 August 1946, 10 March 1947. In November 1946 the RCMP proposed that all members of the NRC be finger-printed so that effective security screening could be carried out. CSIS Documents, folder 141, file 117-91-99, Inspector Parsons to Mackenzie, 29 November 1946.
68 Four years later Pontecorvo, now a scientist at Harwell, defected to the Soviet Union. NRC, vol. 106, Mackenzie to Chadwick, 10 March 1946; Hyde, *Atom Bomb Spies*, 130–42.
69 Mackenzie Diary, 15 February 1946; Cyril James Papers, C-89, Box 10, file 917, James to Boyer, 16 March 1946.
70 Molson not only denounced Boyer's associates, he reserved special scorn for his marriage to a Jewish woman of leftist inclinations. Ibid., Walter Molson to James, 19 March 1946; Interview with Jack Edward, 8 March 1994.
71 James Papers, James to Molson, 22 March 1946.
72 Debates, 20 March 1946; James Papers, Notes, 20 March 1946.
73 In preparing his rebuttal, James was briefed about possible embarrassing incidents, many of which involved economist Eugene Forsey and law professor Frank Scott. He also ordered an investigation of the academic careers of Fred Poland, David Shugar, Matt Nightingale and Harold Gerson. Ibid., James to T.H. Matthews, 23 March 1946.
74 Ibid., Press Release, Principal James, 20 March 1946.
75 James was particularly incensed by an article in the *Chronicle-Telegram* that stated that a student goes 'to McGill as a Canadian democrat, and he graduates as an international communist.' Ibid., Arthur Penny to James, 23 March 1946.
76 Whitaker and Marcuse, *Cold War Canada*, 107.
77 DEA, 2620, file 7-1-5-9, Robinson to Robertson, 15 March 1946.
78 DEA, 2619, 8531-40 C #2, J.A. Corry to King, 3 April 1946.
79 Halperin would teach another seventeen years at Queen's before joining the University of Toronto. Frederick Gibson, *To Serve and Yet to Be Free: Queen's University*, vol. 2: *1917–1961* (Kingston, 1983), 283.
80 Francis, *Frank Underhill*, 114–29.
81 Leopold Infeld, *Why I Left Canada: Reflections on Science and Politics* (Montreal, 1978), 29–30. In August 1946 the RCMP Intelligence Service reported that Infeld, 'a Polish Jew teaching applied science at Toronto university who is a pro-Communist,' had been an active member of Halperin's defence

committee. CSIS Documents, folder 158, file 23-8-46, Report Sgt. Winmill to Inspector Leopold, 23 August 1946.
82 In 1950 Infeld requested a leave of absence to carry out research in Western Europe and Poland. This trip became a cause célèbre when the Catholic weekly *The Ensign* alleged that Infeld would be conveying atomic secrets to the Soviet bloc, a charge that was repeated in the House of Commons by Conservative leader George Drew. Facing intense public pressure, University of Toronto president Sydney Smith gave Infeld an ultimatum: return immediately or be fired. When Infeld decided to remain in Poland, the Canadian government denaturalized him and his family. Ibid., 51–8.
83 RC, Espionage, *Report*, 616–17.
84 Andrew, *Secret Service*, 490.
85 RC, Espionage, *Report*, 617–19.
86 Ibid., 152–6, 298.
87 Ibid., 159.
88 Reference was also made to five documents found in Matt Nightingale's apartment, including a book entitled *R.C.A.F. Landlines Construction and Maintenance* and an old technical manual for telephone equipment issued by the U.S. War Department. None were restricted documents in August 1945. Ibid., 278–9.
89 According to Gouzenko's testimony, David Shugar, an electrical sub-lieutenant in the Royal Canadian Navy between 1943 and 1945, had been recruited by the spy group coordinated by Sam Carr, a communist organizer in Toronto, and given the code name 'Prometheus.' Ibid., 281–317.
90 Ibid., 402, 404.
91 Jack Edward, one of Boyer's McGill colleagues, claims that the RDX formula Rose supposedly conveyed to the Soviets was 'useless.' Edward interview, 8 March 1994; *Montreal Gazette*, 27, 28, 29 March 1947, 5, 6, December 1947.
92 After his arrest in February 1946, May admitted that he had provided Colonel Motinov with 'a slightly enriched sample' of uranium 235 'consisting of about a milligram of oxide. The U 233 was about a tenth of a milligram.' Ibid., 450, 455.
93 On 3 September May sailed for England with instructions to rendezvous with his new GRU handlers on 7 October 'on the street in front of the British Museum.' Ibid., 451–3, 505–25.
94 Conant-Bush Correspondence, vol. 3B, Hoover to Matthew Connelly, secretary to the president, 18 September 1945.
95 NAC, DND, file TS 711-270-16-1, Colonel R.E.S. Williamson to Colonel W.A.B. Anderson, Director of Intelligence, Canadian Army, 14 June 1946.

96 Ibid., Douglas Abbott to Louis St Laurent, 4 July 1946.
97 Ibid., Mackenzie to Solandt, 24 August 1946; ibid., Mackenzie to Solandt, 19 September 1946.
98 Bothwell and Granatstein, *Gouzenko Transcripts*, 244.
99 NAW, MED Records, 333.5, 'Investigation File,' Groves to Senator B.B. Hickenlooper, 12 March 1946.
100 David Holloway, *Stalin and the Bomb: The Soviet Union and Atomic Energy, 1939–1956* (New Haven, 1994), 104–8.
101 Holloway also observes that Soviet scientists independently developed their own thermonuclear bomb. Ibid., 366.
102 In March 1945 Igor Kurchatov, the 'father of the Soviet bomb,' had urgently requested 'several tens of grams of highly enriched uranium.' But May was only able to supply 'a microscopic' amount of uranium 235 and a small sample of uranium 233. Ibid., 105.
103 The primary value of Fuchs's information to the Kremlin was that it ensured that the Soviet bomb would follow the guidelines of the Manhattan Project instead of the unique, and potentially more efficient, approach recommended by Russian physicists Igor Kurchatov and Peter Kapista. Ibid., 108, 163–90; Norman Moss, *Klaus Fuchs: The Man Who Stole the Atom Bomb* (Toronto, 1987).
104 DND, TS 711–270–16–1, Mackenzie to Solandt, 19 September 1946.
105 Ibid., Report of Lieutenant Commander R.N. Battles, 7 September 1946.
106 RC, Espionage, Transcript, Exhibit, D.W.R. McKinley to Supt. G.E. Rivett-Carnac, Intelligence Branch, RCMP, 3 April 1946.
107 Ibid.
108 Ibid. There was concern, however, that Durnford Smith had passed information about the top-secret British 931 radar set, since he had been 'during the relevant time, the chief man working on the K-band field in the Research Council,' and he would appreciate 'the significance of K-band and the 931 set.' Among the ten documents Smith gave the GRU, four came from the Radiation Laboratory at MIT. DEA, 2620, 7-1-5-9, Robertson to Stephen Holmes, Deputy High Commissioner, 10 April 1946.
109 There is no evidence that either Mackenzie or Solandt were concerned that the Soviets had penetrated Canadian signals intelligence research and development, or gained access to Allied SIGINT secrets. Bryden, *Best Kept Secret*, 221-46.
110 D.J. Goodspeed, *A History of the Defence Research Board of Canada* (Ottawa, 1958), 1–37; Douglas Bland, *The Administration of Defence Policy in Canada, 1947 to 1985* (Kingston, 1987), 1–33; Merritt Roe Smith, *Military Enterprise and Technological Change: Perspectives on the American Experience* (Cambridge,

MA, 1985), 203–381; Michael Sperry, *Preparing for the Next War: American Plans for Post-War Defence* (New Haven, 1977), 191–239.

111 Carroll Pursell, 'Scientific Agencies in World War II: The OSRD and Its Challengers,' in Nathan Reingold, ed., *The Sciences in the American Context: New Perspectives* (Washington, 1979), 359–78; Daniel Kevles, 'Scientists, the Military, and the Control of Postwar Defence Research: The Case of the Research Board for National Security, 1944–46,' *Science and Technology* (1975), 24–6.

112 Goodspeed, *Defence Research*, 20–43; Mackenzie Diary, 25 August 1945; NAC, PCO, Series 18, vol. 101, Draft Submission to Defence Committee of Cabinet, 5 October 1945.

113 As superintendent of Army Operational Research Group, Solandt had been involved in planning the Normandy invasion and in deploying proximity fuse artillery attacks against German V-1 rocket attacks on London. University of Toronto Archives, Omond Solandt Papers, vol. 31, Solandt to Brigadier B.F.J. Schonland, 12 October 1944.

114 The British Bombing Survey team included nine researchers from the Ministry of Home Security, one from the Admiralty, two from the Ministry of Supply, one from the Ministry of Aircraft Production, and Solandt, representing the Scientific Advisor to the Army Council. Solandt Papers, vol. 11, Notes on the Atomic Bomb.

115 DRB, vol. 4217, file 700-0-171, Organization Defence Research Board, 29 April 1947.

116 Solandt Papers, Box 33, Address to the Naval Officers Association of British Columbia, 20 June 1947.

117 In addition to Solandt, there were five other ex officio members: the Deputy Minister of National Defence, the Chiefs of Staff of the Army, Navy, and Air Force, and the President of the NRC. There were also six appointed members, three of whom were scientists: physiologist Charles Best of Toronto, and physicists J.H.L. Johnstone of Dalhousie and Gordon Shrum of the University of British Columbia. Goodspeed, *Defence Research*, 39–46.

118 NAC, Cabinet Conclusions, 14 January 1947; Eayrs, *Peacemaking*, 86.

119 NAC, PCO, Series 18, vol. 59, C-10-9, Memorandum for the Cabinet Defence Committee: Arrangements for cooperation with United Kingdom and United States on defence matters, 7 May 1946.

120 NAC, Cabinet Conclusions, 22 May 1946.

121 During the last year of the war Alexander King of the British Commonwealth Scientific Office in Washington took the initiative, along with Henry Tizard, in trying to establish postwar Commonwealth defence science cooperation. After successful meetings in Ottawa in July 1944, the British War Cabinet endorsed the program. Tizard Papers, 399, King to Tizard, 9 Sep-

tember 1944; NRC, 153, 25-1-118, Extract from the Minutes of Meeting War Cabinet, 23 October 1944.
122 Patrick Blackett Papers, vol. 149, Brief C.D. Ellis, n.d.; J.D. Bernal, 'Notes on Future Armament Policy'; Patrick Blackett, 'Strategy and Development.'
123 McGucken, *Scientists*, 307–35.
124 PCO, file C-10–9–D, H.H. Wrong, Memorandum for CDC: Proposed informal Commonwealth conference on defence science, 5 March 1946.
125 CAB 131/6, 58921, Minutes of the 5th meeting of the Commonwealth Advisory Committee on Defence Science, 1947 meeting, 21 November 1947. (National Security Archives, Washington.)
126 Interview with Omond Solandt, 21 March 1989.
127 NAC, Cabinet Conclusions, 22 December 1945.
128 Ibid., 7 May 1946.
129 Musk-Ox was a joint U.S.–Canadian exercise carried out February to May 1946 to determine how troops and air crews adapted to operational conditions in the Canadian Arctic. During a three-month period the Army–RCAF contingent travelled over 3000 miles from Churchill, Manitoba, to Fort Norman, Northwest Territories. Cabinet Conclusions, 22 December 1945; NAC, Defence Research Board Papers, vol. 2474, 716-10-18, Major General J.H. MacQueen, Master-General Ordnance to Defence Headquarters, 5 March 1947.
130 Joseph Jockel, *No Boundaries Upstairs: Canada, the United States, and the Origins of North American Air Defence, 1945–1958* (Vancouver, 1987), 10.
131 Ibid., 15.
132 Joint defence planning was institutionalized in May 1946 with the creation of the Canadian–United States Military Co-operation Committee (MCC), which became the primary vehicle for discussions about continental defence during the next six years. Cabinet Conclusions, 6 May 1946, 27 February 1947; David Bercuson, *True Patriot: The Life of Brooke Claxton, 1898–1960* (Toronto, 1993), 70–135.
133 Herken, *Winning Weapon*, 22–68.
134 Mackenzie Diary, 15 November 1945; Eayrs, *Peacemaking*, 281–2.
135 Howe Papers, 11/23, Truman to Attlee, 16 April 1946.
136 Margaret Gowing, *Independence and Deterrence: Britain and Atomic Energy, 1945–1952* (London, 1974), 5–62.
137 Howe Papers, 13/29, British High Commissioner P.A. Clutterbuck to Mackenzie King, 28 November 1946.
138 Ibid., 11/23, Robertson to DEA, 25 November 1946.
139 DEA, 1946, 429, Memorandum by Head, First Political Division, 26 April 1946.

140 The U.S. government was obviously pleased by the Canadian response, and it made a point of officially inviting a Canadian delegation to witness the June–July Bikini atomic tests. DEA, 1946, 434; T.A. Stone to Dean Acheson, 29 April 1946; Howe Papers, 13/29, Atherton to Pearson, 6 May 1946.
141 With British and U.S. support, Canada took advantage of article 31 of the UN Charter, which allowed countries to attend Security Council discussions if the issues had direct bearing on their national interests. Eayrs, *Peacemaking* 283.
142 Eayrs, *Peacemaking*, 275–85; Herken, *Winning Weapon*, 171–92.
143 DEA, 1946, 443: Memorandum by First Political Division, Provisional Instruction for the Canadian Representative on the Atomic Energy Commission, 7 June 1946. McNaughton served as acting president of the AEC between 14 August and 14 September 1946.
144 Howe Papers, 11/23, McNaughton to DEA, 16 June 1946. Baruch flatly refused to consider a joint Canadian-British-French amendment that would have allowed the USSR to retain some use of its Security Council veto. DEA, 1946, Reid report, 19 December 1946.
145 Interview George Lawrence, Deep River, Ontario, 20 November 1981. Diplomat George Ignatieff, who assisted McNaughton during the UNAEC hearings, provided a vivid recollection of these events in our interview of 13 November 1987. See also George Ignatieff, *The Making of a Peacemonger* (Markham, 1987), 89–106.
146 CAB 122/373, Report of the Joint Planning Staff to the British Chiefs of Staff, 27 February 1946.
147 By the end of the war the capital costs of the Experimental Station amounted to $2,218,346; the Chemical Warfare School (S-11) cost $670,702 and the Toxic Bulk Storage $308,222. HQS 4354-26-4, A. Ross, Deputy Minister, DND to R. Gordon Munro, Office of the High Commission for the United Kingdom, 6 July 1945.
148 HQS, 4354-24-5-3, minutes of the seventh meeting of the special DCWS Committee and the CW Inter-Service Board, 11 September 1945.
149 PRO, DEFE, 2 1252/7827, Minutes of British Defence Committee, 27 November 1945; ibid., memorandum of 6 December 1945.
150 HQS 4354-31-2, Major Taber to National Defence Headquarters, 19 March 1947. During the spring of 1947 the Special Weapons Advisory Committee (SWAC) was created to assist the DRB in its biological and chemical warfare planning. The committee was chaired by Otto Maass. NRC, vol. 7, 3-12-M3-24, Minutes of the SWAC meeting, 8 March 1947.
151 NRC, vol. 7, file 3-12-M3-24, O. Maass, Scientific Adviser to the Chief Gen-

364 Notes to pages 254–64

eral Staff, General Staff Requirements for Research in the Field of Special Weapons, 4 September 1947.
152 NAW, Joint Chiefs of Staff Papers, 'Report ... October 1950.'
153 Bothwell, *Nucleus*, 105, 138–42.
154 The Soviet Union exploded its first nuclear device in the fall of 1949; the British test came in 1952, and the French test in 1960. Jean Lacouture, *De Gaulle: The Ruler 1945–1970* (New York, 1991), 413–30.
155 Of particular note were Canadian efforts to produce advanced air frames and jet engines, first through the Crown corporation Turbo Research Ltd., and then with the Toronto-based A.V. Roe Company. In June 1950 the CF-100, powered by two British-built Avon jet engines, was ready for tests. Two years later plans were made to develop the sophisticated supersonic jet interceptor CF-105, or Avro Arrow. Eayrs, *Peacemaking*, 97–108; Bothwell and Kilbourn, *C.D. Howe*, 253–7, 266–7. See also Julius Lukasiewicz, 'Canada's Encounter with High-Speed Aeronautics,' *Technology & Culture*, 223–61.
156 C.J. Mackenzie retired as NRC president in 1952; his successor was E.W.R. Steacie, who held the position until 1962. Cited in Eggleston, *National Research*, 283, 458.
157 After leaving the NRC, Mackenzie became president of the Crown corporation Atomic Energy of Canada Ltd. (1953–4), while continuing as president of the Atomic Energy Control Board (1948–61). Although nominally retired in 1961, he remained a valuable consultant during the reviews of Canada's science policies in the 1960s.

Conclusion

1 John English and Norman Hillmer, eds, *Making a Difference: Canada's Foreign Policy in a Changing World Order* (Toronto, 1992), vi.
2 Stacey, *Arms, Men and Governments*, 512.
3 Interview with Omond Solandt, 21 March 1989.
4 The first meeting of Canadian, British, and American biological and chemical warfare specialists occurred in the United States. in March 1947; Suffield co-hosted the 1948 meetings. Gradon Carter and Graham S. Pearson, 'North-Atlantic Chemical and Biological Research Collaboration: 1916–1994,' (draft, June 1994), 14–29.
5 Banting War Diary, 17 April 1940.
6 Interview with Harry Thode, 15 May 1982.
7 John Rigden, *Rabi: Scientist and Citizen* (New York, 1987).
8 RDX explosives, the tank, the bomber, the fighter plane, the submarine, and

poison gas were among the most important weapons of the First World War. These, in turn, contributed to the evolution of innovative military strategy and tactics that included strategic bombing, the blitzkrieg, amphibious landings, submarine wolf packs, and the tactical use of chemical weapons. See Williamson Murray and Allan Millett, eds, *Military Innovation in the Interwar Period* (Cambridge, MA, 1966), 6–49, 96–143.
9 Legro, *Cooperation under Fire*, 94–143.
10 Paul Dufour, in his article 'Eggheads and Espionage: The Gouzenko Affair in Canada,' *Journal of Canadian Studies* 16 (304) (1981) was one of the first historians to explore this topic.
11 RC, Espionage, *Report*, 375–409.
12 Maurice Goldsmith, *Sage: A Life of J.D. Bernal* (London, 1980), 92–123.
13 MED, Groves to Secretary of War, 24 March 1947.
14 RC, Espionage, *Report*, 57.
15 Whitaker and Marcuse, *Cold War Canada*, 110.
16 Joseph Rotblat, *Pugwash – The First Ten Years: History of the Conferences of Science and World Affairs* (New York, 1968).
17 By the fall of 1996 some 160 countries, including Canada, had signed and ratified the Convention. The U.S. joined this club in early 1997; Russia has announced it will follow soon.
18 Roberts Brad, ed., *Biological Weapons: Weapons of the Future?* (Washington, 1993), 7–42.
19 Banting Diary, 20 October 1939

Bibliography

Primary Sources

Canada

McGill University Archives
   J.R. Donald papers
   Cyril James papers (office correspondence)
   Otto Maass papers

   Department of Chemistry files
   Department of Physics files
   Faculty of Medicine files

McGill Osler Library
   Wilder Penfield papers

McMaster University Archives
   Canadian Youth Congress papers

National Archives of Canada

   Private Papers

   Adrien Cambron
   Hon. Brooke Claxton
   General H.D.G. Crerar
   Raymond Dearle
   Eugene Forsey

David Keys
J.T. Henderson
Hon. Clarence Decatur Howe
Rt. Hon. William Lyon Mackenzie King
Mackenzie King Diaries (microfilm): 1939–47
Ernest Lapointe
C.J. Mackenzie (wartime diaries)
General A.G.L. McNaughton
E.G.D. Murray
Frank and Libby Park
General Maurice Pope (wartime diaries)
Hon. J.L. Ralston
Frank Scott
Hazen Sise

Government Records

Cabinet Conclusions, 1945–9 (on microfilm)
Cabinet War Committee Minutes and Records (PCO)
Department of National Defence Records
    DND Chemical and Biological Warfare Files on microfilm (C-5001–5019)
    Defence Research Board Records, 1946–52.
Department of External Affairs
National Research Council
    Chemical Warfare files
    Biological Warfare files
    Explosives Research files
    General Correspondence
Privy Council Office
    Solicitor General, Documents Released by the Canadian Security Intelligence Service under the Access to Information Act, folder 141, file 117-91-99 (Security Screening of NRC and the CAScW); folder 158, file 23-8-46 (Civil Liberties Organizations, 1930–1950)
    Solicitor General, Transcript of the evidence presented to the Royal Commission on Espionage (on microfilm)

National Defence Headquarters: Directorate of History
    Chemical Warfare Records: Inter-Service Board

National Research Council
    Frederick Banting papers

Ontario Public Archives
    Canon H.J. Cody papers

Queen's University Archives
    Grant Dexter papers
    Joseph A. Gray papers
    C.G. Power papers
    B. Weldon Sargent papers

University of Manitoba Archives
    Sydney Smith papers (official correspondence)

University of Toronto Archives
    Canon H.J. Cody papers (official correspondence)
    Omond Solandt papers
    George Wright papers

University of Toronto Rare Book Room
    Wartime Diary of Sir Frederick Banting

University of Western Ontario Archives
    Papers of G. Edward Hall

Great Britain

Churchill College Archives, Cambridge
    James Chadwick papers
    John Cockcroft papers
    A.V. Hill papers

Imperial War Museum
    Sir Henry Tizard papers (and diary)

Public Records Office (KEW)
    Atomic Energy files: AB1 series
    Cabinet files: CAB series
    War Office files: WO series

Royal Society, London
    Patrick Blackett papers

United States

Library of Congress
  Vannevar Bush papers
  Robert Oppenheimer papers
  Carl Spaatz papers

National Archives and Records Administration, Washington
  Bush-Conant Correspondence (microfilm)
  Records of the U.S. Joint Chiefs of Staff
  Records of the Joint Committee on New Weapons
  Records of the Manhattan Engineer District
  Records of the Office of Scientific Research and Development
  Records of the War Department: General and Special Staffs,
    New Developments Division (George Merck correspondence
    re: biological warfare)
  State Department Records
  State/War/Navy Coordinating Committee, Lot 57
  U.S. Congress, Joint Committee on Atomic Energy

National Security Archives (Washington)
  Files on Chemical and Biological Warfare

Pusey Library, Harvard University
  Administrative History of the Harvard University Research Laboratory, 1942–1945
  James B. Conant personal papers
  Wendell Furry papers

University of California, Berkeley, Bancroft Library
  Ernest O. Lawrence papers

University of Chicago, Regenstein Library
  American Association of Scientific Workers papers
  Federation of American Scientists papers
  Federation of Atomic Scientists papers

Personal Interviews
(All interviews were conducted in person.)

Raymond Boyer, Montreal, 27 February 1984

Frank Chubb, Montreal, 14 June 1986
T.C. Douglas, Ottawa, 25 May 1983
H.E. Duckworth, Winnipeg, 8 May 1985
Jack Edward, McGill University, 8 March 1994
Eugene Forsey, Ottawa, 26 May 1983
C.G. Gotlieb, University of Toronto, 20 May 1994
George Haythorne, Ottawa, 27 May 1983
Gerhard Herzberg, Ottawa, 31 August 1990
George Ignatieff, Toronto, 13 November 1987
Robert Kenney, Toronto, 23 July 1981
George Lawrence, Deep River, 20 November 1981
Gordon Lunan, Ottawa, 31 August 1990
Robert Nicholls, Merrickville, Ontario, 4 March 1994
J. Carson Mark, Los Alamos, 15 November 1993
Robert Murray, London, Ontario, 25 January 1994
Frank Park, Ottawa, 1 September 1983
B. Weldon Sargent, Kingston, 28 July 1982
Graham Spry, Ottawa, 31 August 1983
R.B. Stewart, Queen's University, 21 May 1993
Harry Thode, Hamilton, 6 May 1992
George Volkoff, Vancouver, 8 June 1983
Kenneth Woodsworth, Toronto, 16 June 1983
Leo Yaffe, McGill University, 16 June 1984; 8 March 1994

Contemporary Bulletins, Newspapers, and Journals

Bulletin, Montreal Branch of the CAScW
Canadian Forum, 1936–1947
The Canadian Scientist, 1944–6
Council Bulletin, National Council for Canadian Soviet Friendship, 1943–5
Montreal Gazette, 1939–47
Nature, 1936–47
New York Times, 1936–47
Science, 1936–47
The Scientific Worker (GB), 1936–47
Toronto Globe and Mail, 1939–47
Winnipeg Free Press, 1939–47

Secondary Sources

Adachi, Ken. *The Enemy That Never Was: A History of the Japanese Canadians.* Toronto, 1977.
Ainley, Marianne Gosztonyi, and Catherine Millar. 'A Select Few: Women and the National Research Council of Canada, 1916–1991.' In Richard Jarrell and Yves Gingras, eds, *Building Canadian Science: The Role of the National Research Council*, 105–16. Ottawa, 1992.
Alperovitz, Gar. *Atomic Diplomacy: Hiroshima and Potsdam: The Use of the Atomic Bomb and the American Confrontation with Soviet Power.* New York, 1965.
– *The Decision to Use the Atomic Bomb and the Architecture of an American Myth.* New York, 1995.
Andrew, Christopher. *Secret Service: The Making of the British Intelligence Community.* London, 1985.
Andrew, Christopher, and Oleg Gordievsky. *KGB: The Inside Story of Its Foreign Operations from Lenin to Gorbachev.* London, 1990.
Avery, Donald. 'Allied Scientific Cooperation and Soviet Espionage in Canada, 1941–45.' *Intelligence and National Security* 8 (3) (July 1993), 100–28.
– 'Atomic Scientific Cooperation and Rivalry among Allies: The Anglo-Canadian Montreal Laboratory and the Manhattan Project, 1943–1946.' *War in History* 2 (3) (1995), 274–305.
Badash, Lawrence. 'The Origins of Big Science: Rutherford at McGill.' In Mario Bunge and William Shea, eds, *Rutherford and Physics at the Turn of the Century.* New York, 1979.
Balawyder, Aloysius, ed. *Canadian–Soviet Relations, 1939–1980.* Oakville, 1981.
Baldwin, Ralph. *The Deadly Fuze: The Secret Weapon of World War II.* San Rafael, CA, 1980.
Barr, Murray. *A Century of Medicine at Western.* London, ON., 1977.
Batchelor, G.K. 'Geoffrey Ingram Taylor, 7 March 1886–27 June 1975.' *Biographical Memoirs of Fellows of the Royal Society of London.* London, 1976.
Bator, Paul, and James Rhodes. *Within Reach of Everyone: A History of the University of Toronto School of Hygiene and the Connaught Laboratories.* Ottawa, 1990.
Baxter, James Phinney, III. *Scientists against Time.* Boston, 1948.
Beardsley, E.H. 'Secrets between Friends: Applied Science Exchange between the Western Allies and the Soviet Union during World War II.' *Social Studies of Science* 7 (1977), 447–73.
Bercuson, David. *True Patriot: The Life of Brooke Claxton, 1898–1960.* Toronto, 1993.
– *Maple Leaf against the Axis: Canada's Second World War.* Toronto, 1995.
Bernal, J.D. *The Social Function of Science.* London, 1939.
Bernstein, Barton. *The Atomic Bomb: The Critical Issues.* Boston, 1976.

- 'America's Biological Warfare Program in the Second World War.' *Journal of Strategic Studies* 11 (3) (Sept. 1988), 292–317.
- 'Origins of the U.S. Biological Warfare Program.' In Susan Wright, ed., *Preventing a Biological Arms Race*, 1–65. Cambridge, 1990.

Birkenhead, Frederick Winston Furneaux Smith, 2d Earl of. *The Prof in Two Worlds: The Official Life of Professor F.A. Lindemann, Viscount Cherwell.* London, 1961.

Black, Edgar C., ed. *History of the Associate Committee on Aviation Medical Research 1939–1945.* Ottawa, 1946.

Bland, Douglas. *The Administration of Defence Policy in Canada, 1947 to 1985.* Kingston, 1987.

Bliss, Michael. *Banting: A Biography.* Toronto, 1984.

Bothwell, Robert. *Eldorado: Canada's National Uranium Company.* Toronto, 1984.
- *Nucleus: The History of Atomic Energy of Canada Limited.* Toronto, 1988.
- 'Weird Science: Scientific Refugees and the Montreal Laboratory.' In Norman Hillmer et al., eds, *On Guard for Thee: War, Ethnicity, and the Canadian State, 1939–1945.* Ottawa, 1988.

Bothwell, Robert, and Jack Granatstein. *The Gouzenko Transcripts: The Evidence Presented to the Kellock–Taschereau Royal Commission of 1946.* Ottawa, 1982.

Bothwell, Robert, and William Kilbourn. *C.D. Howe: A Biography.* Toronto, 1979.

Bowen, E.G. *Radar Days.* Bristol, 1987.

Bower, Tom. *The Paperclip Conspiracy: The Battle for the Spoils and Secrets of Nazi Germany.* London, 1987.

Boyer, Patrick. *A Passion for Justice: The Legacy of James Chalmers McRuer.* Toronto, 1994.

Brad, Roberts, ed. *Biological Weapons: Weapons of the Future?* Washington, 1993.

Brax, Ralph. *The First Student Movement: Student Activism in the United States during the 1930s.* London, 1981.

Brinkley, David. *Washington Goes to War: The Extraordinary Story of the Transformation of a City and a Nation.* New York, 1988.

Broadfoot, Barry. *Years of Sorrow, Years of Shame: The Story of the Japanese Canadians in World War II.* Toronto, 1977.

Brophy, Leo, and George Fisher. *The Chemical Warfare Service: Organizing for War.* Washington, 1959.

Brown, Andrew. *The Neutron and the Bomb: A Biography of Sir James Chadwick.* New York, 1997.

Brown, Frederic. *Chemical Warfare: A Study in Restraints.* Princeton, 1968.

Bryden, John. *Best-Kept Secret: Canadian Secret Intelligence from the Second World War to the Cold War.* Toronto, 1993.

- *Deadly Allies: Canada's Secret War, 1937–1947.* Toronto, 1989.
Buderi, Robert. *The Invention That Changed the World.* New York, 1996.
Bundy, McGeorge. *Danger and Survival: Choices about the Bomb in the First Fifty Years.* New York, 1988.
Burchard, John, ed. *Rockets, Guns and Targets: Rockets, Target Information, Erosion Information and Hypervelocity Guns Developed during World War II by the Office of Scientific Research and Development.* Boston, 1948.
Bush, Vannevar. *Pieces of the Action.* New York, 1970.
Butler, Loren. 'Robert Mulliken and the Politics of Science and Scientists.' *Historical Studies in the Physical and Biological Sciences*, vol. 25, part 1 (1994), 31–6.
Callwood, June. *Emma: The True Story of Canada's Unlikely Spy.* Toronto, 1984.
The Canadian Who's Who, vol. 8: *1958–60*.
Carroll, F. Thomas. 'Immigrants in American Chemistry.' In Jarrell Jackman and Carla Borden, eds, *The Muses Flee Hitler: Cultural Transfer and Adaptation, 1930–1945*, 189–204. Washington, 1983.
Carter, G.B. *Porton Down: 75 Years of Chemical and Biological Research.* London, 1992.
Carter, G.B., and Graham S. Pearson. 'British Biological Warfare and Biological Defence: 1925–1945.' In E. Geissler and John Ellis van Courtland Moon, eds, *Biological and Toxin Weapons 1915–1945.* Stockholm International Peace Research Institute, forthcoming.
Cassidy, David. 'Controlling German Science, I: U.S. and Allied Forces in Germany, 1945–1947.' *Historical Studies in the Physical and Biological Sciences*, vol. 24, part 2 (1994), 197–235.
Caute, David. *The Fellow-Travellers: A Postscript to the Enlightenment.* New York, 1973.
Chartrand, Luc, et al. *Histoire des sciences au Québec.* Montréal, 1987.
Clark, A.L. *The First Fifty Years: A History of the Science Faculty of Queen's University, 1893–1943.* Kingston, 1943.
Clark, Ronald. *Tizard.* London, 1965.
- *Sir Edward Appleton.* Toronto, 1971.
Cochrane, Rexmond C. *History of the Chemical Warfare Service in World War II: 1 July 1940–15 August 1945*, vol. 2: *Biological Warfare Research in the United States.* Historical Section, Plans, Training and Intelligence Division, Office of the Chief, Chemical Corps, November 1947.
Cockburn, Stewart, and David Ellyard. *Oliphant: The Life and Times of Mark Oliphant.* Adelaide, 1981.
Collard, Edgar Andrew. *The McGill You Knew: An Anthology of Memories, 1920–1960.* Don Mills, ON, 1975.
Compton, Arthur. *Atomic Quest: A Personal Narrative.* New York, 1958.

Conant, James Bryant. *My Several Lives: Memoirs of a Social Inventor.* New York, 1970.
Copp, Terry, and William McAndrew. *Battle Exhaustion: Soldiers and Psychiatrists in the Canadian Army, 1939–1945.* Montreal/Kingston, 1990.
Crook, Paul. 'Science and War: Radical Scientists and the Tizard–Cherwell Area Bombing Debate in Britain.' *War and Society* 12 (2) (October 1994), 69–101.
Cuff, R.D., and J.L. Granatstein. *Ties That Bind: Canadian–American Relations in Wartime from the Great War to the Cold War.* Toronto, 1977.
Danchev, Alex. *Establishing the Anglo-American Alliance: The Second World War Diaries of Brigadier Vivian Dykes.* London, 1990.
Davis, Richard G. 'Carl A. Spaatz and the Development of the Royal Air Force–U.S. Army Air Corps Relationship, 1939–40.' *The Journal of Military History* 54 (October 1990), 453–72.
Defries, R.D. *The First Forty Years: Connaught Medical Research Laboratories.* Toronto, 1968.
Dickson, Paul. 'The Hand That Wields the Dagger: Harry Crerar, First Canadian Army Command and National Autonomy.' *War & Society* 13 (2) (October 1995), 113–41.
Dinnerstein, Leonard. *America and the Survivors of the Holocaust.* New York, 1982.
Donald, James Richardson. *Reminiscences of a Pioneer Canadian Chemical Engineer, 1890–1952.* ed. Nicholls, Robert, and Onyszchuk. Montreal, 1989.
Donovan, Robert. *Conflict and Crisis: The Presidency of Harry S. Truman, 1945–1948.* New York, 1977.
Douglas, W.A.B. and Brereton Greenhous. *Out of the Shadows: Canada in the Second World War.* Toronto, 1977.
– *The Official History of the Royal Canadian Air Force,* vol. 2: *The Creation of a National Air Force* Toronto, 1986.
Dower, John. *War without Mercy: Race and Power in the Pacific War.* New York, 1986.
Downing, Doug. 'The RDX Research Program at U of T in World War II.' Unpublished manuscript in possession of author.
Draper, Paula Jean. 'Muses behind Barbed Wire: Canada and the Interned Refugees,' in Jackman and Borden, eds, *Muses Flee Hitler,* 271–82.
Dziuban, Colonel Stanley. *Military Relations between the United States and Canada, 1939–1945.* Washington, 1959.
Eagan, Eileen. *Class, Culture and the Classroom: The Student Peace Movement of the 1930s.* Philadelphia, 1981.
Eayrs, James. *In Defence of Canada,* vol. 1: *Peacemaking and Deterrence.* Toronto, 1972.
– *In Defence of Canada,* vol. 2: *Appeasement and Rearmament.* Toronto, 1981.

Edward, John. 'Wartime Research on RDX: A False Hypothesis Is Better Than No Hypothesis.' *Journal of Chemical Education* 64 (July 1987), 17–29.
– 'The Scientific Career of Raymond Boyer: A Tribute.' July, 1993.
Eggleston, Wilfrid. *Scientists at War*. Toronto, 1950.
– *National Research in Canada: The NRC, 1916–1966*. Toronto, 1978.
English, John. *Shadow of Heaven: The Life of Lester Pearson*, vol. 1: *1897–1948*. Toronto, 1989.
English, John, and Norman Hillmer, eds. *Making a Difference: Canada's Foreign Policy in a Changing World Order*. Toronto, 1992.
Enros, Philip, comp. *Bibliography of Publishing Scientists in Ontario between 1914 and 1939*. Thornhill, ON, 1985.
Estey, Ralph. 'The National Research Council and Seventy-Five Years of Agricultural Research in Canada.' In Richard Jarrell and Yves Gingras, eds, *Building Canadian Science: The Role of the National Research Council*. Ottawa, 1992.
Etzkowitz, Henry. 'The Making of an Entrepreneurial University: The Traffic among MIT, Industry, and the Military, 1860–1960.' In E. Mendelsohn et al., eds, *Science, Technology, and the Military*, vol. 2, 515–40. Boston, 1988.
Eve, A.S. *Rutherford: Being the Life and Letters of the Rt. Hon. Lord Rutherford, O.M*. Cambridge, 1939.
Fawcett, Ruth. *Nuclear Pursuits: The Scientific Biography of Wilfrid Bennett Lewis*. Montreal, 1994.
Feasby, W.R. *Official History of the Canadian Medical Services, 1939–1945*. 2 vols. Ottawa, 1953.
Featherstone, R.C. *McGill University at War*. Montreal, 1947.
Feingold, Henry. *The Politics of Rescue: The Roosevelt Administration and the Holocaust, 1938–1945*. New Brunswick, NJ, 1970.
Ferguson, Ted. *Desperate Siege: The Battle of Hong Kong*. Toronto, 1980.
Ferns, H.S. *Reading from Left to Right: One Man's Political History*. Toronto, 1983.
Flood, E.A. 'Otto Maass, 1890–1961.' *Biographical Memoirs of Fellows of the Royal Society of London*, vol. 8 (1962), 183–204.
Fox, J.P. 'Great Britain and the German Jews 1933.' *Weiner Library Bulletin* 26, (October 1972), 21–33.
Francis, Douglas R. *Frank Underhill: Intellectual Provocateur*. Toronto, 1986.
Frisch, Otto. *What Little I Remember*. Cambridge, 1979.
Frost, Stanley. *McGill University*, vol. 2: *1895–1971*. Montreal, 1984.
Fussell, Paul. *Wartime: Understanding and Behavior in the Second World War*. New York, 1989.
Galison, Peter, 'Physics between War and Peace.' In Everett Mendelsohn et al., eds, *Science, Technology and the Military*, vol. 1, 47–86. Boston, 1988.

Geiger, Roger. *To Advance Knowledge: The Growth of American Research Universities, 1900–1940.* New York, 1986.
- *Research and Relevant Knowledge: American Research Universities during World War II.* New York, 1993.

Geissler, Erhard. 'Biological Warfare Activities in Germany, 1923–1945.' In Erhard Geissler and John Ellis van Courtland Moon, eds, *Biological and Toxin Weapons: Research, Development and Use from the Middle Ages to 1945.* Stockholm International Peace Research Institute, forthcoming.

Gibson, Frederick. *To Serve and Yet Be Free: Queen's University,* vol. 2: *1917–1961.* Kingston, 1983.

Gibson, T.M., and M.H. Harrison. *Into Thin Air: A History of Aviation Medicine in the RAF.* London, 1984.

Gingras, Yves. 'The Institutionalization of Scientific Research in Canadian Universities: The Case of Physics.' *Canadian Historical Review* 67 (2) (June 1986), 181–94.
- *Physics and the Rise of Scientific Research in Canada.* Montreal, 1991.

Gladstone, G.P. 'Paul Gordon Fildes, 1882–1971.' *Biographical Memoirs of Fellows of the Royal Society of London,* vol. 19 (1973), 317–47.

Goldberg, Stanley. 'Groves and the Scientists: Compartmentalization and the Building of the Bomb.' *Physics Today* 48 (8) (August 1995), 38–43.

Goldberg, Susan. 'Inventing a Climate of Opinion: Vannevar Bush and the Decision to Build the Bomb.' *Isis* 83 (September 1992), 429–52.

Goldschmidt, Bertrand. *Atomic Rivals.* New Brunswick, NJ, 1990.

Goldsmith, Maurice. *Sage: A Life of J.D. Bernal.* London, 1980.

Goodell, Rae. *The Visible Scientists.* Toronto, 1977.

Goodspeed, D.J. *DRB: A History of the Defence Research Board of Canada.* Ottawa, 1958.

Gorodetsky, Gabriel. *Stafford Cripps' Mission to Moscow, 1940–42.* Cambridge, 1984.

Gotlieb, C.C., P.E. Pashler, and M. Rubinoff. 'A Radio Method of Studying the Yaw of Shells.' Reprint from *Canadian Journal of Research* 26 (May 1948), 167–98.

Gowing, Margaret. *Britain and Atomic Energy, 1939–1945.* London, 1965 [1964].
- *Independence and Deterrence: Britain and Atomic Energy, 1945–1952.* London, 1974.

Granatstein, J.L. *Canada's War: The Politics of the Mackenzie King Government, 1939–1945.* Toronto, 1975.
- *A Man of Influence: Norman A. Robertson and Canadian Statecraft, 1926–1968.* Ottawa, 1981.
- *The Ottawa Men: The Civil Service Mandarins, 1935–1957.* Toronto, 1982.

- *How Britain's Weakness Forced Canada into the Arms of the United States.* Toronto, 1993.
- *The Generals: The Canadian Army's Senior Commanders in the Second World War.* Toronto, 1993.
Granatstein, J.L., and Desmond Morton. *A Nation Forged in Fire: Canadians and the Second World War.* Toronto, 1989.
Greenhous, Brereton, Stephen Harris, William Johnson, and William Rawling. *The Official History of the Royal Canadian Air Force,* vol. 3: *The Crucible of War, 1939–1945.* Toronto, 1994.
Grenville, David, et al. *Perspectives in Science and Technology: The Legacy of Omond Solandt.* Kingston, 1995.
Gridgeman, N.T. *Biological Sciences at the National Research Council of Canada: The Early Years to 1952.* Waterloo, ON, 1979.
Groom, A.J.R. *British Thinking about Nuclear Weapons.* London, 1974.
Grove, J.W. *In Defence of Science: Science, Technology and Politics in Modern Society.* Toronto, 1989.
Groves, Leslie. *Now It Can Be Told: The Story of the Manhattan Project.* New York, 1962.
Gruber, Carol. *Mars and Minerva: World War I and the Uses of Higher Learning in America.* Baton Rouge, LA, 1975.
- 'The Overhead System in Government-Sponsored Academic Science: Origins and Early Development.' *Historical Studies in the Physical and Biological Sciences* 25 (2) (1995), 185–240.
Guerlac, Henry. *Radar in World War II.* Philadelphia, 1987.
Gwynne-Timothy, J.R.W. *Western's First Century.* London, ON, 1983.
Haber, L.F. *The Poisonous Cloud: Chemical Warfare in the First World War.* Oxford, 1986.
Hadley, Michael. *U-Boats against Canada: German Submarines in Canadian Waters.* Montreal/Kingston, 1985.
Haldane, J.B.S. *Callinicus: A Defense of Chemical Warfare.* New York, 1972
Hannant, Larry. 'Inter-War Security Screening in Britain, the United States and Canada.' *Intelligence and National Security* 6 (4) (1991), 711–35.
- *The Infernal Machine: Investigating the Loyalty of Canada's Citizens.* Toronto, 1995.
Harris, Robert, and Jeremy Paxman. *A Higher Form of Killing: The Secret Story of Gas and Germ Warfare.* London, 1982.
Harris, Sheldon. *Factories of Death: Japanese Biological Warfare, 1932–45, and the American Cover-Up.* New York, 1994.
Harris, Stephen. *Canadian Brass: The Making of a Professional Army, 1860–1939.* Toronto, 1988.
Hartcup, Guy. *The Challenge of War: Britain's Scientific and Engineering Contributions to World War Two.* New York, 1970.

Hartcup, Guy, and T.E. Allibone. *Cockcroft and the Atom*. Bristol, 1984.
Harvison, C.W. *The Horsemen*. Toronto, 1967.
Hecht, Gabrielle. 'Political Designs: Nuclear Reactors and National Policy in Postwar France,' *Technology and Culture* 35 (1994), 657–85.
Heilbron, J.L. 'Physics at McGill in Rutherford's Time.' In Mario Bunge and William Shea, eds, *Rutherford and Physics at the Turn of the Century*. New York, 1979.
Heilbron, J.L., and Robert W. Seidel. *Lawrence and His Laboratory: A History of the Lawrence Berkeley Laboratory*, vol. 1. Berkeley, 1989.
Herken, Gregg. *The Winning Weapon: The Atomic Bomb in the Cold War, 1945–1950*. New York, 1980.
Hershberg, James G. *James B. Conant: Harvard to Hiroshima and the Making of the Nuclear Age*. New York, 1993.
Hewlett, Richard, and Oscar Anderson, Jr. *The New World, 1939–1945: A History of the United States Atomic Energy Commission*. University Park, 1962.
Hirschfeld, Gerhard. *Exile in Great Britain: Refugees from Hitler's Germany*. NJ, 1984.
Hoch, Paul. 'The Crystallization of a Strategic Alliance: The American Physics Elite and the Military in the 1940s.' In E. Mendelsohn et al., eds, *Science, Technology and the Military*, vol. 2. Boston, 1988.
Holloway, David. *Stalin and the Bomb: The Soviet Union and Atomic Energy, 1939–1956*. Toronto, 1987. (New Haven, CT:, 1994)
Hooks, Gregory. *Forging the Military-Industrial Complex: World War II's Battle of the Potomac*. Urbana, IL, 1991.
Horn, Michael. '"The Mildew Discretion": Academic Freedom and Self-Censorship in Canada.' *Dalhousie Review* 72 (4) (Winter 1992–93), 439–66.
Howard, Victor. *The Mackenzie–Papineau Battalion: Canadian Participation in the Spanish Civil War*. Toronto, 1969.
Huxley, Julian. *A Scientist and the Soviets*. London, 1932.
Hyde, H. Montgomery. *The Atom Bomb Spies*. London, 1980.
Ignatieff, George. *The Making of a Peacemonger: The Memoirs of George Ignatieff*. Markham, ON, 1985.
Infeld, Leopold. *Why I Left Canada: Reflections on Science and Politics*. Montreal, 1978.
Jockel, Joseph. *No Boundaries Upstairs: Canada, the United States, and the Origins of North American Air Defence, 1945–1958*. Vancouver, 1987.
Jones, Greta. 'The Mushroom-Shaped Cloud: British Scientists' Opposition to Nuclear Weapons Policy, 1945–57,' *Annals of Science* 43 (1986), 1–26.
Jones, Martin Hugh. 'Wickham Steed and German Biological Warfare Research.' *Intelligence and National Security* 7 (4) (1992), 379–402.

Jones, R.V. *Most Secret War: British Scientific Intelligence, 1939–1945*. London, 1978.
– *Reflections on Intelligence*. London, 1989.
Kapp, Richard. 'Charles Best, the Canadian Red Cross Society, and Canada's First National Blood Donation Program.' *Canadian Bulletin of Medical History* 12 (1) (1995), 28–42.
Kennedy, J. de N. *History of the Department of Munitions and Supply*. Ottawa, 1950.
Kevles, Daniel. *The Physicists: The History of a Scientific Community in Modern America*. New York, 1978.
– 'Scientists, the Military, and the Control of Postwar Defence Research: The Case of the Research Board National Security, 1944–46.' *Science and Technology* (1975), 20–47.
Keyserlingk, Robert. 'Breaking the Nazi Plot: Canadian Government Attitudes towards German Canadians, 1939–1945.' In Norman Hillmer et al., eds, *On Guard for Thee: War, Ethnicity, and the Canadian State, 1939–1945*, 53–70. Ottawa, 1988.
King, M. Christine. *E.W.R. Steacie and Science in Canada*. Toronto, 1989.
King, Jonathan, and Harlee Strauss. 'The Hazards of Defensive Biological Warfare Programs.' In Susan Wright, ed., *Preventing a Biological Arms Race*, 120–32. Cambridge, 1990.
Koch, Eric. *Deemed Suspect: Interned Wartime Refugees in Canada*. Toronto, 1980.
Kochan, Miriam. *Britain's Internees in the Second World War*. London, 1983.
Kohn, Richard. 'The Scholarship on World War II: Its Present Condition and Future Possibilities' *The Journal of Military History* 55 (3) (July 1991), 365–93.
Kolasky, John. *The Shattered Illusion: The History of Ukrainian Pro-Communist Organizations in Canada*. Toronto, 1979.
Kuznick, Peter. *Beyond the Laboratory: Scientists as Political Activists in 1930s America*. Chicago, 1987.
Lacouture, Jean. *De Gaulle: The Ruler 1945–1970*. New York, 1991.
Lanouette, William. *Genius in the Shadows: A Biography of Leo Szilard, the Man behind the Bomb*. Chicago, 1992.
Lawren, William. *The General and the Bomb: A Biography of General Leslie R. Groves, Director of the Manhattan Project*. New York, 1988.
Legro, Jeffrey. *Cooperation under Fire: Anglo-German Restraint during World War II*. Ithaca, NY, 1995.
Leslie, Stuart. *The Cold War and American Science*. New York, 1993.
Levering, Ralph. *American Opinion and the Russian Alliance, 1939–1945*. Chapel Hill, NC, 1975.
Lewis, Jefferson. *Something Hidden: A Biography of Wilder Penfield*. Toronto, 1981.

Lewis, Julian. *Changing Direction: British Military Planning for Post-War Strategic Defence, 1942–1947.* London, 1988.
Lewis, W.B. 'George Hugh Henderson.' *Royal Society, Obituary Notices.* London, 1950.
Li, Alison. 'Expansion and Consolidation: The Associate Committee and the Division of Medical Research of the NRC, 1938–1959.' In R. Jarrell and Y. Gingras, eds, *Building Canadian Science: The Role of the National Research Council.* Ottawa, 1992.
Livesay, Dorothy. *Right Hand, Left Hand.* Erin, ON, 1977.
Lowell, Sir Bernard. 'Patrick Maynard Stuart Blackett, Baron Blackett of Chelsea,' *Biographical Memoirs of the Royal Society,* 1–107. London, 1975.
Lukasiewicz, Julius. 'Canada's Encounter with High-Speed Aeronautics,' *Technology and Culture* 27 (1986), 223–61.
Lunan, Gordon. *The Making of a Spy: A Political Odyssey.* Montreal, 1995.
McGucken, William. *Scientists, Society, and State: The Social Relations of Science Movement in Great Britain, 1931–1947.* Columbus, OH, 1984.
McKillop, A.B. *Matters of Mind: The University in Ontario, 1791–1951.* Toronto, 1994.
Mackenzie, Hector. 'Finance and Functionalism: Canada's War Effort as Justification for Its Participation in Wartime and Post-War Organizations.' Paper presented at the International Second Quebec Conference: A 50th Anniversary Commemoration. Quebec City, October 1994.
Macleod, Roy. '"All for Each and Each for All": Reflections on Anglo-American and Commonwealth Scientific Cooperation, 1940–1945.' *Albion* 26 (1994), 79–112.
Macleod, Roy, and Kay Macleod. 'The Contradictions of Professionalism: Scientists, Trade Unionism and the First World War.' *Social Studies of Science* 9 (1979), 1–32.
Macrakis, Kristie. *Surviving the Swastika: Scientific Research in Nazi Germany.* New York, 1993.
Mark, J. Carson. 'A Maverick View.' Unpublished recollections, in possession of author.
Mellor, D.P. *Australia in the War of 1939–45: The Role of Science and Industry.* Canberra, 1958.
The Merck Manual of Diagnosis and Therapy. Rathway, NJ, 1992.
Middleton, W.E.K. *Physics at the National Research Council of Canada, 1929–1952.* Waterloo, ON, 1979.
– *Radar Development in Canada: The Radio Branch of the National Research Council of Canada, 1939–1946.* Waterloo, ON, 1981.
Mikesh, Robert. *Japan's World War II Balloon Bomb Attacks on North America.* Washington, 1973.

Milner, Marc. *North Atlantic Run: The Royal Canadian Navy and the Battle for the Convoys*. Toronto, 1985.
- 'The Implications of Technological Backwardness: The Canadian Navy, 1939–1945.' *The Canadian Defence Quarterly* 19 (3) (Winter 1989), 46–52.
- *The U-Boat Hunters: The Royal Canadian Navy and the Offensive against Germany's Submarines*. Toronto, 1994.
Moon, John Ellis van Courtland. 'Project SPHINX: The Question of the Use of Gas in the Planned Invasion of Japan.' *The Journal of Strategic Studies* (September 1989).
- 'United States Biological Warfare Planning and Preparedness: The Dilemmas of Policy.' In Erhard Geissler and John Ellis van Courtland Moon, eds, *Biological and Toxin Weapons: Research, Development, and Use from the Middle Ages to 1945*. Stockholm International Peace Research Institute, forthcoming.
Morton, W.L. *One University: A History of the University of Manitoba*. Toronto, 1972.
Moss, Norman. *Klaus Fuchs: The Man Who Stole the Atom Bomb*. Toronto, 1987.
Mott, Sir Nevill. *A Life in Science*. London, 1986.
Murray, Williamson, and Millett, Allan, eds, *Military Innovation in the Interwar Period*. Cambridge, MA, 1996.
Noyes, W.A. *Chemistry: A History of the Chemistry Components of the National Defence Research Committee, 1940–1946*. Boston, 1948.
Oliphant, Mark. *Rutherford, Recollections of the Cambridge Days*. London, 1972.
Owens, Larry. 'The Counterproductive Management of Science in the Second World War: Vannevar Bush and the Office of Scientific Research and Development.' *Business History Review* 68 (Winter 1994), 515–76.
Pattison, Michael. 'Scientists, Inventors and the Military in Britain, 1915–19: The Munitions Inventions Department.' *Social Studies of Science* 13 (1983), 8–26.
Paul, Diana. 'A War on Two Fronts: J.B.S. Haldane and the Response to Lysenkoism in Britain.' *Journal of History of Biology* 16 (1) (Spring 1983), 1–37.
Peierls, Rudolf. *Bird of Passage: Recollections of a physicist*. Princeton, NJ, 1985.
Penfield, Wilder. *The British-American-Canadian Surgical Mission to the U.S.S.R.* Ottawa, National Research Council, 1943.
- *No Man Alone: A Neurosurgeon's Life*. Toronto, 1977.
Penner, Norman. *The Canadian Left: A Critical Analysis*. Toronto, 1977.
Phillipson, Donald. 'International Research Co-ordination in Wartime: The Ango-American Alliance, 1939–1945.' Unpublished paper, 1985.
Pickersgill, J.W., and D.F. Forster. *The Mackenzie King Record*, vol. 3: *1945–1946*. Toronto, 1968.
Pincher, Chapman. *Too Secret Too Long: The Great Betrayal of Britain's Crucial Secrets and the Cover-Up*. London, 1984.

Pope, Maurice. *Soldiers and Politicians: The Memoirs of Lt. Gen. Maurice A. Pope*. Toronto, 1962.
Powers, Richard. *Secrecy and Power: The Life of J. Edgar Hoover*. New York, 1987.
Price, Matt. 'Roots of Dissent: The Chicago Met Lab and the Origins of the Franck Report.' *Isis* 86 (2) (June 1995), 222–44.
Pullen, Hugh Francis. 'The Royal Canadian Navy between the Wars.' In James Boutilier, ed., *The RCN in Retrospect, 1910–1968*. Vancouver, 1982, 42–73.
Pursell, Carroll. 'Scientific Agencies in World War II: The OSRD and Its Challengers.' In Nathan Reingold, ed., *The Sciences in the American Context: New Perspectives*, 359–78. Washington, 1979.
Ramirez, Bruno. 'Ethnicity on Trial: The Italians of Montreal and The Second World War.' In Norman Hillmer et al., eds, *On Guard for Thee: War, Ethnicity, and The Canadian State, 1939–1945*, 71–84. Ottawa, 1988.
Ranger, Robin. *The Canadian Contribution to the Control of Chemical and Biological Warfare*. Toronto, 1976.
Reingold, Nathan. 'Choosing the Future: The U.S. Research Community, 1944–46,' *Historical Studies in the Physical and Biological Sciences*, 25 (2) (1995), 301–28.
Richardson, F.D. 'Charles Frederick Goodeve, 21 February 1904–7 April 1980.' *Biographical Memoirs of Fellows of the Royal Society of London*. London, 1981, 307–49.
Rigden, John. *Rabi: Scientist and Citizen*. New York, 1987.
Rivett-Carnac, Charles. *Pursuit in the Wilderness*. Toronto, 1965.
Roland, Alex. 'Technology and War: The Historiographical Revolution in the 1980s.' *Technology and Culture* 34 (1) (January 1993), 117–35.
Roland, Charles G. 'Allied POWs, Japanese Captors and the Geneva Convention,' *War & Society* 9 (2) (October 1991), 83–101.
Romano, Terrie. 'The Associate Committee on Medical Research and the Second World War.' In Richard Jarrell and Yves Gingras, eds, *Building Canadian Science: The Role of the National Research Council*, 71–87. Ottawa, 1992.
Roskill, Stephen. *Hankey: Man of Secrets*, 2 vols. London, 1974.
Rotblat, Joseph. *Scientists in the Quest for Peace: A History of Pugwash Conferences*. Cambridge, MA, 1972.
Rothwell, Victor. *Britain and the Cold War, 1941–1947*. London, 1982.
Royal Commission appointed under Order in Council PC411 of 5 February 1946, *The Report of the Royal Commission to investigate the facts relating to and the circumstances surrounding the communication, by public officials and other persons in positions of trust, of secret and confidential information to agents of a foreign power*. Ottawa, 1946.
Sapolsky, Harvey. 'Academic Science and the Military: The Years since the Sec-

ond World War.' In Nathan Reingold, ed., *The Sciences in the American Context: New Perspectives*, 379–99. Washington, 1979.
– *Science and the Navy: The History of the Office of Naval Research*. Princeton, 1990.
Sargent, B.W. 'Joseph Alexander Gray, 1884–1966.' *Proceedings of the Royal Society of Canada*. Fourth series, vol. 6. Ottawa, 1968.
– 'Nuclear Physics in Canada in the 1930s.' Kingston, ON, n.d.
– 'Recollections of the Cavendish Laboratory Directed by Rutherford.' Mimeo. Kingston, ON, Queen's Nuclear Physics Lab, n.d.
Sawatsky, John. *Men in the Shadows: The RCMP Security Service*. Toronto, 1980.
– *Gouzenko: The Untold Story*. Toronto, 1981 [1984].
Selye, Hans. *The Stress of My Life: A Scientist's Memoirs*. Toronto, 1979 [1977].
Shephard, Ronald. 'The Influence of Solandt on the Development of Early Operational Research in Britain.' In C.E. Law et al., eds, *Perspectives in Science and Technology: The Legacy of Omond Solandt*, 30–58. Kingston, 1995.
Sherman, A.J. *Island Refuge: Britain and Refugees from the Third Reich, 1933–1939*. London, 1973.
Sherwin, Martin. *A World Destroyed: The Atomic Bomb and the Grand Alliance*. New York, 1977.
Shore, Marlene. *The Science of Social Redemption: McGill, the Chicago School, and the Origins of Social Research in Canada*. Toronto, 1987.
Shortt, S.E.D., ed. *Medicine in Canadian Society: Historical Perspectives*. Montreal/Kingston, 1981.
Simpson, John. *The Independent Nuclear State: The United States, Britain and the Military Atom*. New York, 1983.
Sinclair, Andrew. *The Red and the Blue: Intelligence, Treason and the Universities*. London, 1986.
Smith, Alice Kimbell. *A Peril and a Hope: The Scientists' Movement in America, 1945–47*. Chicago, 1965.
Smith, Bradley. *The Ultra-Magic Deals and the Most Secret Special Relationship, 1940–1946*. Novato, CA, 1993.
Smith, Denis. *Diplomacy of Fear: Canada and the Cold War, 1941–1948*. Toronto, 1988.
Smith, John K. 'World War II and the Transformation of the American Chemical Industry.' In E. Mendelsohn et al., eds, *Science, Technology and the Military*, vol. 2, 307–22. Boston, 1988.
Smith, Merritt Roe. *Military Enterprise and Technological Change: Perspectives on the American Experience*. Cambridge, MA, 1985.
Sperry, Michael. *Preparing for the Next War: American Plans for Post-War Defence*. New Haven, 1977.
Spiers, Edward. *Chemical Warfare*. London, 1986.

Spinks, John. *Two Blades of Grass: An Autobiography.* Saskatoon, SK, 1980.
Stacey, C.P. *Official History of the Canadian Army in the Second World War,* vol. 1: *Six Years of War. The Army in Canada, Britain, and the Pacific.* Ottawa, 1966.
– *Arms, Men and Governments: The War Policies of Canada, 1939–1945.* Ottawa, 1970.
– *Canada and the Age of Conflict: A History of Canadian External Policies,* vol. 2: *1921–1948.* Toronto, 1981.
Stafford, David. *Camp X: Canada's School for Secret Agents, 1941–45.* Toronto, 1986.
Stent, Ronald. *A Bespattered Page? The Internment of His Majesty's 'Most Loyal Enemy Aliens.'* London, 1980.
Stevenson, William. *Intrepid's Last Case.* New York, 1984.
Stewart, Irwin. *Organizing Scientific Research for War: The Administrative History of the Office of Scientific Research and Development.* New York, 1948.
Stewart, Roderick. *Bethune.* Markham, ON, 1980.
Stokes, Lawrence. 'Canada and an Academic Refugee from Nazi Germany: The Case of Gerhard Herzberg.' *Canadian Historical Review* 57 (2) (June 1976), 150–70.
Suvorov, Victor. *Soviet Military Intelligence.* London, 1984.
Swettenham, John. *McNaughton,* 2 vols. Toronto, 1968.
Szasz, Ferenc. *British Scientists and the Manhattan Project: The Los Alamos Years.* New York, 1990.
Tardif, H.P. *Recollections of CARDE/DREV, 1945–1995.* Valcartier, Quebec, 1995.
Taylor, A.J.P. *Beaverbrook.* London, 1972.
Technical Manual, U.S. War Department, Military Explosives. Washington, 29 August 1940.
Thistle, Mel, ed. *The Mackenzie–McNaughton Wartime Letters.* Toronto, 1975.
Thomas, Hartley. *U.W.O. Continent C.O.T.C.* London, 1956.
Thomas, Jerry. 'John Stuart Foster, McGill University, and the Renascence of Nuclear Physics in Montreal, 1935–1950.' *Historical Studies in the Physical Sciences,* vol. 14, part 2 (1984), 357–77.
Thompson, John Herd. *Ethnic Minorities during Two World Wars.* Ottawa, 1991.
Travers, Timothy. *How the War Was Won: Command and Technology in the British Army on the Western Front, 1917–1918.* London, 1992.
Unsworth, Michael E. '"A Threat Well within the Capabilities": Japanese Balloon Bombs and Biological Warfare.' Paper presented at the American Military Institute Annual Meeting, Bethesda, Maryland, 5 April 1986.
Villa, Brian. *Unauthorized Action: Mountbatten and the Dieppe Raid.* Toronto, 1994.
Walker, Mark. *National Socialism and the Quest for Nuclear Power, 1939–1949.* Cambridge, 1989.
Ward, Peter. 'British Columbia and the Japanese Evacuation.' *Canadian Historical Review* 57 (3) (September 1976), 289–309.

- *White Canada Forever: Popular Attitudes and Public Policy toward Orientals in British Columbia.* Montreal, 1978.
Wark, Wesley. 'Cryptographic Innocence: The Origins of Signals Intelligence in Canada in the Second World War.' *Journal of Contemporary History* 22 (1987), 642–58.
Weart, Spencer. *Scientists in Power.* Cambridge, 1979.
 - *Nuclear Fear: A History of Images.* Cambridge, MA, 1988.
Weisskopf, Victor. *The Joy of Insight: Passions of a Physicist.* New York, 1990.
Werskey, Gary. *The Visible College: A Collective Biography of British Scientists and Socialists of the 1930s.* New York, 1978.
Wheeler-Bennett, John. *John Anderson: Viscount Waverley.* New York, 1962.
Whitaker, Reg. 'Official Repression of Communism during World War II.' *Labour/Le Travail* 17 (1986), 135–66.
Whitaker, Reg, and Greg Kealey, eds. *R.C.M.P. Security Bulletins, The War Series, 1939–1941.* St. John's, 1989.
Whitaker, Reg, and Gary Marcuse. *Cold War Canada: The Making of a National Insecurity State, 1945–1957.* Toronto, 1994.
Williams, Peter, and David Wallace. *Unit 731: The Japanese Army's Secret of Secrets.* London, 1989.
World's Who's Who in Science. Chicago, 1968.
Wyden, Peter. *Day One: Before Hiroshima and After.* New York, 1984.
Wyman, David. *Paper Walls: America and the Refugee Crisis, 1938–1941.* New York, 1968.
Zachary, G. Pascal. *Endless Frontier: Vannevar Bush, Engineer of the American Century.* New York, 1997.
Zimmerman, David. 'The Canadian Navy and the National Research Council, 1939–1945.' *Canadian Historical Review* 69 (2) (June 1988), 203–26.
 - *The Great Naval Battle of Ottawa.* Toronto, 1989.
 - *Top Secret Exchange: The Tizard Mission and the Scientific War.* Montreal/Kingston, 1996.
Zuckerman, Solly. *Scientists and War: The Impact of Science on Military and Civil Affairs.* New York, 1967.
 - *From Apes to Warlords: The Autobiography of Solly Zuckerman.* London, 1978.

Illustration Credits

Donald Avery
Grosse Île; Black Brant missile

F.G. Banting Papers, Thomas Fisher Rare Book Library, University of Toronto
Banting and Pavlov; Banting prepares for scientific mission

National Archives of Canada
University of Toronto physicists, PA-200697; commemorative opening of the NRC, PA-198202; C.J. Mackenzie, PA-198203; Otto Maass, PA-171607; E.G.D. Murray, PA-198207; McNaughton and Churchill, PA-119399; Canadian Military Mission, PA-188987; Nikolai Zabotin, PA-116421; Breadner and Sokolov, PA-200695; Raymond Boyer, PA-129633

National Archives, Washington. Still Picture Branch, Department of Energy
Oak Ridge Collection, RG 434-OR Box 2: British atomic pioneers; American defence science mandarins, 434-OR-7-37; Chicago metallurgical lab, 434-OR-7-31, binder 7
U.S. Army Signal Corps, WWII: examining Chinese civilian, 111-SC-247375

University of Toronto Archives
Harold Ireton Collection: radar technicians, B92-0030/002P
Omond Solandt Collection, courtesy Canadian Department of National Defence: Solandt and Tizard, B94-0020/003 (07); commonwealth defence scientists led by Tizard, B94-0020/003 (08)

York University Archives, *Toronto Telegram* Collection
Courtesy Canadian Department of National Defence: unused poison gas, Canadian Army photo 4407-6; flame-throwing tank, B99F674 #1536, Canadian Army overseas photo 37469-N
Courtesy Associated Press: atomic bomb tests in Nevada, B85F556 #1534

Index

Abbott, D., 240
ABC-1 defence arrangement, 41, 65
ABC-22, 65, 250
Acheson-Lilienthal Report, 237
Adams, E., 220, 230
Advisory Committee on Atomic Energy (ACAE), 198–9, 250
aeronautical labs, 310
aeronautical research/design, 284
Agent 'W,' 137, 150
Air Defence Research and Development Establishment (ADRDE), 99–100
Air Defence Research Sub-Committee, 75
Akers, W., 178, 181, 185–9, 202, 335
Allied War Supplies Corporation, 114
Allison, S., 191
Alperovitz, G., 196
American Association for the Advancement of Science (AAAS), 38
American Association of Scientific Workers (AAScW), 34, 38, 224
American Radar Development Planning Committee, 88
American RDX Committee, 112

Anderson, J., 178–81, 186–7, 199, 265
Anglo-American alliance, 249, 257
Anglo-American atomic energy, 192, 251
Anglo-American gas warfare policy, 141
Anglo-Soviet Technical Accord, 214–17, 223, 233
anthrax, 157, 159–60, 163, 165, 170, 172, 174, 210; GIN project, 160–2
anti-submarine weapons, 90–1; depth charges, 64; hedgehogs, 115; torpedo technology, 115
Appleton, E.V., 74, 178
Argonne National Laboratory, 208, 213, 246; heavy water reactor – Chicago Pile 3 (CP-3), 190, 193
arms control, and role of scientific groups, 225
Associate Committee for Aviation Medicine, 69
Associate Committee on Explosives, 109, 118–19
Associate Committee for Medical Research (ACMR), 33, 59, 154
atomic research, 12, 30, 95, 139, 148, 176, 264; Allied cooperation, 177,

181, 185–8, 194, 201, 252; American research, 235, 252; American superiority, 201, 252; Bikini atomic tests, 363; British dependence on United States, 252; British research, 180; Canadian research and development, 198, 255, 260–1; control of, 225, 253; health risks associated with, 193; secrecy and security, 178, 182, 188, 199, 212, 219, 229, 235–6, 244; use of bomb against Japan, 195
Attlee, Clement, 198, 252
Auger, P., 182
Australia, 21; experimental station, 323
Avro Arrow, 364

Bachmann, W., 110–12
Bachmann process, 112, 115–16, 217, 244, 314
Baer, E., 205
Baer, R., 26
Bain, W.J., 26
Baker, N., 213
BAL, 146
Banting, F.: activities during war, 47, 55–6, 124–5, 149, 152–7, 174–5; death of, 59, 152; Nobel prize for discovery of insulin, 31–2; postwar activities, 256, 260, 262, 266, 272, 325; prewar career, 6–7, 24, 33–5, 39
Banting–Rabinowitch Mission, 124–5
Barrett, H.M., 132, 164, 254, 263, 272
Baruch, B., 253, 363
Baruch Plan, 253
Battle of the Bulge, 106
Battles, R.N., 247
Battle of Stalingrad, 214
Bavin, E.W., 208
Beaverbrook, Lord, 79

Benbow, Major H., 193, 213
Bennett, R.B., 18
Bernal, J.D., 223, 225, 265, 272, 352
Best, C., 32, 39, 53
Bethune, N., 35, 272
Bevin, E., 232
biological and toxin warfare (BTW), 23, 98, 140, 150–1; Allied cooperation, 152, 156–9, 161, 163, 165–7, 175, 254; American research, 163, 165, 254; anti-animal agents, 172; anti-plant agents, 165, 171; biological agents, 153, 158–9, 164–5; British research, 165; Canadian research and development, 152, 158, 161–5, 254, 258, 260; countermeasures, 153; defensive uses of, 155, 173; ESS 4-lb bomb, 163; fear of German use of, 156, 168; 500-lb cluster bomb, 163; health risks, 162, 164, 253; immunization of Allied troops and civilians, 168; Japanese research, 173; offensive uses of, 155; role in Cold War, 174–5; secrets and security, 209, 226–7, 249; Soviet research, 226
Biological Warfare Committee (USBWC), 165
Biological Warfare Sub-Committee, 154–5, 157, 175
Biological Weapons Convention (1972), 265
Blackett, P.M.S., 21, 98, 215, 223, 225, 272
Bletchley Park, 12
Bliss, Michael, 7–8
Blomquist, A., 114
Bohr, N., 29
Borden, Robert, 16
Bothwell, R., 8

Bottomley, Air Marshal N., 165
botulinum toxin ('bot tox'), 160, 163, 167–9, 174, 331; cluster bombs, 164; Type A, 163; Type B, 163
Bowen, E.G. (Taffy), 73, 81, 93
Bowen, Vice-Admiral H.B., 76–7
Bowles, E.L., 80
Boyer, R., 36, 112, 114, 119, 208, 217, 220, 223, 232–5, 239–40, 243–4, 263–4, 272
Boyle, R.W., 7, 29, 85, 239, 273
Bracken, J., 221
Bragg, L.W., 79, 103
Breadner, Air Marshal L.S., 140
Bren gun contract, 22
Britain, 4–5, 12–16; alliance with USSR, 214, 223; atomic policy, 198; atomic research, 176–7, 179, 184, 186–7, 190, 194–5, 214; atomic security, 188; biological warfare research and policy, 151, 154–5, 157, 159, 167, 174; Biological Warfare Sub-Committee, 154–5, 157, 175; Bletchley Park, 12; chemical research, 122–3, 126, 146, 154; Committee of Imperial Defence, 14, 23, 152, 154; cooperation with Allies, 41, 58, 76–7, 78, 89, 98, 124, 215, 250; decline of scientists' movement in, 238; espionage, 232, 234, 238, 242, 246; explosives research, 115; fear of enemy use of biological and chemical warfare, 123–4, 130, 145, 154–5, 157, 169; fear of Soviet nuclear capabilities, 231–2; postwar defence, 194, 198–9; radar research (RDF), 19, 20–1; RAF, 18, 20, 54, 73, 115; rearmament policies, 20, 37, 52, 257; Royal Navy, 20, 83, 115; Scientific Advisory Committee, 58; secret intelligence, 5, 55, 206, 213; Telecommunications Research Establishment (TRE), 73, 81, 88; threat of German V weapons, 146
British Air Ministry, 69, 73, 75, 78–9, 86; Air Defence Research and Development Establishment (ADRDE), 99–100; Air Defence Research Sub-Committee, 75; Committee for the Scientific Study of Air Defence (CSSAD), 22, 52
British Armed Forces, 19, 22, 55–6, 82–3, 134; Army gas school at Porton, 125
British Army Staff College, 20
British Association of Scientific Workers (BAScW), 34, 37–8, 222, 225, 264
British Atomic Energy Bill, 238, 252
British Atomic Scientists Association (BASA), 226, 238
British Central Commonwealth Scientific Committee, 116
British Central Radio Bureau, 88
British Central Scientific Office (BCSO), 61, 67, 111
British Chiefs of Staff (BCOS), 88, 105, 123, 127, 145–7, 165, 169–70, 174, 323
British Empire Air Training Scheme, 305
British Foreign Office, 53
British Inter-Services Sub-Committee for Biological Warfare (ISSCBW), 174
British Medical Research Council (BMRC), 124, 152, 154, 158, 335
British Ministry of Aircraft Production (MAP), 79, 99–100,

392 Index

British Ministry of Supply, 99, 102, 109, 131, 140, 215
British Purchasing Commission (BPC), 108
British Radar Sub-Committee, 88
British Special Committee on Biological Warfare, 24
British Supply Council, 143, 209
British Tube Alloys, 182–5, 189, 201–2, 258; Directorate of Tube Alloys, 178–9
British War Cabinet, 54, 58, 126, 157, 177–9, 214
British War Office, 20, 22–3, 90, 148, 153–4, 164, 168–9, 245
British Woolwich Explosives Station, 109
Bruceton, Pennsylvania, test laboratory, 113, 118
Brunskill, Major General G., 144, 148–9
Bukharin, N., 291
Bureau of Preventive Medicine (U.S. Navy), 158
Burns, Colonel E.L.M., 20
Burton, E., 47, 83–4, 104, 273
Bush, Vannevar: postwar role, 257–61, 273; prewar role, 3, 5, 39; wartime role, 57, 59–61, 63–4, 67–70, 72, 77, 93, 100–1, 105–6, 112, 116, 140, 143, 173, 179–80, 182, 186, 188–90, 194, 201–2, 211, 239, 248
BW-CW Tripartite Agreement, 129, 261

California Institute of Technology, 72
Cambridge University, 29, 37
Cambron, A., 114
Camp Detrick, Maryland, 152, 162–71, 174–5, 254, 257, 261

Campbell, G., 57
Canada: alliance with Soviet Union, 214, 216–17, 221, 226; atomic research, 176–7, 180, 184, 186–7, 192, 196, 200–1; atomic security, 188, 209; biological warfare research, 155–6, 173–4; chemical weapons research, 138–9, 143, 149; cooperation with Allies, 176–7, 263; defence preparedness, 259; fear of Soviet nuclear capabilities, 231; internment of Japanese Canadians, 205–6; military intelligence, 203; Official Secrets Act, 204, 228, 233, 236, 239; operational research, 167; postwar nuclear research, 177, 194–5; prewar military capabilities, 15–16, 18, 257; role in scientific information distribution, 225; Soviet espionage in, 218–19, 226–7, 230, 236, 239, 242, 265
Canada-Deuterium-Uranium (CANDU), 177
Canada Evidence Act, 233
Canada Inquiries Act, 230
Canada–U.S. Permanent Joint Board on Defence, 66
Canada–U.S. rinderpest research station, 158–9
Canadian–American RDX Committee, 114, 116, 119, 208
Canadian Armament Research and Development Establishment (CARDE), 120, 249
Canadian Armed Forces, 15, 18–19, 22, 82, 96–7, 113, 115, 203–4, 208, 247–8, 251, 258; dependence on British technology, 14, 20, 25, 89, 94; RCAF, 9, 20, 74, 91–3, 130, 140, 210, 243; RCN, 9, 20, 50, 83, 89–94, 119,

140, 222, 233, 258; Royal Canadian Army Medical Corps, 7
Canadian Army, 6, 19, 90, 130; biological and chemical warfare, 12, 16, 127; Bren gun contract, 22; Directorate of Mechanization and Artillery, 109; radar, 16; RDX explosives, 8, 12, 16; role in security 209–10; role in WWI, 122
Canadian Army Technical Development Board, 118
Canadian Association of Professional Physicists (CAPP), 239
Canadian Association of Scientific Workers (CAScW), 222–6, 232, 234, 236–40, 242, 264–5
Canadian Cabinet Defence Committee (CDC), 248–9
Canadian Cabinet War Committee, 65, 135, 137, 152, 159, 168–9, 172–3, 216, 256
Canadian Chiefs of Staff, 75, 103–4, 135, 142, 150, 154, 168, 172
Canadian Civil Liberties Union, 37
Canadian Committee to Aid Spanish Democracy, 36
Canadian Commonwealth Federation (CCF), 206, 221
Canadian Communist party, 35, 219
Canadian Cryptographic Unit (Examination Unit), 11
Canadian CW Inter-Service Board, 131–2, 136, 140, 144, 148, 169, 254
Canadian Industries Ltd. (CIL), 108
Canadian Institute of Chemistry (CIC), 26–7
Canadian Joint Staff Mission (CJSM), 65–7, 141
Canadian League for Peace and Democracy, 36

Canadian Military Headquarters (London), 168
Canadian Military Mission in Washington, 197
Canadian National Research Council (NRC), 3, 11, 39, 56, 115, 117, 120, 218; Allied cooperation, 256, 260; asdic/sonar research, 90, 97; atomic research, 98, 181–2, 184; biological warfare research, 152, 155, 157, 160; chemical warfare, 151; cooperation with British, 12, 23–5, 42, 53, 86, 91, 98; cooperation with United States, 50, 54, 59, 63, 67, 80, 129; explosives research, 47, 98, 107, 109–11, 113–14, 118–20; founding of, 5–6, 16–17; funding of research, 84–5, 96–7; operational research, 124; prewar defence technology, 14–15, 44; proximity fuse research, 98; radar research, 15, 47, 62, 69, 74–5, 78–9, 82–91, 257–8; Radio Field Station, 19, 63, 74–8, 80–1, 88–9, 92, 102, 208, 220, 247; recruitment of scientists, 8–9, 42–3, 82–3, 86, 182–3; relations with military, 6, 11, 42, 80, 88, 91, 93–4, 97–8, 131, 258; relations with universities, 28, 30, 44–6, 49, 80, 82–3, 94, 129, 152, 158; role of C.J. Mackenzie, 61, 68–9, 80, 85, 87 89, 98, 113, 181, 184–6, 195, 198, 206–8, 239, 255–6, 260; secrets and security, 203, 206, 208–9, 216, 227, 232, 242; war medicine, 15, 69
Canadian Pacific Railway, 132
Canadian RDX Committee, 112
Canadian Signal Research and Development Establishment, 249
Canadian–U.S. Chemical Warfare Advisory Committee (CWAC), 142

Carr, S., 219, 232
Cave, H.M., 84
Cavendish laboratory, 29, 31, 224
CF-105 (Avro Arrow), 364
Chadwick, J., 179, 188–9, 191–2, 195, 199, 201, 213–14, 240, 257, 273
Chalk River nuclear laboratories, 176–7, 195, 199–201, 218, 237, 239, 244–6, 255, 259, 261, 343; reactor project, 199
chemical warfare, 62; Advisory Sub-Committee, 148; Allied cooperation, 133, 142, 144, 149, 181, 254; Allied policy controlling use of, 254; American research, 254; Canadian research, 254, 258, 260; effects of, 253; KB-16, 138–9; operational research, 142, 148; role of Otto Maass, 133–4; screening of Canadian scientists, 208; secrets and security, 209, 226, 249; Soviet research, 226
Chemical Warfare Convention on the Prohibition of the Development, Production, Stockpiling and Use of Chemical Weapons, and on their Destruction (1992), 265
Chicago Metallurgical Laboratory, 186–7, 191–4, 202, 212–13, 220; cooperation with Montreal Laboratory, 191
Chief of the General Staff (CGS), 130, 136, 157, 168, 178, 198
China, chemical weapons used against, 136
Chipman, R.A., 84
Churchill, Winston, 5, 22, 56, 71, 127, 136, 145–6, 149, 170, 178–80, 188, 194, 196, 198, 223–4, 232, 265
C-I Committee, 166

Civil Defence Canada, 134
Claxton, B., 240, 251
Clinton (Oak Ridge), 193
Cochrane, R., 169
Cockcroft, J., 77, 79, 99, 102, 179, 190–1, 194–5, 199, 201–2, 213, 215, 246, 273, 336, 341
Cody, H.J., 118, 205–6, 325
Cold War: Allied defence policy, 253; American policy, 228, 251; American–Russian nuclear agreement, 230, 254; British policy, 228, 251; Canadian policy, 228, 249–50, 255; end of, 266; espionage, 265; United Nations' role, 251
Coldwell, M.J., 221
Collip, J.B., 59, 152, 157, 327
Columbia University, 94, 183
Combined Chiefs of Staff (CCOS), 65–6, 111, 141, 166
Combined Policy Committee (CPC), 188–9, 191–2, 196
Committee of Imperial Defence, 14, 23, 152, 154
Committee for the Scientific Study of Air Defence (CSSAD), 22, 52
Commonwealth Advisory Committee on Defence Science, 250, 261
Communist activity, 205, 208
Compound Z, 129, 137, 150
Compton, A., 189
Compton, K., 77, 80, 88
Conant, J., 5, 42, 60, 63, 100–1, 129, 138–41, 143, 177, 185–92, 194, 201–2, 258, 273, 333
Conant mission, 67, 95, 100
Connaught Laboratories, 156, 327
Connor, R., 114
Cook, S.J., 50
Cornell University, 112, 115

Corson, D., 81
Craigie, J.G., 152, 156, 263
Crerar, Major General H.D.G., 20–4, 51, 130–1, 209, 273
Croil, Air Vice-Marshal G.M., 74

Dale, H., 158–9, 189
Dalhousie University, 6, 26, 28, 32, 85
Darwin, C., 61, 223
Davies, E.L., 129–30, 140, 142, 163, 166, 274, 332
Davies, J., 221
Dearle, R.C., 83, 85–6, 274
Defence of Canada Regulations, 205
Defence Industries Ltd. (DIL), pilot plant at Valleyfield, 108, 118
Defence Research Board (DRB), 228, 248–9, 255, 259, 261; British-American cooperation, 249; cooperation with military, 249
de Gaulle, General Charles, 214
Department of Agriculture, 154, 172
Department of Defence Production, 255
Department of Justice, 230, 232
Department of Munitions and Supply (DMS), 96–7, 107–9, 112, 115, 181, 217–19; Explosives and Chemical Branch, 107, 114; Shawinigan Falls pilot plant, 112, 209, 218. *See also* C.D. Howe
Department of National Defence (DND); 6, 10, 14, 18, 42, 67, 97, 114, 131, 152, 155, 162, 197, 209, 216,218, 220, 249; funding of chemical warfare research, 131; support of biological warfare research, 158–9
Department of National Health and Welfare, 172

Department of Pensions and National Health, 154
Department of Scientific and Industrial Research (DSIR), 5, 16, 178, 181
DesRosiers, Colonel (Deputy Minister of Defence), 50
Dill, J., 126–7, 141, 188
dinitrodiethanol nitramine (DINA), 117–20
Dirac, P., 223
Directorate of Chemical Warfare and Smoke (DCW&S), 104, 124, 131–2, 134–5, 137–44, 146, 148–51, 160, 163, 172, 174, 209, 226, 258; C-1 Committee, 161; cooperation with military, 140
Directorate of Mechanization and Artillery, 109
Directorate of Technical Research, 94, 131
Dominion Arsenals, 113
Dominion–Provincial Conference in Ottawa (1945), 197
Donald, J.R., 107–8, 110, 113, 115
Douglas, W.A.B., 9
Drew, G., 287
DuBridge, L., 77, 86–7, 92
Duncan, J.S., 47, 56
Dunning, C., 241
Du Pont Company, 186
Dykes, V., 65

Eaton, J.D., 221
École Polytechnique, 50
Edgewood Arsenal, 128, 257
Edward, J., 359
Eggleston, W., 8
Einstein, A., 28–9
Eldorado Mining Company, 341
Element '49,' 186

Embick, Major General, 21
Esoid, 168
Exercise Musk-Ox, 251, 362
explosives research: Allied cooperation, 107–13, 181, 260; American research, 107, 113; Bachmann process, 112, 115–16, 217, 244, 314; British research, 107; Canadian research, 260; cooperation with Soviet Union, 217; explosive polymers, 117; Explosives Research Laboratory (ERL), 113; HMX, 115; NENO, 118–19; Ross-Scheissler process, 110–11; secrets and security, 219, 243–5; TNT, 115, 117. *See also* RDX explosives
Explosives Research Laboratory (ERL), 113

Fawcett, G., 210
Federation of American Scientists (FAS), 225–6, 234–9, 264
Fermi, E., 182
Fildes, P., 157, 160–1, 164, 167, 330, 332
Fischer, H.O., 26
flame-thrower weapons, 128, 137, 142, 148; Barracuda model, 143; Crocodile, 143; Rattlesnake Mark II, 143, 148; research, 124, 322; Ronson, 143
Flavelle, E., 221
Flood, Colonel E.A., 23–4, 129, 138–9, 142
Ford, R., 231
Foreign Enlistment Act, 36
Forsey, E., 206, 292, 294
Foster, J.S., 7, 30 , 81, 86–7, 263, 274
Fothergill, Captain L., 164

Foulkes, Major General C.H., 20, 122, 248
Fowler, R.H., 54–6, 59, 63, 78–9, 84, 93
Fowler, R.L., 99
France, 155; nuclear research, 200
Fraser, D., 156
Fred, E.B., 156–7, 162
Free French scientists, 182, 190, 200, 214; security threat, 190
Fuchs, K., 190, 205, 246, 360
Fu-Go balloon project, 173
Furer, Rear Admiral J.A., 93
Furlong, Admiral, 57
Furry, W., 237, 357

Gallie, W.E., 7
Geneva Protocol (1925), 122–4, 126, 152–3, 171; rules on use of chemical and biological weapons, 16
German High Command, 145; threat of biological and chemical weapons use, 158, 163, 165, 169–70
Germany, 4, 37, 256; aerial bombing of Britain, 72, 99; atomic research and development, 178; biological warfare research, 153, 155, 167, 328; blitzkrieg, 204; chemical research, 128, 142; espionage, 212; 'fifth column,' 204; Luftwaffe, 146; poisonous gas, 16, 122–3, 128, 146, 149, 169; proximity fuses, 106; takeover of explosives research operations by USSR, 244; threat of chemical and biological weapons, 23, 136, 149, 158, 163, 165, 169, 170; threat of U-boats, 71, 73, 89, 91–2, 94; threat of V weapons, 105, 146; Treaty of Versailles, 16; unconditional surrender (May 1945), 170, 195

Gibbons, N.E., 153
GIN project, 160–2
GIR project, 160
GL Mark III C, 216, 309
Goering, Reich Marshal H., 106, 312
Goldschimdt, B., 182, 187, 193, 200, 214, 274
Goodeve, C.F., 274, 288
Gordon, A.R., 113, 132, 263, 274
Gotlieb, C.C., 104, 274
Gousev, F., 221
Gouzenko, I., 203, 218–19, 227, 230, 232, 236, 239; defection of, 229, 242
Graham, D., 7
Granatstein, J.L., 9
Granite Peak, 163
Gray, J.A., 83–4, 213, 274
Greey, P.H., 156
Greig, J.W., 64–5
Grosse Île project (GIR), 152, 158–63, 174, 209, 258, 329; protection of research and weapons, 210
Groves, General L.R., 177, 180, 187–93, 198–202, 212–14, 227, 235, 239–40, 245–6, 257, 261, 265
Gruinard Island, 160
Grundfest, 206
Gulf War, 266
Gusev, F., 215

Hackbusch, R.A., 91, 308
Hafstad, L.R., 100
Hahn, O., 29, 39, 176
Haldane, J.B., 153
Halifax, Lord, 61
Hall, G.E., 125, 156
Halperin, I., 205, 218, 220, 232–3, 241–2, 274, 350, 357
Hanford, Washington, 176, 180, 246
Hankey, Lord, 71, 155, 157, 178

Harbin (Manchuria), Units 100 and 731, 173
Harris, Air Marshal A.T., 111
Harris, Colonel J.P., 113
Harvard University, 6–7, 38, 60, 112, 237
Haskins, C., 63–4
Heath, F., 81
Hebrew Committee of National Liberation, 149
Henderson, G.H., 275
Henderson, J.T., 19, 74–5, 78, 83–4, 275
Henderson, W.J., 85
Herzberg, G., 183, 275, 288
Hibbert, H., 45
Higinbotham, W.A., 235
Hill, A.V., 21, 34, 52–4, 66, 70–1, 76, 79, 223, 275
Hill, G., 62
Himmler, H. (SS leader), 169
Hiroshima, 196–8, 225, 248
HMX, 115
Holland, 155
Holloway, D., 246
Honorary Advisory Council on Scientific and Industrial Research, 17
Honshu (Operation Coronet), 148
Hoover, J. Edgar, 237, 244, 354
Howe, C.D.: postwar role, 260, 264, 275; prewar role, 8, 11; wartime role, 42, 56 , 96, 107, 111–12, 118, 181, 186–8, 192, 196, 198, 217, 255
Howlett, L.E., 19, 62
Huffman, J., 193, 213
Hull, C., 59
Hyde Park Agreement, 41, 52, 88, 108, 194
Hyde Park Aide-Mémoire, 251

Ignatieff, G., 363

Imperial Conference (1937), 18
Imperial Defence College, London, 20
Industrial Chemical Industries (ICI), 178, 184
Infeld, L., 205–6, 222, 241–2, 275, 358–9
infra-red research, 22, 72
Inglis, J., 22
Innis, H., 224
Inspection Board of the United Kingdom and Canada, 109, 134
Inspection Board's Explosives Proof Establishment, 118
Inter-Services Sub-Committee for Biological Warfare (ISSCBW), 165–6, 175
Inventions Board (Canada), 49
Iraq, nerve gas and biological weapons acquired by, 266
Ismay, General H.L., 127
Italy, 16, 256; biological warfare research, 153; use of poisonous gas, 23, 122–3, 149
Iwo Jima, 147

James, C., 206, 240–1
Japan, 16, 256; atomic bomb used against, 195–8; biological warfare research, 157–8, 167, 172–3, 227; chemical weapons used against China, 136; fear of proximity fuse, 106; invasion of, 172, 324; poisonous gas use, 134, 149; treatment of POWs, 135; use of humans for experiments, 173. *See also* Hiroshima; Nagasaki
Japanese Canadians, internment of, 135, 205
Johns Hopkins University, 94, 101, 104, 257

Johnstone, J.H.L., 85
Joint Committee on New Weapons (JCNW), 70–2, 105, 248, 258. *See also* Bush, Vannevar; United States
Joint Intelligence Committee, 66
Joint Technical Warfare Committee, 174, 250
Joint War Production Committee of Canada and the United States, 108
Joliot-Curie, F., 214, 225, 240, 347, 352

Kapitza, P., 224
Kellock, R.L., 228, 231, 233
Kellogg-Briand Pact (1928), 264
Kelly, W., 227
Kempton, A.E., 191
Kevles, D., 95
Keys, D., 83
King, A., 361
King, William Lyon Mackenzie: funding for defence science, 47, 84; postwar role, 260, 265; prewar role, 17–18, 20–1, 24; pro-Soviet policies, 221; rearmament policies, 20, 22; recruitment of immigrant scientists, 182–3; role during war, 43, 47, 49, 51, 57, 135, 180–1, 187, 197, 204, 215, 221, 229, 232, 236, 251–2
Kingan, R., 131
Kistiakowsky, G.B., 107, 112, 118
KL-16, 150
Kowarski, 191, 200, 214, 336
Kremlin, 215, 217, 223, 231–2, 242, 246
Krotov, I., 216
Kyushu (Operation Olympia), 148

Labour party, 198
Labour Progressive Party of Canada, 217
LaFlèche, Major General L.R., 154

Index 399

Laird, E., 306
Lathe, G., 225
Lauritsen, C.C., 100
Laval University, 32, 50
Lawrence, G., 183–5, 253, 275
League of Nations, 16
Leahy, Admiral W., 230
Letson, Brigadier H.F.G., 61
Levine, S., 205–6, 241
Lewis, W.K., 129
Lindemann, Brigadier C., 276
Lindemann, F.A. (Lord Cherwell), 5, 7, 21–2, 54–9, 66, 70, 178, 194, 257, 276
Linstead, P.B., 209
Linstead, R.P., 111–12
Litvinoff, M., 349
Llewellin, Colonel J.J., 188–9
Loomis, A., 77, 80, 87
Los Alamos, 176, 184, 190–1, 195, 212, 246, 265
Lothian, Lord, 53, 57
Low, S., 240
Lucas, C., 125
Luftwaffe, 146
Lunan, G., 218, 220, 230, 232–3, 354–5

M-1000 Committee, 152, 157, 207, 210
Maass, Otto: postwar career, 256, 258, 260–2, 276, 294, 320, 322, 330, 363; prewar role, 6–8, 26–7; role during war, 45–6, 53, 56–7, 59, 94, 103, 108–9, 115, 125, 127–9, 130–4, 137–43, 149–50, 152, 154, 159–61, 164, 166, 168–9, 172, 174–5, 184, 207, 210, 226, 240, 254
Macdonald, M. (British High Commissioner), 180–1
Mackenzie, C.J.: activities during war, 42–4, 46–50, 53, 56–9, 61–3, 66, 68–9, 77–80, 85–6, 92–4, 97, 100, 103, 112–14, 116, 118–19, 125, 131, 133, 152, 155, 182–6, 188–9, 191–2, 194–6, 199–202, 206–9, 214, 239–40, 245–8; postwar career, 255–8, 260–2, 264, 276, 364; prewar career, 3, 6–8, 11–12, 18–19, 40. *See also* Canadian National Research Council
Macklin, C., 207
Macklin, M., 263
Macreary, G., 65
Madsen, J., 76
Manhattan Engineer District (MED), 179, 187, 191, 194, 200, 213–14
Manhattan Project, 31, 176–7, 180–81, 186–91, 196, 199–200, 207, 235, 245–6, 253, 257; recruitment of scientists, 183; secrets and security, 200, 212–14
Mark, J.C., 184, 190
Marsh, L., 206
Marshall, General G., 57, 141, 147
Massachusetts Institute of Technology (MIT), 38, 63–4, 80, 92–4, 129, 208; Radiation Laboratory, 257
Massey, V. (Canadian High Commissioner), 83, 124, 131, 155
Maud Committee, 177–8
May, A.N., 208, 220, 223, 226, 232, 234, 236, 239, 244–6, 276, 357, 359; conviction for espionage, 238
May-Johnson Bill, 236, 238, 356
Mazerall, E., 218, 220, 232–3, 239, 247, 350
McGill University, 6–7, 16, 26, 28–30, 32, 36, 44, 46, 80, 83, 86–7, 107, 109, 113–14, 116, 129, 132, 150, 152, 154, 156, 206–7, 217, 222, 224, 240–1, 256–7; War Advisory Board, 45

McKinley, D.W.R., 79, 102, 210, 247, 263
McLennan, J.C., 28, 84, 289
McMahon, B. (Senator), 356
McMahon Bill, 237
McMaster University, 119, 183–4
McNaughton, General A.G.L.: activities during war, 40, 42–3, 51, 74-75, 79, 84–5, 100, 103, 114, 117–18, 124, 127, 130–1, 152–5, 253; postwar activities, 276, 286, 304, 325; prewar career, 6, 9, 14, 17–20, 22–4, 27, 33
McTaggert-Cowan, P.D., 239
Mellanby, E., 124–5, 154, 326
Merck, G., 159–60, 165–6, 173, 175, 329, 332
Military Co-operation Committee (MCC), 362
Millard, J.R., 89
Mills, W.G., 198
Mitchell, C., 156
Moffat, P. (U.S. ambassador), 51–2
Molson, W., 240
Montreal Atomic Laboratory, 12, 31, 176–7, 180–7, 189, 190–6, 199–202, 207–9, 212–14, 220, 222, 225, 255, 257, 263; cooperation with Chicago Metallurgical Laboratory, 191; dependence on American nuclear research, 193; European émigré scientists, 207; security risks, 212
Montreal Neurological Institute (MNI), 32, 69, 222
Morrison, Colonel G.P., 134
Motinov, Colonel, 218, 220, 349
Mott, N.F., 223
Mountbatten, Lord L., 265
Munich crisis of 1938, 37, 153
Murchie, Major General J.C., 136–7, 169, 172

Murray, Everitt George Dunne (E.G.D.), 7, 32, 152, 156, 158–63, 166–9, 174–5, 260–2, 277, 330, 332, 335
Murray, W.C., 15
mustard gas, 129, 133, 135, 137, 143, 145, 150, 170, 209, 285, 287. *See also* Poisonous gas
Mutual Aid Board, 215–17, 233, 244

Nagasaki, 196–8, 225, 248
napalm weapons, 143
National Conference of Universities (NCU), 43
National Council for Canadian–Soviet Friendship (NCCSF), 217, 219, 221–2, 242
National Defence Act, 249
National Defense Research Committee (NDRC), 5, 39, 57, 59–63, 77, 87, 93–4, 100–2, 104, 107–8, 112, 150, 258; atomic research, 185; chemical research, 138; Committee on Explosives and Chemical Warfare, 129; explosives research, 118, 120
National Research Council. *See* Canadian National Research Council
National Research X-perimental (NRX), 177, 194, 199
National Resources Mobilization Act (1940), 39
Nazi-Soviet Pact (August 1939), 37
NENO, 118–19
nerve gases: sarin, 146, 323; soman, 146; tabun, 128, 146
Newall, Air Marshal C., 53
New London naval testing laboratory, 64
Nicholls, R., 114

Nightingale, M., 210, 220, 233, 309, 350, 359
Norden bombsight, 55, 297
Normandy landing (D-Day), 144, 169
NRC. *See* Canadian National Research Council
nuclear research. *See* atomic research

Oak Ridge, 176, 180, 190–1, 193, 212–14
Office of Naval Intelligence, 12, 93, 99
Office of Scientific Research and Development (OSRD), 4, 10, 63–4, 67–8, 72, 93, 116, 156, 172, 258, 312; atomic research and development, 186; relations with NRC, 185, 248; role of Bush, 68, 70, 80, 179, 182, 257; role of Conant, 201
Office of Strategic Studies (OSS), 167–8
Official Secrets Act (Canada), 204, 228, 233, 236, 239
Ogdensburg Agreement of 1940, 41, 65
Okinawa, 147
Oliphant, M., 76, 179, 190, 199, 340
Ontario Public Inquiries Act, 233
Operation Olympia, 148
Operation Overlord (D-Day), 106, 145, 168
Oppenheimer, J.R., 183, 189, 265, 357

Padlock Law, 36
Paget, Colonel H., 165–6
Paterson, Major J.C., 132
Pavlov, V., 219
Pearl Harbor, 135–6, 149, 157
Pearson, D., 230–1
Pearson, Lester B., 355–6
Peierls, R., 179, 190

Penfield, W., 222, 277
People's Commissariat of State Security (NKVD), 218–19, 226, 242
Permanent Joint Board on Defence (PJBD), 41, 51, 65–7, 250
Perry, Major General H.M., 154–5
Petawawa, 205
Petroleum Warfare Development Department (PWDD), 143
Pew, A.E.H., 90
Phillips, Colonel W.E., 78, 308
phosgene, 129, 133, 137, 145, 170
Pirie, Air Vice-Marshal, 53
Pitt, A., 81, 84, 100, 102–4, 261, 263, 277
Placzek, G., 182, 184
plutonium, 187, 193–4, 201–2, 212, 220, 245–6, 255, 261; element '49,' 186
poisonous gas: Agent 'W,' 137, 150; blood, 123; choking, 123; Compound Z, 129, 137, 150; HS mustard gas, 137; HT mustard gas, 129, 133, 135, 137, 143, 145, 150, 170, 209, 285, 287; KL-16, 150; phosgene, 129, 133, 137, 145, 170; possible use against Japan, 135; sarin, 146, 323; soman, 146; tabun, 128, 146
Pontecorvo, B., 182, 239–40, 277, 358
Pope, Major General Maurice, 9, 20–1, 24, 65, 141, 258, 277, 301–2
Porsild, A.E., 224
Porter, General W., 137–8, 141–2, 147, 150, 159, 166, 175, 261, 333
Porton, 127–30, 141, 157, 160–1, 164–7, 257, 261
Porton Down, 152, 175
Power, C.G. (Minister of Defence for Air), 42
Pratt, J.D., 131, 138, 140, 149

402   Index

Princeton University, 82
Progressive Conservative party, 221
protective ointment (BAL), 146
proximity fuse, 8, 12, 127, 139; Allied cooperation, 100, 103–4; American research, 100–2; British research, 84, 99–100; Canadian research, 101–3, 260; early use of, 105–6; military requirements, 101; security of, 105, 243
Purvis, A.B., 108

Quebec Agreement, 188, 196, 201, 251–2
Quebec Conference (1943), 66, 177, 187–9, 215, 340
Queen's University, 16, 26, 28–9, 32, 39, 43, 83, 113, 119, 150, 152, 161, 174, 213, 233, 241

Rabi, I., 77, 263
Rabinowitch, Major I.R., 124–7, 131, 154, 318
radar research (RDF), 72–5, 82, 95, 139, 172, 216, 264; Allied cooperation, 58, 69, 80, 87, 95, 181; American research, 71–2, 74, 77, 81, 86, 92; British research, 69, 74–5, 78–9, 82–8; Canadian research, 260; screening of Canadian scientists, 208; secrets and security, 76, 209, 216, 219, 247
Radiation Laboratory (Rad Lab), 63–4, 73, 77, 81–2, 86–7, 92–3, 95, 208. *See also* Massachusetts Institute of Technology
Radio Field Station, 19, 63, 74–8, 80–1, 88–9, 92, 102, 208, 220, 247
radio research: American, 72; British, 71–2; Radio Locators GL 3, 216

Ralston, Colonel J.L. (Minister of Defence), 56, 130–1, 136, 155, 168
Rasetti, F., 182, 295, 337
Rawling, William, 9
RDX explosives, 98, 107, 109, 113, 115–16, 120, 313, 316; Allied cooperation, 113; patent issues, 116–17; secrecy and safeguarding of, 116, 209, 217; Type B campaign, 115–16
Red Army Intelligence (GRU), 214–17, 219–21, 223, 226–7, 229, 234, 242–5, 247, 265
Reed, G., 35, 152, 158, 160–1, 163, 168, 174, 263, 277, 328, 334–5
Reed, R.W., 210
Research Enterprises Ltd. (REL), 78, 82, 86, 88, 91, 216
Rice, C., 263
Richardson, Air Marshal A.V., 124
rinderpest, 154, 156–9, 174, 210, 331; GIR project, 160; vaccine, 161
Rivett-Carnac, C., 230
Robertson, N., 229, 252
Rogov, Colonel V., 218, 232, 242
Roosevelt, F.D., 5, 21, 51, 57, 60, 136, 145, 147, 149, 173, 179–80, 188, 194, 211, 258, 260, 265
Rose, D.C., 89
Rose, F., 217–19, 232, 236, 243, 353
Ross, J.R., 107
Ross-Scheissler process, 110–11
Royal Canadian Mounted Police (RCMP), 203, 205–6, 208, 210, 213, 227, 230, 234; Intelligence Service, 208; role in espionage, 241; security of chemical warfare research, 132
Royal Canadian Navy Directorate of Electrical Engineering, 247

Royal Commission on Espionage (1946), 13, 36, 208, 219, 228–9, 231–7, 239, 241–2, 245, 264
Rutherford, Ernest, 15, 28, 30–1

St Laurent, Louis (Minister of Justice), 229
Sargent, B.W., 184, 213, 263, 346
sarin, 146, 323
Scheissler, R., 314
Schild, A., 206
Schrader, G., 128
Scientific Advisory Committee, 58
Scientific Committee, 183
Scientific War Service (University of Toronto), 46
Secret Intelligence, 203; Allied cooperation, 210; British intelligence, 245; Canadian espionage, 208, 214; Canadian policies, 233; effects of security measures on, 227; role of scientific groups, 225; Soviet intelligence, 212, 215, 246
Security Panel, 228
Selye, H., 224
Shawinigan Falls pilot plant, 112, 209, 218
Shenstone, A.G., 61, 82, 300
Shope, R., 161
Shugar, D., 208, 218, 220, 222, 232–4, 243, 350, 359
Simon, F., 190
Sinclair, A., 54
Sivertz, C., 132, 222, 238, 277
Slotin, L., 190, 277, 357
Smith, A.H.R., 81–2
Smith, Durnford (code named Badeau), 92, 208, 218, 220, 232–3, 239, 243, 247, 350, 355, 360
Smith, S., 359

smoke weapons research, 124, 142, 148
Social Credit party, 240
Society for Cultural Relations with Foreign Countries (VOKS), 226
Sokolov, Colonel V., 218, 349
Solandt, O.M., 120, 245, 248, 254, 261, 278, 360–1
soman, 146
sonar/asdic research, 90–1; secrets and security, 219, 233, 243, 247. *See also* Anti-submarine warfare
Soviet Academy of Science, 223–4, 238
Soviet Information Bureau, 224
Soviet Union (USSR), 12; agent recruitment, 219–20, 242; alliance with British Commonwealth, 206; Allied–Soviet scientific cooperation, 216, 223–4; atomic capabilities, 212, 265, 231, 246, 251, 360, 364; espionage in Canada, 204, 214, 217–18, 220–1, 226, 228–9, 233–6, 242, 244–5, 265; fear of chemical weapons, 136; Soviet intelligence, 3, 216, 220; support of Spanish republican government, 35
Spaatz, Colonel C., 76
Spain, 35–6; Civil War in, 37, 122
Speakman, B.H., 222
Stacey, Colonel C.P., 8–9, 259
Stalin, Joseph, 246
Stamp, Lord T.C., 165–6
Steacie, E.W.R., 27, 191, 194, 202, 278, 364
Stedman, Air Vice-Marshal E.W., 58, 310
Stephenson, Captain C.S., 158
Stevenson, J.W., 210
Stimson, H.L. (Secretary of War), 57,

136, 147, 159, 166, 173, 188, 258, 321
Strategic Air Forces, 145
Strategy and Policy Group, 66
Strong, Major General G.V., 66
Stuart, Lieutenant General K., 58, 134–5, 168
Suffield Experimental Station (SES), 9, 124, 130–3, 137, 140–4, 146, 149, 150, 152, 161–8, 174–5, 209, 249, 254, 257–8, 261, 266, 319, 322, 335
Synge, J.L., 206, 278, 310
Szilard, L., 227

tabun, 128, 146
Taschereau, R., 228, 231, 233
TASS, 218
Taylor, G. (U.S. senator), 238
Technical Development Board (TDB), 94
Technical Sub-Committee on Chemicals and Explosives, 108, 137
Telecommunications Research Establishment (TRE), 73, 81, 88
Tennessee Eastman (Kodak) pilot plant, 116
Thode, H., 183–4, 263, 278
Thompson, G.P., 86, 179, 183
thorium, 193, 246
Thorvaldson, T., 132
Tizard, Henry, 7, 21–2, 42, 54–9, 61–2, 66, 69–71, 75–9, 88, 93, 99, 127–8, 174, 177, 189, 214–15, 223, 248, 250, 257, 278, 298–9; mission of August 1940, 12, 52, 77, 93, 95, 127, 156, 214, 257; Moscow mission (1943), 215
Tolman, R., 100, 253
Torr, Colonel W.W.T., 20–1
toxoid (Esoid), 168
Treaty of Versailles, 16

Tripp, B.M.H., 224
Truman, H.S., 196, 211, 230, 244, 252
Tuve, M., 100–4, 278, 311

U.K. military mission (Washington), 143
Underhill, F., 206
United Nations, 225, 232, 251, 262; role in control of nuclear research, 253, 262
United Nations Atomic Energy Commission (UNAEC), 252–3; Scientific and Technical Committee, 253
United States, 4, 20–1, 53; atomic bomb used against Japan, 195–8; atomic research, 30, 176, 179, 185–8, 190, 198, 257; Biological Warfare Committee (USBWC), 165; biological warfare research, 156, 159, 174; capability for war, 53, 55, 259; chemical warfare research, 127, 129, 136, 138, 148–9, 170–1; cooperation with Allies, 50–1, 176–7, 200, 215, 250; cooperation with USSR, 223, 230; decline of scientists' movement in, 238; entry into WWII, 42, 68; explosives research, 261; fear of atomic espionage, 246; fear of Soviet nuclear capabilities, 231, 236, 255, 265; immunization of troops, 168; Joint Intelligence Committee, 66; McMahon Bill, 237; military research and development, 3, 11–12, 55; military superiority, 189–90, 253, 261; Office of Naval Intelligence, 12, 93, 99; rearmament policies, 55; recruitment of immigrant scientists, 183–4; security of military information, 179, 193, 206, 211, 217, 226, 244;

Soviet espionage in, 234, 242, 244–5; war aims, 60, 66; War Department Industrial Mobilization Plan, 21; WBC, 156–9, 162
University of Alberta, 17, 26, 119
University of British Columbia, 26, 28, 119, 183
University of Chicago, 139, 176, 180
University of Manitoba, 26, 28, 33, 119
University of Michigan, 110, 112, 257
University of Montreal, 184
University of Pennsylvania, 112
University of Rochester, 77, 263
University of Saskatchewan, 6, 15, 18, 26, 119, 132, 183
University of Toronto, 7, 16, 25–6, 28, 31–2, 44, 46, 69, 80–4, 100, 102–4, 109–13, 116, 118, 120, 129, 132, 150, 152, 155–6, 205–6, 222, 224, 241, 257; Scientific War Service, 46
University of Western Ontario, 26, 28, 30, 83, 85–6, 119, 132, 150, 207, 222, 238, 263
University of Wisconsin, 156
uranium 233, 191, 193–4, 201–2, 220, 245–6
uranium 235, 177–8, 181, 220, 245–6
Urey, H., 183, 339
U.S. Army Chemical Warfare Service (CWS), 127–9, 133, 137–8, 144–5, 147, 149–50, 156, 159, 163–4, 170, 172, 226, 254
U.S. Atomic Energy Commission, 255
U.S. Biological Warfare Committee, 261
U.S.–Canada Sub-Committee on Flamethrowers, 143
U.S. Federal Bureau of Investigation, 229, 244

U.S. Interservice Electronic Research Supply Agency (ERSA), 88
U.S. Joint Chiefs of Staff (JCS), 66–7, 72, 102, 105–6, 141, 145, 148–9, 167–8, 171, 211
U.S. Navy, 66, 93, 119, 158, 161
U.S. War Department, 81, 173, 198; Army, 66, 115; Ordnance Board, 113, 116

Valcartier explosives and propellants complex, 113, 118–19, 209
Valdee, M.V., 156
Veall, N., 204, 222, 226, 234–7, 278
V-J Day, 259
Volkoff, G., 183–4, 279
von Halban, H., 176, 179–82, 184, 188, 190, 202, 214, 274, 336

Waitt, Brigadier General A.H., 144–5
Wallace, Colonel F.C., 78, 91–2, 207
Wallace, P., 184
Wallace, R.C., 233
Wansborough-Jones, Lieutenant-Colonel C.H., 148
War Advisory Board, 45
War Department Industrial Mobilization Plan, 21
War Disease Control Station, 140, 152, 159, 161
War Measures Act (Canada), 228, 230
War Research Services (WRS), 159, 161
War Technical and Scientific Development Committee (WTSDC), 47–9, 78, 132
Washington Agreement (1945), 250–1
Watson, H.W., 193, 213
Watson, W.H., 91
Watson-Watt, R., 69, 74, 76, 225

Webster, W.L., 61, 64, 116, 299
Wedemeyer, Brigadier General A.C., 66
Wehrmacht, 145; Army division, 127–8
Weil, G., 200
Weir, Lord, 131
Welles, S., 51–2
Wenner-Gren, A., 84
Wilgress, D., 355
Williams, E.K., 230
Williamson, Colonel R.E.S., 245
Wiseman, W., 143
Woods Hole Oceanography Institute, 64, 251
Woolpert, Lieutenant-Colonel O., 170–1
Woolwich process, 118–19, 217, 244
Woonton, G.A., 85

World Federation of Scientific Workers, 225
Wright, C.S., 83
Wright, G., 109–14, 116–20, 132, 261–2, 279, 317
Wrong, H., 229

Yaffe, L., 184–5, 207, 279, 322
Young, J.V. (Master General of Ordnance), 168

Zabotin, Colonel N., 216, 218–19, 220, 229, 231–2, 242–3, 265
Zaroubin, G., 218
Zero Energy Experimental Pile (ZEEP), 199–200, 244
Zimmerman D., 9, 89, 94, 296
Zinn, W., 213